Re-Engineering Philosophy for Limited Beings

Cartoon by Sydney Harris, *Physics Today,* February 1988. Courtesy of Sidney Harris and ScienceCartoonsPlus.com.

Re-Engineering Philosophy for Limited Beings

PIECEWISE APPROXIMATIONS TO REALITY

WILLIAM C. WIMSATT

HARVARD UNIVERSITY PRESS
Cambridge, Massachusetts, and London, England 2007

Library of Congress Cataloging-in-Publication Data

Wimsatt, William C.
Re-engineering philosophy for limited beings : piecewise approximations to reality / William C. Wimsatt.
p. cm.
Includes bibliographical references and index.
ISBN-13: 978-0-674-01545-6 (alk. paper)
ISBN-10: 0-674-01545-2 (alk. paper)
1. Philosophy. I. Title.
B29.W498 2007
191—dc22 2006052988

To my generators and support—

past: my parents, Ruth and Bill Wimsatt
present: my partner in life, Barbara Horberg Wimsatt
future: our son, William Samuel Abell ("Upski") Wimsatt

Preface

In the early 1990s, when I finally decided it was time to put together the papers that would become this volume, the tides of rationalism were near their high-water mark. Game theory increasingly dominated economics and political science, and optimizers in economics were talking to those in biology. Rationalism had been strong in philosophy for my entire professional generation, beginning in the mid-1960s. Naturalism's hold seemed less certain. Realism seemed on the retreat, not only within philosophy but also with the rise of science studies.

Now we see a growing and refreshing empiricism about many things "philosophical," even rationality itself. Traditional idealizations seem less satisfactory on all sides. New disciplines have emerged, intersecting many of our knottiest problems of the biological and human sciences—those problems often called methodological or conceptual or philosophical, and increasingly described as complex, in the senses of chapters 8, 9, and 10 of this volume. Behavioral decision theory has begun to have its day. Gigerenzer, Hutchins, and others project a more heuristic, collective, and contextually imbedded image of our intellectual powers. An explosion in the past decade of works on human biological, cognitive, social, cultural, and above all collective evolution have made it only natural to look for our human natures through a historically and socially informed multidisciplinary scientific perspective. The *human sciences* and the *Darwinian sciences,* cluster terms emerging in the past decade, are both appropriately named and increasingly interpenetrating. Perhaps Herbert Simon and Donald Campbell's different but consilient evolutionary visions of our kind will come to pass (see more on this below).

But there is also a lively antiscientism abroad opposing this—perhaps largely in response to the excessive reductionism of the past generation. Philosophers like John MacDowell see scientific perspectives on human nature as too coarse, rigid, and insensitive to capture our intentional worlds of mind and culture. Past materialisms have regularly promised urban renewal of these neighborhoods to make room for the latest seemingly spare materialism, bringing a bulldozer when a cultural liaison mission was called for. This is not the form of integrative materialism now emerging. Scientific-cultural liaisons now blossom, and new progress in all of the human and Darwinian sciences could result from richer and more appreciative interdisciplinary interactions. Hopefully, enough practitioners of traditional disciplines will recognize that it is again time for new infusions for the health of all and will resist the temptation to erect defensive conceptual trade barriers. The ironies of the frontispiece simultaneously document the hubris of a too-simple physics-inspired reductionism, and my commitment to "everything in between." It is thus a singularly suitable way to begin this book. I thank Sidney Harris for permission to use his inspired wry insights on the human scientific condition.

The oldest of these chapters goes back more than 30 years; the youngest, only a few months. It has been about 25 years since Lindsay Waters suggested that I put a book of them together, and I'm happy now to surrender my seniority as his longest-standing project though it has evolved from a collection of papers to a more systematic work. He has maintained an ongoing conversation and stream of suggestions (readings as well as modulations) all along. For years I didn't have the right mix, and others developed alongside. These are, I hope, the first of more "right mixes" to come. This collection combines older papers, newer ones (several never before published and written especially for this volume), and substantial new introductory essays.

With the older papers (chapters 4–6 and 8–11) I have made only minor stylistic revisions, removed old misprints, and occasionally added new bibliographical references. Chapter 5's section on group selection was reorganized to eliminate redundancies with Chapter 4. These papers have aged well—often anticipating newer directions and still relevant as they stand, though references and descriptions of other positions are occasionally dated. When not central to the argument, I have left them unchanged. I apply a naturalistically conceived rationality to the analysis of the complex world we live in, and to what kinds of beings we are that we do so. I hope the collection has the right hooks and balance to interest scientists and philosophers alike. It is a naturalistic and realist philosophy of science for limited beings studying a much less limited world. It mines science for philosophical lessons more commonly than the reverse. The key concepts here (robustness, heuristic strategies, near-decomposability, levels of organization, mechanistic explanation, and generative entrenchment) are applied to explore the cognitive tools with which we approach the world and our parts in it. Reductionism—the source of past and continuing threats of assault from below on the human sciences and humanities—is both target and resource, as much for its methodology as for its results. The result is a softer, richer vision of our world and our place in it than promised by both sides in the history of the warfare between mentalisms and materialisms. It is a more appropriate philosophy of science, I argue, than we have been given so far.

Acknowledgments are spread through the chapter endnotes. The new materials (chapters 1–3, 7, most of 12, and 13; the part introductions; and the appendixes) emerged during and between several "writing retreats." I thank the staff of Villa Serbelloni, the Rockefeller Foundation Center at Bellagio (in a warm, dry, beautiful, and amicable March 1997), the Franke Humanities Center at Chicago (for another part of that year, and of 2004–2005), and finally the staff and fellows of the National Humanities Center in Research Triangle Park of Raleigh-Durham, North Carolina, for a glorious year in 2000–2001. Their catalyses were real, though nominally directed toward other projects, now further along and better than they would have been without their help. Some chapters benefited from the support of the National Science Foundation and the System Development Foundation, and my students, undergraduate and graduate, played a continuing role as midwives and sounding boards, many even as they developed their own careers. Particularly among these (in temporal order) have been Bob Richardson, Bill Bechtel, Bob McCauley, Jim Griesemer, Sahotra Sarkar, Jeffrey Schank, and Stuart Glennan. Schank and Griesemer coauthored multiple papers (and Schank, software) with me in the process, and were particularly productive for my own thought. Deeply formative influences of Herbert Simon, Donald Campbell, Richard Lewontin, Richard Levins, and Frank Rosenblatt all began for me in the mid-1960s, and have endured. Michael Wade's example and research since 1975 has been a rich source of inspiration. Virtually everything consistently combined from their perspectives is found here—as well, perhaps, as a few that aren't! Productive tensions arise from their less common points of disagreement and my attempted projections of their viewpoints to new domains. Eclecticism has given me more productive influences than I can remember. I hope that any people left out here will forgive my lack of global insight.

Those commenting on newer materials include Bob Batterman, Bill Bechtel, Chris DiTeresi, Stuart Glennan, Jim Griesemer, Jessica Heineman-Peiper, Elizabeth van Meer, Tom Nickles, Trevor Pearce, Matt Schockey, Sahotra Sarkar, Jeff Schank, Eric Schliesser, Howard Stein, J. D. Trout, Jon Tsou, Julio Tuma, Barbara Wimsatt, and Billy Wimsatt. Joe Williams and Nathan Reich made productive organizational and editorial suggestions. Barbara Wimsatt helped design and compile the index. Barbara and Billy have traversed the long road to this book with me, observing, helping, and exhorting. To all, a very big and extended thanks!

Contents

Figures and Tables

Re-Engineering Philosophy for Limited Beings

PART ONE

Introduction

CHAPTER ONE

Myths of LaPlacean Omniscience

Why is it that academics who claim to seek the truth want to pretend that they have always had it? What are they paid for, anyway?

This behavioral paradox is found throughout academia, but is nowhere stronger than in the sciences. We preach with pride that science is fallible and testable—contrasting it with religion and other dogmas. And we should. But we hide with embarrassment papers that are undercut or shown to be erroneous by later work. (Try to get tenure with such a paper!) We commonly don't report negative results. Errors are okay only if they are someone else's or belong to prior generations. Less reflective scientists teach undergraduates only those parts of their science that are "established fact," deferring until graduate school anything disputed or under active investigation. These same instructors then complain that beginning graduate students don't seem to notice when something is wrong or missing, or how to find questions worth researching. But has anyone tried to teach this skill? We publicly celebrate fortuitous discovery, as if to convince humanists and artists that science too is a creative liberal art, but discoveries to be celebrated are those that required a prepared mind and laboratory. We ought to celebrate the well-crafted research plan even if it failed, and not when dumb luck let us succeed without preparation. We should thereby *reinforce what our students can do* to help them to succeed.

Many of the contradictions emerge from our idealizations of science and philosophical views that see knowledge only in certainty. Even when these viewpoints have been given up explicitly, their residue remains. I address these issues in overview in the first three short chapters. In the rest of the book I elaborate an alternative picture.

The focus on truth is understandable. The sciences are the best-crafted repositories of truth on this planet. Their results are important for our technology, and for the public validation of our craft. At least as important are the cognitive technologies, material procedures, and social structures we have crafted for seeking and checking our results. But the way we talk about science and teach it not only fails to communicate this fact, but actively misrepresents it. An alien power seeking to undercut our civilization could do no better than to teach most of our science and its history and philosophy as we do now. This could have been written in 1962 by Thomas Kuhn, or Stephen Brush in 1974. What is striking is that it needs saying now more than ever.

These chapters are primarily about method in science; particularly, the strengths, weaknesses, and proper interpretations of reductionist approaches. This is not as limiting as it might seem. Reductionism is the dominant methodology of "big science" today, percolating widely through the sciences, with methodological strategies applying still more broadly to areas where analytic methods are found. But our analytic training leaves us with a set of bad reactions—that we don't really understand something until we have broken it down into the smallest possible pieces. This broader methodology of reductionism has important implications, but not those that many philosophers have assumed. Philosophers once put these methods at the core of a formalistic foundationalist philosophy of science. But this view is fundamentally flawed by idealizations we make about how we reason. As real beings, we "deviate" from these idealizations about computation, cognition, inference, decision making, and the structure of our conceptual schemes.

One important mistake is the belief that reductionist or analytic methods eliminate or analyze away what is being analyzed or reduced. This produces a false opposition between those working at different levels of organization, or between those using "humanistic" and "scientific" approaches to the phenomena. The aim of what are called reductionist explanations in science is not to atomize phenomena to the lowest possible level, but to be able to articulate and understand entities, events, and processes at diverse levels, and to give explanations involving heterogeneous relations among them in producing complex phenomena. We can get a robust and lasting appreciation of processes at higher levels of organization in their own terms that is not compromised by having lower-level accounts. And large parts of our reasoning are profoundly heuristic at our working levels, as we craft and modify our procedures. It is here that we must start.

Many of these paths were first sketched four decades ago by Herbert Simon. I have explored them often, and philosophers now traverse them in increasing numbers.[1] We also need a context of real problems, keeping us honest in ways that idealizations and toy examples common in philosophy often fail to do. In practice we use surprisingly powerful heuristic tools rather than such idealizations to structure our search. These new tools are sources of inspiration for an alternative, robust, naturalistic, and scientifically motivated realist philosophy of science. With this focus, scientific processes are seen as end-directed or teleological activities, generating a cognitive engineering of science, with natural resonance with design analyses in engineering (Chapter 13)—a fruitful path so far ignored in our profession.

Realism for Limited Beings in a Rich, Messy World

I seek methodological tools appropriate to well-adapted but limited and error-prone beings. We need a philosophy of science that can be pursued by *real* people in *real* situations in *real* time with the kinds of tools that we *actually* have—now or in a realistically possible future. This must be a central requirement for any naturalistic account. Thus I oppose not only various eliminativisms, but also overly idealized intentional or rationalistic accounts. In these chapters I advocate an approach that can provide both better descriptions of our activities and normative guidance based on realistic measures of our strengths and limitations. No current philosophy of science does this fully, though increasing numbers are moving in that direction. A philosophy of science for real people means real scientists, real engineers, historians or sociologists of real science and engineering, and real philosophers interested in how any of the preceding people work, think about their practice, think about the natural worlds we all inhabit, and think about what follows reflectively and reflexively from these facts.

This view involves a species of realism, though not of the usual sort. It fits current scientific practice and illuminates historical cases better than other approaches, and it has implications for how to do philosophy.[2] It neither has nor seeks the stark simplicities of current foundationalist theories. This philosophy must be based from top to bottom on heuristic principles of reasoning and practice, but it also seeks a full accounting of other things in our richly populated universe—including the formal approaches we have sought in the past.

This project is a philosophy for messy systems, for the real world, for the "in-between" (see the frontispiece), and for the variegated ecologies of reality supporting and increasingly bent to our science and technology. *Pace Quine,* this is ontology for the tropical rainforest. The "piecewise approximations" of the book title is unavoidable: we are, must be, and can be realists in our science and much of our practice. But our realism, like our practice, and even our inferential consistency, must be piecemeal and usually satisfied with a local rather than a global order. We aren't God and we don't have a God's-eye view of the world. (In this piecemeal world, we don't even have a *gods' eyes* view.)

But then the first part of the title is only half-truth: to re-engineer the whole of philosophy in a human image is still ambitiously global. I don't do all this. I sketch how to do it for significant parts of philosophy of science and closely connected areas of science. This captures new phenomena and reconceptualizes old in ways that fit more naturally with how we proceed. I hope that others find these results sufficiently suggestive to use, extend, and add to the tools I describe here to employ them elsewhere.

"Re-Engineering" appears in the title as a verb: this view of science and nature is constructed largely (as with all creative acts) by taking, modifying, and reassessing what is at hand, and employing it in new contexts, thus *re*-engineering. *Re*-engineering is cumulative and is what makes our cumulative cultures possible. And any engineering project must be responsive to real world constraints, thus realism. Our social, cognitive, and cultural ways of being are no less real than the rest of the natural world, and all together leave their marks. But putting our feet firmly in the natural world is not enough. Natural scientists have long privileged the "more fundamental" ends of their scientific hierarchy, and pure science over applied—supposing that (in principle) all knowledge flowed from their end of the investigative enterprise.

Not so: *Re-engineering* also works as an adjective, and has a deeper methodological role. Theorists and methodologists of the pure sciences have much to learn about their own disciplines from engineering and the study of practice, and from evolutionary biology, the most fundamental of all (re-)engineering disciplines. Our cognitive capabilities and institutions are no less engineered and re-engineered than our biology and technology, both collections of layered kluges and exaptations. We must know what can be learned from this fact about ourselves to better pursue science of any sort.

Social Natures

Philosophers were once deaf to claims that cognitive science, evolutionary biology, and issues of practice were centrally important to *our own* methodology. In those days, psychology led only to "psychologism," the study of history only encouraged committing the "genetic fallacy," and "sociology" seemed worse than socialism. Knowledge after all was mind-independent and timeless, and minds were individual and both abiological and asocial. Today such opinions seem quaint; such philosophers find themselves increasingly embattled, circling the wagons in defense of Reason. We show our own disciplinary biases and force them on others: the various "philosophies of X" often seem to be more about arguments internal to philosophy than "of" anything. In much of our broader humanistic intellectual community, the social rootedness of our activities is taken for granted, the real world is seen as a construction, and reason is but a chattel of interests. But this course also swings too far. How can we bridge these divergent perspectives?

I want to re-psychologize, re-socialize, and re-embed us in the world, where we reason about that world as well as about how we interact with and reflect upon it. Can we still be recognizably philosophical while letting the subjects of "philosophies of" shine through much more clearly and inspire new philosophies, rather than merely exporting our same old "philosophical" disputes to these new territories?

Contrary to some philosophical views, "empirical contingencies" are crucially important to philosophy. We are embodied socialized beings: evolved and developing in a world conditioned by our sociality and technology. We steer, often unreflectively, through it with values that are uneasy combinations of history, religion, science, pseudo-science, and the latest fads—most recently the ideological preachments and short planning horizons of a "free" market. Any adequate account of reason must see it as the adaptation that it is: fallible, but self-correcting. And self-correcting not just through reason alone, but in the way that DNA is self-replicating—when embedded in a larger supporting complex that is both of the world and self-continuing in the world.

Those who want to recognize our social natures rarely see biologists as useful allies. But Levins and Lewontin (1985) argue that organisms often actively construct their environments, and urge a rich developmental and evolutionary *relationalism,* treating organism-environment interactions as real (and often primary) without succumbing to relativism.[3] And Paul Ehrlich (2000) speaks of "human natures" because culture is part of human nature, and different cultures make different natures. He thus recognizes that the Hobbesian state of nature never existed. We were social throughout our primate history—long before language or tool use. The social contract is a creation myth of western political science. Our culture is midwife to and product of our extended juvenile period of learning, before we mature and our habits become less plastic. Yet this richly relational world is still a natural world, and we must study both biology and cultures to understand it.

But how do we study nature? How do we organize it? How do we avoid chaos, disorder, or unfathomable confusions in our account of natural complexities?

Heuristics as Adaptations for the Real World

Philosophers have established ways of creating order working from a few key idealizations about decision making and rational or logical inference. These principles crop up continually; normative frameworks for almost everything else they do, and (supposedly) everything else *we* do. These assumptions are deeply *generatively entrenched,* widely used pivotal assumptions that play a role in generating our philosophical theories of almost everything else.[4]

So is this a problem? Unfortunately, yes: these particular idealizations demand unrealistic degrees of knowledge, unlimited inferential powers, or both (see Appendix D on LaPlacean demons). They don't fit our behavior. As normative rather than descriptive principles, that's not necessarily a problem. But they make assumptions that we *cannot* satisfy, and that *is* a problem. How can these provide goals toward which we should orient ourselves? How could they be correct even as normative claims? Should we give them up? We hear that to do so would be to open the doors to irrationalism, relativism, or other unnamed horrors precipitating total cognitive collapse and chaos. *But the very success of our cognitive and social adaptations in the real world belies those fears. It is an existence proof that there are better idealizations or models to be made.*[5]

Where should we look instead? I study the heuristic techniques scientists use to explore, describe, model, and analyze complex systems—and to "de-bug" their results. These are plausibly more carefully tuned and evaluated descendants of more broadly used inferential practices. So we start with our actual practice—but seeing these practices for their strengths as evolved cognitive adaptations rather than as compromised attempts to pursue our ideals. *These "deviations" are not failings, but the source of our peculiar strengths in this uncertain world of complex, evolving beings, technologies, and institutions.*

Considering our cognitive powers and limits and the social contexts and embodiments of our decision processes yields many benefits.[6] As philosophers we can analyze new scientific activities left unexplored or dismissed because our overly idealized models have lacked the resources to analyze them. Theories of science and other human activities can be both broader in scope *and* more sensitive to detail while giving better—more effective, detailed, and realistic—normative guidance for how to proceed. As scientists, more realistic accounts of these activities help us to sidestep avoidable errors, find and fix them when they do occur, and pursue our goals more effectively. Thus my intended audience is philosophers—to get them to change their practice—and scientists—in the hope that these ideas might be genuinely helpful in their practice. There *is* room for a theory of practice: a genuinely philosophical theory with lessons in how to do philosophy as well as for philosophy of science.[7] I present case studies in and of this practice. I will return to some of the ways we could expect changes in Chapter 3.

Nature as Backwoods Mechanic and Used-Parts Dealer

The worldview of logical empiricism was built with minimalist tastes, axiomatizations, on the inspiring constructive triumphs of the new symbolic logic, and on the more rigorous standards of proof that accompanied them. The computational worldview descending from it builds on algorithms—powered and inspired by the marvelous and increasingly powerful engines we have built to execute them quickly and reliably. The algorithmic view adds to the axiomatic an appreciation of problems of computational complexity, and procedures and subroutines make explicit the advantages of a natural modular and hierarchical organization. But the algorithmic view, an improvement, supposes or tries to construct a well-defined structure to the problem-space (Simon, 1973), and an abstract crispness in the engines for exploring it.

These views are each appropriate to many tasks, though neither is as general as commonly supposed. There is order in our practice, raw material for a generalizable account. These *in principle* claims deflect our attention, and don't help here. So I will systematically ignore the kinds of *in principle* claims often arising in these two contexts and much beloved by philosophers to see how we do—and should—deal with situations we face *in practice.*[8]

Our ancestors didn't adapt in this complex world with simpler but still formally respectable deduction systems. Biology is full of diverse kinds of inductive pattern-detectors. The first problem was finding order, not testing it, and finding it according to rough error-prone procedures. If we look at ourselves as biological social cognitive beings, we see that our responses to problems of adaptation in a complex world are crafted with heuristics. Insofar as these are products of selection processes, biological and cultural dimensions of our reason must also be heuristic. *Heuristic principles are most fundamentally neither axioms nor algorithms, though they are often treated as such. As a group, they have distinct and interesting properties.*[9] *Most importantly, they are re-tuned, re-modulated, re-contextualized, and often newly reconnected piecemeal rearrangements of existing adaptations or exaptations, and they encourage us to do likewise with whatever we construct.* We are just now beginning to come to grips with that picture: Nature as a reconditioned parts dealer and crafty backwoods mechanic, constantly fixing and redesigning old machines and fashioning new ones out of whatever comes easily to hand.[10] So much of engineering—even of "new" designs—is *Re*-engineering. This has profound implications for the character of evolutionary products (and of evolutionary change; see Chapter 7).

Are heuristics merely an interesting subset of the class of all possible algorithms, as Dennett (1995) suggests? If we imagine a list of all executable procedures, all truth-preserving procedures as well as all heuristics would be on it. This broader class is sometimes called the class of algorithms. This picture has a misleading air of necessary truth. It encourages us to treat heuristics as derivative from algorithms—a mistake. (We might as well treat texts as derivative from alphabetic sequences.) The truth is quite the reverse. Heuristics are both causally and phylogenetically prior to our algorithmic abstractions. Evolution opportunistically favors results, using any available method, no matter how crude, that is cost effective in its context relative to other available alternatives. Recognizing the heuristic character of our reasoning solves many otherwise intractable problems (Aloimonos and Rosenfeld, 1991; McClamrock, 1995).

The context-free elegance we are taught to value is of no broader foundational moment than the contexts these methods were designed for. Traditional foundationalisms combined the need for systematization and the reliability of socially cross-checkable proof processes and mistook them for an account of the internal character of reasoning (Hutchins, 1995; Gigerenzer, 1993). But formalistic foundationalism fails miserably as a process model for cognition, for the construction of theories or devices, or for any biological adaptive processes. It has done little better for philosophy. Working practices with axioms and algorithms can be seen as limiting special kinds of heuristic methodologies. Well adapted to appropriate problems (the deductive exploration of formal models), the success of formal methods in those instances has no generalizable foundational significance. A naturalistic worldview of the genesis and operation of functionally organized evolving systems is heuristics all the way down—as far down as entities or configurations of them are products of selection. (Biological adaptations share the central features of heuristic procedures—see chapters 4 and 5; appendixes A and B; and Wimsatt, 1980b.)

Error and Change

So should we just substitute heuristics for axioms as new fundamental rules, acting on whatever comes easily to hand, and proceed as before with a neo-classical foundationalism? Not quite! We also need adjustments elsewhere: a change in the scope of (and expectations from) acceptable inference rules; a new conception of foundations, which must become dynamical and potentially changeable; a more error-tolerant conception of the organization of our conceptual structures; and a new treatment of error. Our adaptive mechanisms must be capable of detecting and responding to—nay, feeding on—errors at different levels and across varied contexts, and exploiting parallelisms and redundancies eschewed by formalists to detect errors and fine-tune responses. One of the most common and effective uses of deductive arguments in science is to detect and localize errors and refine concepts.[11]

With a new focus on error and change, a richer theory must address not only how generative structures work, but how they are changed or modified in midstream, and what patterns such changes should follow—*for there are such patterns, important ones, and they make such an endeavor worthwhile.* These are totally ignored in traditional approaches, but are central to a fuller reading of the nature of developmental and evolutionary processes—whether biological, cognitive, or cultural, including processes of scientific change. *Generative entrenchment* provides a heuristic and dynamical reading of an attractive feature of foundational theories: the remarkable generative power of a few well-chosen assumptions, structures, or processes, and the consequences of this power for freezing-in the assumptions that engender it. This crucial idea and its implications are sketched in Chapter 7, suggesting the form of a foundationalism for non-foundationalists—scientists, engineers, and students of adaptive complexity across the disciplines.

The roots of the views offered here are diverse. The rich backwoods of evolution (Darwin's "tangled bank") is a heterogeneous, multi-level tropical rainforest, with converging overlapping branches, and patterns of intersecting order, residents, and connections at a variety of levels, but no *single* stable foundational bedrock that anchors everything else. Yet this multiple rootedness need not lead to "anything goes" perspectival relativism, or an anti-naturalist worship of common sense, experience, or language. It yields a kind of multi-perspectival realism anchored in the heterogeneity of "piecewise" complementary approaches common in biology and the study of complex systems (chapters 9 and 10). Here an overlapping diversity of roots, assumptions, approaches, and methods is fruitful (Chapter 4). But tracing foundations, sensible for more simply rooted systems, becomes both more difficult and less important. One could argue that multiple rootedness only poses *practical* problems for tracing foundations; and, *traditionally,* as philosophers, we are only interested in matters of principle! This is a serious mistake. Several chapters consider arguments from principle, and give reasons for rejecting, deflecting, or reinterpreting them in specific contexts (especially chapters 5, 9, 11, and 13). Like false models and qualified generalizations, *in principle* arguments are often honored more in the breach than in the observance. But we need to consider the role of *in principle* arguments more generally—especially where these spring from idealizations concerning our rationality. We start this in Chapter 2.

Organization and Aims of This Book

These essays, written between 1971 and 2002, are about a fifth of my published work. This chapter and the next two, chapters 7 and 13, and the introductory materials to parts II–IV are published here for the first time. Chapter 12 has been substantially expanded from earlier versions with new scientific examples.

Chapters 2 and 3 elaborate in more detail the fallibilist, realist, and heuristic perspective I urge. After the introductory section, the following chapters form three groups.

First (in Part II) is a series of general methodological essays on some deep and pivotally important heuristic principles of inference, principles that are used over and over again in the chapters that follow. They are, in effect, both heuristic and foundational (if "heuristic" is unfamiliar or problematic, Chapter 5, on heuristics, might be the best way to start). These chapters provide perspectives and principles that should be of much broader use in philosophy. They have relevance to the analysis and construction of diverse complex evolved structures—from macromolecules, to musical composition, to social networks and methodologies.

Next (in Part III) is a collection of chapters on the nature of reduction and emergence—a specific set of methodological problems central to modern science and particularly relevant to the study of complex systems, which is a constitutive theme of my work generally. These topics are strongly informed by the methodologies advocated in the first section, and provide further examples of their application. (Indeed, grappling with these issues reinforced for me their centrality.) Reductionist methodologies commonly take for granted the perspectives or levels of organization they utilize and focus on how they are related. These chapters—particularly 9, 10, and 12—also deal with how we recognize or individuate perspectives, levels, and decompositions in the first place.

The introduction to Part IV and Chapter 13 look backward to theoretical physics and engineering simulation in a pre-computer age. Incidents there led me away from my naive reductionism to this heuristic-based realist view of science and engineering. It deals with relations between pure and applied science, and what philosophers have to learn from a closer look at the engineering perspective.

Substantial introductions to the different sections track the most important themes in the chapters, and attempt to create a new order among them. For some major elaborations of their arguments, you might wish to return to the introductions after reading the chapters. This chapter is itself the introduction to the next one, which expands on two themes: the nature and differences between our uses of idealizations in science and in philosophy, and the fundamental reorientation toward error required in this fallibilist and heuristic view of adaptation and knowledge. In Chapter 3 I advance proposals for how to view the philosophical fallout. Further meta-reflections are found particularly in Chapter 10 and in the final sections of chapters 12 and 13.

The first two appendixes concern heuristics, elaborating Chapter 5, and complementing the rest of the book. Appendix A lists important properties of heuristics crucial to explaining characteristics of their use. Appendix B lists 20 different heuristics used in conceptualizing, model-building, experimental design, and other activities in the analysis of systems and their behavior. These heuristics are reductionistic or have that effect under specified circumstances. They may be helpful in analyzing a system, but carry with them characteristic biases associated with ignoring or downplaying context. They flesh out specific portable methods in how a reductionist might approach a system or problem—a "toolbox" for methodological reductionists. Appendix C is a limited glossary for key concepts or assumptions used in the text. Appendix D reviews unrealistic "in principle" idealizations found in philosophical or scientific arguments, assuming powers that are infinitely (or sometimes just indefinitely) greater then we could possibly achieve. Their obvious falsity points to the need to re-engineer philosophy for limited beings.

CHAPTER TWO

Normative Idealizations versus the Metabolism of Error

Inadequacies of Our Normative Idealizations

Scientists use idealizations, often conflicting ones, for various ends. Planets and stars are point masses in deriving Kepler's laws, as are molecules for the ideal gas law, while some mid-sized bodies (springs) are extended but massless in the simplest treatment of the harmonic oscillator. Geophysicists, stereochemists, and race-car designers, respectively, use other idealizations for these same objects sharing none of these assumptions. Organisms may be ageless, identical, and sexless in some ecological models, sexed and variably aged in others; discretely variable in specified ways in genetic models; and continuously variable in a small set of specified directions in evolutionary optimization models. Different sets of false assumptions—idealizations—play crucial roles in teasing apart different aspects of the causal structure of our world and permeate model-building, which is the cutting edge of theory construction. They are implicit in and motivate experimental design. They are used in acquiring, describing, and analyzing data. Models or idealizations increasingly replace theories as foci for studying interactions among these three central scientific activities.

Idealizations commonly play different roles for scientists and philosophers. Scientists treat models as simplifications of or useful counterfactual transformations of nature—visions ultimately to be corrected and elaborated in an ongoing theoretical, experimental, and observational dialectic with our world. This dialectic is socially, culturally, and technologically arranged, like other activities in our "carpentered" world. But it is clear that when the model doesn't fit reality, as it commonly doesn't, we should make changes in the model to do better. We don't expect reality to change to fit the model!

Philosophers sometimes seem to want the reverse: for human agents, we assume that they should change to fit the models. How can this be? Philosophers use idealizations too—to simplify and generalize analyses, to abstract away from particular details peripheral to the point at hand, and often to urge certain norms of behavior. This *normative* role gives idealized models of rationality and inference an ambiguous status. As implicit accounts of the scientist as rational agent—as logical thinker and utility maximizer—these models often play as presuppositions in the background while the analysis targets something else: decision under uncertainty, the idea of a conceptual scheme, theory reduction, supervenience, causal explanation, or the nature of language. These idealizations too are models, though we rarely act as if they were.[1] Calling them "constitutive ideals" or suggesting a special normative or generative role doesn't hide the fact that they embody assumptions and conditions. It is sensible to ask whether and how well these assumptions are realized in that part of the world they are applied to. While they are sometimes inspected directly, more commonly these assumptions are invisible. They aren't treated as if they were part of the analysis. Problems with them are often misidentified or incorrectly localized as due to some failure in the proposed analysis.

Why this strange behavior? I assume that models are kinds of abstract structures or sets of assumptions that can have either a broadly descriptive use, as in science, or to specify a standard for behavior, in which case it is used normatively. (Scientific models are actually used in various ways—see Chapter 6—but here they will all be called "descriptive.") Unlike these uses of models, *a model treated as normative is not usually compromised by our failure to act accordingly.* We can *choose* to follow a normative rule or not—the usual contrast between rules that guide action and causal laws. So violations of a rule may indicate our own failures rather than problems with the rule. But this is not enough: falsely describing human behavior is too weak a test for normative status! Many norms are commonly followed, often better than our best predictions for systems of comparable complexity. Consider Dennett's (1971) rationale for the "intentional stance," to assume that a potential opponent is a rational agent with reasonable beliefs is a powerful predictor of their behavior. Blaming the agent *is* the most likely response to deviant behavior in many contexts. For a well-designed set of norms, this is appropriate. But blaming the agent just places the initial burden of proof; else we could never question or reject a normative model. And sometimes we must. The right rules? Sometimes these are not even *good* rules for us to follow.[2]

Philosophers treat rationality as a normative ideal deeply rooted in our conception of ourselves and all that we do. But in our complex world of limited and fallible agents, with imperfect, noisy, and incomplete information, is it rational to try to be *perfectly rational?* Herbert Simon (1955) first shocked decision theorists four decades ago with the claim that it wasn't. He urged "satisficing" (discussed below) rather than "maximizing" as a basis for rational choice, according to his "Principle of Bounded Rationality." Darwin would have liked his views.[3] Philosophers and economists have resisted this move, favoring extreme idealizations and often deifying them as necessary or conceptual truths. We are still far from realizing the full impact of Simon's revolution. *The normative status of common philosophical idealizations of the scientist-as-agent interacts with various common cognitive biases to serve us poorly. This status hides how badly these idealizations perform as descriptions of our actual behavior and also give misleading advice about what we should do about the deviations we do detect.*

Consider an example of the cobbled together but adaptive way heuristics can act to our benefit. Festinger (1957) found that the perceived utility of a chosen alternative increased after making the choice, apparently just through the act of choosing. *This would give a bias towards answering "yes" to "Did you maximize utility in your choice?" if the question were asked after the fact, even if he or she had not maximized utility.*[4] This behavior seems more delusional than rational, but this "irrational bias" should normally improve the performance of individuals who have it over those who don't.

It may be advantageous to change utility assessments after the fact if there is some cost associated with changing your mind after making or announcing your decision. There often is. A subconscious biasing process favoring a chosen alternative could act as a lock-in device—useful if preferences are psychological variables fluctuating in time.[5] We notice decisions if the choice is important and problematic, so the alternatives being weighed are generally close. In these cases it may cost more to change after committing than to go with a suboptimal choice. Festinger's example involved vacillation between two potential mates—something likely to scare both away! Could this cunning unreason be meta-rational? Such a "utility distorting" device in our cognitive apparatus may serve us better, even if we are "misled" in our assessments.[6] Indeed, *is* this being misled? How (and when) *should* we evaluate the "goodness" of choices? Traditional decision theory just assumes that utilities are fixed. Perhaps calling it a "distortion" is an artifact of a static view of choice. Other such cognitive biases and their effects are discussed in chapters 5, 4, 7, 11, 13, and Wimsatt (1980b).

The status of idealizations is clearer for scientists. When we describe planetary orbits as elliptical we ignore deviations caused by the planets' gravitational attraction for one another, and a host of smaller effects required for more accurate prediction and explanation (Peterson, 1993). Human population growth is logistic (a simple model of exponential growth with resource limitations) if we ignore the age-structure of the population, and treat individuals of different ages as if they were alike. But they aren't: limited resources and other biological and cultural variables have effects that change with age. Women of 2, 32, and 62 years have different birth and death rates responding differently to changing resources. Some changes are intrinsic to the natural life cycle and others are modulated by culture. We accept these inaccuracies when they serve our purposes. For more detail and accuracy, we modify models, predictions, and collect more data on age-structure and other things as required. Modeling is discussed further in chapters 6 and 7.[7]

When we describe decision makers as maximizing expected utility, is it equally clear that, except rarely, they cannot possibly do so, that this is an incomputable task? Such models are not even in the right ballpark for the structure of most of our decisions. Our world is too complex, and our abilities too limited for such a model: we have neither the knowledge nor the computational facility required even to formulate correctly all the terms in the equation for expected utility, much less to determine the parameter values. Nonetheless we are given the common philosophers' de facto idealization that we are LaPlacean demons—omniscient and computationally omnipotent computers.[8]

Consider other common idealizations: virtually no real scientific theories are globally free from contradiction, and few real scientific inferences are truly deductive or truth preserving. Approximations in the theory and in its derivations belie both of these claims, and real scientific theories are laced together with approximations.[9] Sloppy idealizations about birth and death rates don't raise a hair, but what could be more important than birth and death? In science, we at least try to identify where these approximations will get us into trouble; yet contradictions in theories and approximative inferences are skeletons in almost all philosophical closets. *Philosophers internalize their denial and have built such magnificent normative edifices upon them that they don't seriously consider the possibility that they are false—not just in little ways, but in ways fundamentally compromising their approach to the rationality of human behavior and the practice of science.* I'm not against the rationality of science, I'm all for it. But this just is a bad conception of it.

Satisficing, Heuristics, and Possible Behavior for Real Agents

What other options are there? Consider Herbert Simon's (1955, 1996) "satisficing" account. An agent has a level of aspiration or of satisfaction and evaluates alternatives by whether they fall above or below this level. The level of aspiration is set and modified through experience. Alternatives aren't given in advance, but generated, or acquired (and sometimes lost) sequentially in real time. Decisions are made according to heuristic rules—rules as simple as "choose the first alternative which meets or exceeds your level of aspiration, unless you already know one to be probably better." This model demands much less in computational and cognitive resources than rational decision theory, and fits plausibly with decision mechanisms we actually use. Unlike the maximization model, one need not compare—or even generate—all choices (Goldstein and Gigerenzer, 1996). Satisficing also integrates learning: aspiration levels are constantly modified by experience. There's no need to determine an exhaustive and mutually exclusive set of choices, or the comparative judgments required within it, to determine the chosen alternative. Giere (1988) applies satisficing to the decisions of scientists, arguing its advantages over more traditional Bayesian models. *The heuristic principles of reasoning ("heuristics") urged here are natural extensions of satisficing, or of Simon's broader conception of "bounded rationality."*

But with satisficing, wouldn't you accept lower standards and performance? You need not. And higher standards aren't always better. "Reasonable goal setting" commonly contributes to better performance (McClelland, 1973). "Who knows just *how* well we can do?" Indeed. But are these realistic standards? Can we try to high jump 20 feet—or a quarter of a mile? What improvements in diet, training regime, or genetic engineering are supposed to result in that? Another million years of hominid evolution? Not a prayer! Yet our implicit models of decision making are orders of magnitude more demanding of our grey matter, and less plausible. We seem not to want to recognize limitations on our powers of thought. Standards like that we don't need.

Nor is this a benign failure. Idealizations about reasoning enter surreptitiously into our technology as well, and cause real problems. Those who worry about "human engineering" have known and studied this concern since the Second World War. Cognitive psychologist Donald Norman puts it this way:

> Today most industrial accidents are blamed on human error: 75 percent of all airline accidents are attributed to "pilot error." When we see figures like these it is a clear sign that something is wrong, but not with the human. What is wrong is the design of the technology that requires humans to behave in machine-centered ways, ways for which people are not well-suited. . . .
>
> When technology is not designed from a human-centered point of view, it doesn't reduce the incidence of human error, nor minimize the impact of the errors that do occur. Yes, people do indeed err. Therefore the technology should be designed to take this well-known fact into account. Instead, the tendency is to blame the person who errs, even though the fault might lie with the technology. (Norman, 1993, p. 11)

What kinds of machines are *we* supposed to be to run this technology? Machines who, in an airliner or nuclear power plant for example, keep track of hundreds of meter readings, detect any discrepancies, and decide correctly what to do about it. And even if and when we detect anomalies, idealized scenarios of "deciding what to do about it" fail to notice that we may treat aberrant readings as problems with the instruments rather than with the monitored processes. Instrument failures are far more common, and we may resist recognizing more baleful possibilities. Idealized Cartesian assumptions are again to blame: we will treat instrument readings as infallible (just when we should, of course!) *rather than adding dangerous delays as we try to check up on them or grapple with the implications of what they are telling us.* These are both *prima facie* reasonable reasons for delay if immediate action has a significant cost for which we may be blamed if it wasn't required.

These are common patterns of failure in after-the-fact analyses of major disasters. There is seldom a single error, but a series of compounding factors: intersecting errors of design, manufacture, maintenance, training, knowledge, and use. Peterson (1995) has revealing case studies of diverse software errors—errors that often led too much larger accidents. Often the disaster develops over time and could have been better contained if operators had made the "right" decisions;

judged by hindsight, of course. Medvedev (1991) is eloquent on Chernobyl, as are Petroski's many essays celebrating the role of failure in engineering design. Idealizations of our cognitive powers ignore not only our humanity, but also our biology—both our cognitive limits and our strategies for dealing with error. This is not just a biology of constraint, invoking biology to explain and apologize for what we cannot do. Evolution empowers us with reliable and error-tolerant solutions, and we can learn by looking more closely at nature's designs.

So what should we learn for normative idealizations? Kant's "[O]ught implies can" should be a constraint on all of our obligations. Given our unrealistic idealizations, this is perhaps the deepest principle of a heuristic approach: a test that idealizations must pass before we can accept them as providing normative guidance. We aren't talking here about apocalyptic descriptivism, fine-tuning our normative idealizations to adapt to all of the pebbles on the beach. What if no one can *ever* provide what they ask? By Kant's maxim, such idealizations are not good norms for us even to *try* to follow. *If we try to follow methods that require things far beyond our capacities, we may miss more effective tools appropriately tuned to our true abilities.* We have two such sets of tools—one sharpened and directed in hundreds of millions of years of animal, vertebrate, mammalian, primate, and hominid evolution. The other, a social and cultural evolution that has given us a second overlapping and largely complementary set. A broader formulation of the problems posed by Festinger's "utility distorter" may reveal better solutions than narrower "rationalistic" formulations that ignore or oversimplify more of our context and biological and psychological nature. Why not study these biological, psychological, social, and cultural tools for inference and problem solving, and include at least the more robust and less "accidental" (or avoidable) ones in our conception of the scientist as agent? We use more than logic and rational decision theory to get around in this world—even when it appears that in principle, we could get by with them alone. Epistemologists and ontologists like to get by with the minimum required, with clear and simply stated boundaries and no hard cases. But be wary of such promises: any claim that something can be done *in principle* is a tacit admission that it hasn't yet been done in practice!

The Productive Use of Error-Prone Procedures

Our common modes of inference aren't perfect. As with all inductive methods of the sciences, heuristic principles don't guarantee results.

And, through error or accident, we can *mis*use a truth-preserving rule of inference, thereby "voiding the warranty." So one could be a skeptic about any of our inferences. Exactly this line was used to argue that the senses don't give us infallible knowledge of the external world, and served various forms of idealism from the eighteenth to the early twentieth centuries. Philosophers often seem never to get past this point; they get hung up on certainty. Why certainty? Since Descartes, it has seemed to be the best way to avoid errors. We have engaged for over 350 years in futile searches for guaranteed ways. But what if a search for reliable knowledge is *not* best pursued as a search for guarantees?

Maybe errors are okay—or tolerable—as long as they aren't too frequent, and if we have good ways of remembering, detecting, and recovering from common ones; learning from the patterns of our mistakes; and effectively maintaining, refining, and teaching what we have learned. *More remarkable than our occasional failures is the fact that these common methods work so well as often as they do.* But with the above capacities, this isn't surprising: we can localize faults, fix them, and gain generalizable knowledge about how, when, and where they are likely to occur. Over time—evolutionary, cultural, and ontogenetic—this knowledge and these reliable practices accumulate.[10] Why not develop a philosophical theory that makes *these* capacities or similar ones the fundamental facts that we idealize from: the generative powers of evolved, developed, socialized, and acculturated human agents?

The most general maxim for those who study functionally organized systems is that we come to understand how things work by studying how, when, and where they break down. This is true for those who seek to test theories or models and, more basically, for those who wish to figure out how they work. We can't see how to test a theory until we *do* understand how it works. In the real world, knowledge of how to use or test a theory does not come packaged with its axioms! We who are surrounded by technical systems often forget or underestimate the learned wizardry embedded in the knowledge and practices of those who work on them—heuristic knowledge of breakdowns; their likely causes; and how to find, fix, and prevent them. Auto mechanics, doctors, engineers, programmers, and other students of mechanisms learn how to debug their preferred systems by localizing and fixing the faults that occur; both in their machines and in their procedures for working on them. *This works for the mind no less than for any of our other tools.* With that understanding, we can analyze, calibrate, and debug both our reasons and our reasoning.

Now we have come to the crux of the issue: *we can't idealize deviations and errors out of existence in our normative theories because they are central to our methodology.* We are *error-prone* and *error tolerant*—errors are unavoidable in the fabric of our lives, so we are well adapted to living with and learning from them. We learn more when things break down than when they work right. *Cognitively speaking, we metabolize mistakes!* Feedback mechanisms provide pervasive and often automatic means of error correction at all levels of biological and cognitive activity. We are particularly tuned to detecting and working with violated expectations, and more generally, with differences.

Errors are often sources of creative elaboration. Model building, experimental design, and software design exhibit this lesson in rich detail. "Bugs," incidental features of programs, models, designs, or things designed for another purpose often stimulate or initiate a useful result or capacity. Mistakes or non-adaptive features may get special attention because they aren't functional, and then inspire new ideas. The creative role of errors and "neutral" traits also emerges from a closer look at how theories and experimental designs are actually applied and implemented.[11] (Published reports of experiments would lead you to believe that everything was planned in advance!) Gould and Vrba's (1982) "exaptations" are features co-opted to serve new functions in new contexts, to provide new sources for evolutionary innovation and selection. They argued the need to replace the older term "preadaptation," which they thought inappropriately suggested a kind of prescience. *Exaptations* are fortuitous—not error exactly, but not intended either. Most organic features have been exaptations—the more entrenched ones many times over, each time midwifing a new kind of adaptive complex, and acquiring new layers of functional significance. Changes that make traits *vestigial* signal a change in design direction, and are often the mark of past exaptations. Histories of design are littered with changes in direction, both in nature and in our record of technological and cultural artifacts.

Our normative models should reflect how we learn, and we learn from our mistakes. Some errors are important, some aren't. Which ones are important may depend on your question. Some are easier to find, and some are very informative—errors with riches to be found. Heuristic principles leave tracks or footprints—signs of their application—in their systematic errors, failures, or biases through which we can recognize them (see Chapter 5 and appendixes A and B). Some tracks are diagnostic for their causes and others relatively non-specific, with alternative possible causes. These failures can be positive tools but they're not always easy to find. We can be quite resistant to seeing them. They may be too local or too global. Some tracks may be very small, or only found under rocks, or on the underside of leaves, or only visible in ultraviolet light; so if you don't know what to look for, and how, and where, you may not find them. That may take a specialist's knowledge of the area—thus our dependence on experts. So it is with the biases of most heuristics.

For a global perspective, there's the Charles Addams cartoon of the early 1960s: Two pith-helmeted types have pulled into a large shallow depression, and parked their jeep in a southwestern desert landscape punctuated with saguaro cactus and occasional mesas. They are now taking a rest break. One says to the other: "I guess you're right. It was just one of those wild reports." Your viewpoint is back a way, and *you* can see that they have just parked in a giant footprint. Hard for them to see, but just what we don't want to miss: so fundamental that it affects everything in sight, but hard to find a place to stand to view it—or to see where it is not. Here a specialist's knowledge may not help, because it may lead us to focus too closely. It is just this kind of mistake we make with our models of rationality. As with the cartoon, it may sometimes be better to stand back from the search.

What we really need to avoid is not errors, but significant ones from which we can't recover. Even significant errors are okay as long as they're easy to find. With enough time, or enough of us working on the problem from enough different perspectives, or all of the above, we assume we will find them—maybe even all of them, though we'll never know that for sure. This last qualification slips out effortlessly, like a mumbled incantation. But let's not get distracted. We've been sitting in the skeptics pew for 350 years, so it is time to move on to more productive activities.

This positive view of error yields a richer and more natural model of our rationality and decision making than those recently used by decision makers, economists, and philosophers (Nozick, 1993, and Williams, 1996, offer other arguments to this conclusion). This new picture is full of context-sensitive heuristic rules of procedure efficiently tuned to generate solutions for specific situations, with learned (and often *labeled*) warnings as to when breakdowns may occur. Less elegant, but it usually works better. More accurate, not only in characterizing the decisions we make, but also in describing and analyzing the processes by which we reach them. And well tuned to the adaptive complexity we find in nature. The recommendations of this new model are different, though often refreshingly familiar from other "real world" contexts. They have been and should continue to be a lot more fruitful, and this model *doesn't* lose its normative force by being descriptively accurate. That's good for a change.

CHAPTER THREE

Toward a Philosophy for Limited Beings

The Stance and Outlook of a Scientifically Informed Philosophy of Science

An adequate philosophy of science should have normative force. It should help us to do science or, more likely, to find and help us avoid sources of error, since scientific methodologies are by nature open-ended. Without being normative it is not a philosophical account. Mere descriptions of scientific practice, no matter how general or sensitive to detail, will not do. Without normative force, studies of methodology, however interesting, would translate as a catalogue of fortuitous and mysterious particular accidents, with no method at all. So the "special sciences" (Fodor, 1974) can't be *too* special, as I'll argue below.

These lessons should be keyed much more closely to actual scientific practice than is common in philosophy. We should be prepared to do field work in the sciences. We need to learn to speak the language, but even more to understand and to value their practices as a native, while still retaining the conceptual and methodological interests of the philosopher. We will sometimes need *knowing-how* (and *knowing-why,* a comparable knowledge of, access to, and sometimes even identification with their internalized values) as well as *knowing-that,* to be able to write about it competently.[1] And we should be activists, reapplying these lessons to the practice of science. Don't be afraid to *do* science, if that seems to be called for to get the necessary perspective!

We shouldn't be afraid to give advice to scientists, but we must be humbler about it. This is the price of privilege. We are not "keepers of the logic of science," revealing foundational conceptual truths prior to any possible science. Our advice should be contextual and sensitive to feedback, not a priori pronouncements offered *ex cathedra*. (No one ever asks for such help, or takes it willingly!) Such an aprioristic view is fundamentally incompatible with the values of science, and self-compromising for anyone committed to adopting coherent and compatible views of scientific and philosophical method. To use the two together, we need a different view.

If we can understand the science from the inside while retaining a philosophical perspective, we can gain a new and important viewpoint on scientific practice. However, this is ultimately just another kind of empirically informed and thus empirically infirmable theoretical perspective. We will be occupationally more self-reflective, but not immune to error. Statisticians are uncommon among mathematicians for their extensive attention to real-world cases and reasoning processes. Like them, we should be *theorists* of reason, and at the same time among the most applied *therapists* of reason.[2] Like them also, we should be as interested in the detection of pattern as in what inferences we draw from it. We should explore the apparently unsystematic kluges and patches as well as paradigmatic cases of good science to find and criticize new handles for the phenomena. In this, statisticians are finally discoverers or inventers of reason—or at least the first taxonomists of new reasoning processes that they find.

This new attention to scientific practice suggests the image of many philosophies for the many sciences. But for all their particularity and context sensitivity—hallmarks of "the special sciences"—and the new *disunity of science* movement, these new lessons *are* widely projectible across the sciences. In our flight from a monolithic and exceptionless logic of science we should not miss the many techniques that are wide but not universal in scope—a "toolbox of science." We advocates of the former unity of science movement have just sought too much projectibility in terms of all-too-simple and context-free unqualified universals. That's not to be had. But there are many sound *ceteris paribus* generalizations cross-cutting traditional disciplines. *Maxims of reductionistic problem-solving* and *maxims of functional inference* apply generally to these kinds of problems, with qualifications appropriate to different areas. Neither set captures all inferences in given disciplines. The historical sciences share many methodological approaches. The problems and methods arising from the conjoint complementary use of individual and populational (or qualitative and quantitative) data yield insights across many disciplines. Other disciplinary special knowledge includes how tools commonly used together interact for that specific subject matter.

We need a new organization of methodological knowledge around families of heuristics that are applied to characteristic kinds of problems across disciplines, and how they interact with one another, with specific natural constraints, and with other practices for managing our subject matters. This yields breadth greater than that of the special sciences, but only via a new taxonomy. Classifying together families of related heuristics should simplify the task of going from footprints to heuristics, and then back to the probable other biases and corrective therapy (complementary heuristics) required. (A beginning list and classification, particularly of reductionistic heuristics, is found in Appendix B.)

Ceteris Paribus, Complexity, and Philosophical Method

The projectible maxims should tell us new things about *philosophical* methods as well. Studies of scientific methodology can be philosophically innovative. Witness the revolutionary systems of great past philosopher-scientists or "natural philosophers": Aristotle, Galileo, Descartes, Newton, Leibniz, Hume, Kant, and Darwin. An important lesson for philosophical methodology—with seeds sown by Darwin in his move from essentialistic to populational thinking—is that in complex systems you should expect essentially all of your interesting generalizations to be of the *ceteris paribus* kind. They will have an indefinitely large but unspecified class of exceptions, with a good set of conditions and heuristics for when to expect them. Complex systems have too many degrees of freedom, too many ways both for accomplishing something and for failing to do so to expect simple, exceptionless generalizations to work. However, unless they *characteristically* worked, they wouldn't be here. (That is, *we* wouldn't!) Characteristic behavior provides a basis for generalization, but for such generalizations, exceptions are the rule. Cartwright (1983, 1989) and Wimsatt (Chapter 9; 1972, 1976b) have both written about this from complementary perspectives.

These *ceteris paribus* clauses are not detachable, as they might be for simpler systems where we only worry about proper system closure—the idealized unreal images of economic theory, for example. They are intrinsic to the generalizations and to the systems they describe, which makes them characteristic of any patterns produced by a selection process and operating in a population: not only will individual members of that population differ, but evolution feeds opportunistically on outcomes, manipulating probabilities by diverse paths so that *multiple realizability and multiple exceptions* are inevitable. We hide this with in principle assumptions that if we knew everything, the (complex) unqualified universals would emerge. Thus, our attitudes toward reductionism are flawed by failing to correct appropriately for the incompleteness of our knowledge and the nature of our heuristics (Chapter 10). We don't yet have adequate philosophical and conceptual tools for dealing with uncertain messy situations. I provide a "starter set" (see chapters 4, 5, 6, 7, and 10), but more is required.

My intended audience is philosophers, to get them to change their practice, and also scientists, in the hope that these ideas might be genuinely helpful in their practice. There *is* room for a theory of practice—a genuinely philosophical theory with lessons in how to do philosophy as well as philosophy of science.[3] These are case studies in and of this practice. "But where's the theory?" one might say after reading this collection. "I see some localized conceptual analysis, a lot of methodological exegesis, a little history of science, some thinly disguised theoretical science, and probably a bit too much preaching. There's some pretty applied philosophy in there, but there's no theory of practice." To understand why this complaint is ultimately misplaced, note that philosophers sometimes use *theory* interchangeably for theory and for *meta*-theory. We distinguish the two clearly in principle, but not always very well in practice. Thus the so-called identity theory of mind was not a specific account of what specific kinds of mental things were identical with what specific kinds of physical things. That would have been a scientific theory. Philosophers may be interested in such theories, but don't normally propose them. Rather, it was a philosophical theory—a meta-theory for scientific identity theories of mind—of what kinds of constraints apply to and what kinds of consequences would follow from any specific scientific theory that made such an identification.

Philosophers tend to presume that one could have true and interesting things to say about this in general, even in advance of actually seeing the theory. I don't believe it! This is another kind of slippery inference from in principle thinking, an assumption of a kind of cognitive omniscience or omnipotence. *General meta-theories are particularly vulnerable to discovery of new ways to reframe the question.* Such attitudes occasion justifiable suspicion among scientists! They shouldn't if such judgements were advanced as the quite abstract and theoretical but still empirical hypotheses that they are, and defended accordingly. They are aids in exploring the space of theories of a given kind, and thus productive for both philosophy and science. But historically philosophers have done no better at these kinds of projections than scientists. Neither of us have done very well, at least by standards encouraged by a search for certainty! The history of scientific progress and the evolution of our conceptual categories is littered with one generation's projects and category mistakes that have become the next generation's impossibilities and conceptual truths. What has happened to the supposed fixity of species? The supposed incoherence of infinitesimals? The "sensible" program for the reduction of mathematics to logic? The "natural" affinities between determinism and predictability given the claims of chaos? We should be more careful and modest about meta-theoretical claims.

I have no complete and systematic meta-theory of practice to offer here, though these chapters point often in the direction of one. There are specific theories of practice for different important practices in different circumstances; there are many more to be studied. As normative theories of reasoning—of problem solving in various kinds of scientific contexts—they are still of philosophical interest. I believe that these theories—offered piecemeal, but articulating well together—are both descriptively and normatively more adequate to the practice of science than anything you will find in your philosophy. Often, there is no competing philosophical theory of them (or a scientific one either) since I urge a conception of philosophy reaching beyond its current boundaries. There is also meta-theory, offered in justification of specific types of practices, but not yet a *systematic* meta-theory. When it does arrive, it should be as full of *ceteris paribus* qualifiers as its subject matter. Meanwhile, I have neither aimed for completeness nor marked all meta-theoretic remarks as such. Adequate meta-theory should flow from such specific theories of practice after we have made them ours—to realize what they can do so we can generalize appropriately. But that is another book for another time and perhaps for another person.

My motivation here is different. I am by choice a conceptual engineer, not a pure theoretician. If what I have to say is not useful to scientists I have not yet done my job right. This additional constraint makes a harder task than philosophers usually set for themselves. I may not have succeeded. I may be wrong. More often, despite my best efforts, what I offer will still be too abstract to be immediately useful for practicing scientists. This is not yet a tried-and-true recipe book of laboratory formulas for conceptualization, model-building, criticism, and experimental design. For concrete paradigms of analogous texts within the sciences, see the illuminating and entertaining but useful laboratory preparations and formulas in the venerable yearly editions of the *Handbook of Chemistry and Physics,* or the classic *Procedures in Experimental Physics* of John Strong (already in its seventeenth edition since 1938 when I bought my copy in 1959). These authors recognized well the need for piecemeal elaboration. To move in that direction is a worthy aim for any philosopher of science. The last section of Robert Nozick's (1993) *The Nature of Rationality* had some of this character for philosophical investigations.

I have worked mostly in the biological sciences, so my examples are closer to home there. But I am also moved by experience and continuing interests in mathematical modeling in engineering and the less foundational of the physical sciences, and in psychology and the social sciences. (I usually describe myself as a "philosopher of the inexact sciences"—see chapters 13 and 10.) Biology is often a sufficiently variegated paradigm to provide lessons for those areas too. The sciences I seek are the "everything in between" of the frontispiece.

Our real world is complex, and we are faced—as biological and social beings—with the need to make an ongoing stream of decisions that serve us well *here.* We surround ourselves with idealizations. We think we understand these idealized laboratory and conceptual worlds of our own construction, and it is all-too-tempting to refer any questions to one of them. But I am deeply impressed with the often successful strategies we have for dealing with complexity in the real worlds we inhabit. To study these strategies, we must study them *there,* not in one of the idealized constructions we have made for other purposes. We certainly will need new idealized constructions to study this cluster of problems, but idealizations made specifically for these purposes! Controlled and simplified laboratory worlds and idealizations are designed by cognitive scientists or decision theorists to reduce the response of an agent to a few—ideally to one—degrees of freedom in order to simplify theory testing. Such designs try to eliminate interaction effects so we rarely see any. They give few clues, often misleading ones, for how we do or should respond in worlds with many degrees of freedom and interacting constraints. Similarly, we get too easily enamoured of the tidy conceptualizations we model and claim logically necessary consequences from (when practicing "possible worlds semantics") as philosophers. For these problems, such model-worlds do not make enough discriminations among relevant kinds to be adequate models of our world.

Our Present and Future Naturalistic Philosophical Methods

Consider an analogy: suppose our current philosophical methods are like the ideal gas law and phenomenological thermodynamics—rooted in deeper theories, phenomena, and mechanisms of which we are currently unaware, and can normally ignore while confidently and reliably producing useful knowledge using these methods. They constitute idealized phenomenological theories for reasoning abstractly to and from first principles—based in logic, to be sure (what could be deeper than that?), but humor me. Suppose they are *sui generis,* and have no deeper foundation in other things. We take them as foundational to our methodology. Inspired by logic and sometimes by obsolete psychology, these idealizations are at best first-order approximations: *exceptionless generalizations; the search (always and exclusively) for logically valid arguments; analyses in terms of necessary and sufficient conditions; knowledge or belief structures that are supposedly free of inconsistency and closed under entailment; crisp inner-outer and subjective/objective distinctions and other structures from a view of natural order as a compositional hierarchy of logical schemata or computational algorithms; crisp fact-value distinctions, and the separate operation of modular and non-interacting cognitive, conative, and affective processes; the perceived gulf between issues of meaning and empirical fact (and the low status accorded the latter in argument!); the special attention paid to claims that hold in principle but not in practice. . . .* This list could go on for a long time. We have used these tools well when they work well, but now we push them too widely. They have become entrenched, masking their status in a naturalistic philosophy as merely tools.

For some parts of the world these tools and techniques can produce compact, simple, and elegant answers. There, world-pegs fit category-holes so closely that we don't notice the small shavings left over. There, philosophical analysis with present tools can and has made real progress. Like phenomenological thermodynamics, these answers are immensely useful, practically irreplaceable, and work well most of the time. But the world is also very complex, and our concepts and language are constantly changing. The orders that philosophy produces in this way are local approximations to temporal strobe-pictures of changing conceptual relations, end-directed cognitions, and social norms—modulated by affect, and differently seen and constituted by networks of embodied and socialized cognitive beings. Too many problem areas are not well illuminated by our philosophical lampposts—at least where they are currently placed.

This sounds problematic. Is this attacking logic? Logic doesn't break down by being approximately right. It is a reflection of the standing of logic in our scheme that we wouldn't think of violations in this way. I'm not attacking logic, but some of the tools we have crafted with it and other time-honored idealizations that seem almost part of it. These tools "break down" by failing to organize areas compactly and with closure when less exact methods do so much more simply.

The alternative methods I propose aren't necessarily empirical because they are "messy." Insisting on an exact, precise, complete, exceptionless description can hide important order that is there. Sometimes, if we are willing to tolerate a few exceptions, context-dependencies, and approximate truths—if we're willing to "defocus" a little—we can get nice, compact, *ceteris paribus* qualified but robust generalizations. That's still worth doing in the vast regions of our conceptual world—most of it—where that's the best we *can* do (Cartwright, 1983). It's also worth doing because it opens our world to a deeper explanation, for constrained fuzzy order is the general case: *on our time and size scale, and for several orders of magnitude in each direction, it is exceptionless crisp precision that is the degenerate special case. Nor are we—detecting and assessing this order—disembodied cognitive engines: we yoke cognition and affect quasi-independently in a noise-tolerant manner in the service of local and more global ends.*[4]

Ceteris paribus clauses are nice. They allow us the regularities and modularities we know are there while reminding us of the exceptions—fluctuations or deviations from a macroscopic order that point the way to a deeper understanding. This cognitive form (general pattern + exceptions) and its relatives (broad similarities + attendant differences, models + qualifications, etc.) are deeply anchored in the structure of our case-based organization of knowledge: "This is like that (which you already know) but with the following differences." It wears the microstructure of cumulative learning on its face. Similar schemata appear again and again in the chapters that follow. This form is intrinsic to multi-level or multi-perspectival theories (Chapter 8), to the use of models (Chapter 6), and to the broader robust generalizations that form the architecture of our real science (Chapter 4).

Brownian motion was the pivotal imperfection in the picture of macroscopic continuous inert fluids that let us see the inner workings of liquids and gases. The new statistical micro-theory didn't eliminate these macroscopic continua, despite what some have said! It explained their properties, and revealed other dimensions that enriched our view of them immensely. The micro-theory in condensed matter physics in which my kitchen table is mostly empty space between electron orbitals did not make it any less solid, real, or impenetrable to my finger. It can be both, once the different qualities of the *detectors* appropriate to these two levels are taken into account (Wimsatt, 1976a). Physical continua behave as continuous to macroscopic detectors in a *scale-dependent* way that changes and breaks down at lower levels. There is a conflict only if we try to see them on the analogy of the mathematical continuum of the real numbers, which is *scale-independent,* showing the properties of continua on all size scales. This point is nicely made for physical fractal properties by Mandelbrot (1982). This bi-level perspective on the table or other macro-level objects enriches our view of the world too. It shows how multiple legitimate (macro- and micro-) views of our objects are deeply imbedded in our physical world—not just a product of our subjectivities. (Brownian motion and a broader realist multi-perspectivalism both reappear in Chapter 10.)

I seek for philosophical methods the naturally rooted analogues to statistical mechanics; the "deeper" principles that yield and explain them in the limits, and explain when, how, and where fluctuations and deviations from those idealized pictures are possible and to be expected. From them, a naturalistic epistemology and a naturalistic methodology, extending to metaphysics and the valuational dimensions of human experience, should be both possible and inevitable. Naturalism need not be fundamentally eliminative or destructive of traditional views and methods. The essentials of philosophical methods can and must survive a thoroughgoing naturalism, but the naturalism includes societies, cultures, and ecosystems, with essentially all of our cognitive and cultural structures and regularities—descriptive, affective, and normative—intact as "phenomenological" laws and objects: reference groups, ideologies, and markets are as real as neurons, genes, and quarks. This is the vision of the aims of science and its objects exploited in Chapter 8. The tools we already have will look and act only a little differently than we thought—and mostly under unusual circumstances—but we will have a richer understanding of them. We will also have to learn how to use some powerful new tools, but for this minor inconvenience we get rich, deep, and robust connections with the world.

Some older tools of lesser importance should achieve a new centrality, and we may need to recognize or create some brand new ones. At least two topics should have a more central importance in this new epistemology: The first would be heuristics (their general nature, their variety, how we calibrate and debug them, the warnings appropriate to their use in various areas, their relations to counterfactuals and *ceteris paribus* clauses, and how more complex heuristics are articulated into methodologies or customized for specific applications). In many ways, heuristics could come to occupy the methodological position for theories of the middle-range of laws for fundamental theory; these theories will no longer be less important because they are less "fundamental."

The second topic would fit a metaphysics course, an epistemology course, or even an ethics course, depending upon how it was developed. We used to talk blithely about what objects to allow, and on what grounds—the topic of ontology—in traditional metaphysics courses. But objects have boundaries—why not study them?[5] We need to understand how we place or locate boundaries, and how they can be generated or changed over time, particularly with the kind of overlapping real-world complexities discussed in chapters 9 and 10. We need a metaphysics seminar on the nature, detection, and consequences—or as I like to say, the "care and feeding"—of *functional localization fallacies*. This term is common in methodological discussions in neurophysiology; close to the surface in the units of selection controversy in evolutionary biology; and crying out for recognition in the methodological individualism/methodological holism dispute in the social sciences, and in discussions of reductionism generally. It is such a general problem that it infects all areas of philosophy. What else is G. E. Moore's "Naturalistic fallacy" (if it is one) but a kind of *mis*-localization claim? (Moore should have been half as careful elsewhere: his "isolation test" for determining "intrinsic goodness" is one of the more glaring functional localization fallacies in the history of philosophy, since it assumes that intrinsic goodness cannot inhere in the relation of an entity to other entities.) To generalize: *when a system can be described at a variety of levels of organization, or from a variety of perspectives, how do you recognize when a property is attributed to it at the wrong level, or from the wrong perspective? And how do inferences go wrong when this happens?* This study is most naturally pursued through examples, which should lead to deeper characterizations of the problem.

I haven't provided all of the necessary tools in this work—or completed drawings of the new edifice. Neither have the eliminativists, and they want you to make do with nothing at your level in the meantime! Or deconstructionists, who in their haste to leave no stone unturned have left nothing more than word games. Other systematic philosophies—neo-Wittgensteinian, neo-Kantian, or others—have done no better. I have more faith in this multifaceted, multi-level world of intersecting appearances and real interactions than in any of theirs. Here are fairly finished drawings of some important rooms, and still incomplete sketches of the whole thing. I hope that this doesn't strain too many current conventions for architectural rendering for you to be able to see the inferential and presumptive modesty, and the power, coherence, and integrative promise of what I've got so far. A rain forest is a rich place after all, still far richer than we know.

PART TWO

Problem-Solving Strategies for Complex Systems

The five chapters in Part II deal with more general themes for real-world problem-solving strategies. Each reflects the particular fallibilist but ameliorative and cumulative perspective I urged in the introduction. They must, for real-world problems are inevitably complex. Chapters 4, 5, and 6 represent old principles or techniques largely ignored by philosophers, although they are of deep methodological importance in the sciences. They are up front because they inform everything else. They can be caricatured as follows: Chapter 4 is on what's real—or artifactual—and the criteria we use to decide that; Chapter 5 is about how we find out what's real or anything else with methods that are error-prone, and the characteristics and nature of those tools; and Chapter 6 is a plea for falsehoods, especially deliberate ones, exploring their centrality in model-building and theory-construction. Chapter 7 introduces the concepts of *generative entrenchment* to capture the unavoidable influence of history and *inertia* in how we deal with new complexities. And Chapter 8 applies the tools of these four chapters to the problem and nature of evidence in this real and complex world.

Chapter 4: Robustness, Reliability, and Overdetermination

Chapter 4 (first published in 1981) considers robustness analysis—*the use of multiple independent means to detect, derive, measure, manipulate, or otherwise to access entities, phenomena, theorems, properties, and other things we wish to study*. These methods figure centrally in defining our criteria for reality, and for telling trustworthy results from flawed or artifactual ones.[1] Objects have many properties that change at their boundaries, so boundaries and objects are paradigmatically robust. But properties and theorems can be robust too. There's more: robustness is the primary criterion for reality and for error detection used by all of us—swineherd, student, and scientist alike. It has rich and far-reaching implications for a kind of "local realism." No analysis of realism in science can afford to ignore it. *Robustness analysis* is central to problem-solving strategies across the sciences, from math and physics through biology to social psychology and linguistics, and from the construction of logical arguments to the evaluation of good fiction. It illuminates methods of proof, theory construction, mathematical modeling, experimental design, measurement, and observation, and is used widely in contexts of discovery, justification, and the under-studied processes of theory articulation and calibration.[2]

Virtually all scientific activities intersect issues of robustness. It must be a central feature of any adequate reliability based theory of knowledge. In different contexts, robustness analysis can be used both to assess the reality of the target "object" and to determine the characteristics and biases of procedures used to access it. It thus seemingly simultaneously bootstraps the reality of the object while correcting for potential idiosyncrasies of perspective. But robustness analysis is not foolproof. Like any heuristic procedures (Chapter 5), these methods sometimes break down and lead us astray. Sources of error—commonly failures of the independence assumptions—are illustrated as well. Deeper methodological implications of robustness are explored in Chapter 10, where it is applied to philosophical method. I also discuss it in two other papers: Wimsatt (1980a) has a detailed case study of the robustness of chaos across different models of ecological systems (this was also the first paper on chaos by a philosopher), and Wimsatt (1991) uses robustness in the comparative analysis of multiple images in scientific visualization.

Chapter 5: Heuristics and the Study of Human Behavior

The nature and biases of heuristic strategies are discussed in Chapter 5 (first published in 1985). Errors are again critical here, since biases of heuristics are systematic, and leave characteristic footprints through which they can often be identified. We can systematically study the biases to discover more about our reasoning processes, delimiting their use accordingly, and devising local patches or general techniques (other heuristics) to correct for them. Heuristic methods are here characterized in terms of four central properties:[3] *fallibility, relative efficiency, the systematic character of their biases, and the sometimes hidden way in which they can transform problems.* Heuristics are the closest to fundamental inferential tools we have, so this chapter is central to every other one in the book. Our richer methods are multiply connected, complementary, and elaborative layered networks of heuristics. Heuristics are the Lego Blocks for our construction kits for methodologies and technologies alike, so their nature and consequent uses demand close study (see appendixes A and B). If Chapter 4 is ontological for its focus on our criteria for what is real, then this chapter is epistemological and methodological for its focus on the tools through which we come to know.

The specific heuristics discussed here are the rich and diverse strategies associated with reductionist problem solving; approaches reflecting the dominant methodology in most high-profile and high prestige science today. So this chapter also anticipates the next section (chapters 9–12), which focus on reductionist methods. Heuristics help to explain why reductionist methodologies are as powerful as they are, but also when, why, and how they can go awry. This work complements discussions of functional localization fallacies in chapters 10 and 12, and more recent studies by Bechtel and Richardson (1993) and McClamrock (1995). These heuristics give useful new handles to characterize reductionism across the disciplines; to understand its strengths and weaknesses; and to predict when, and how, reductionist approaches should fail. Indeed (Chapter 4 and Wimsatt, 1980b), these heuristics also give us *meta*-handles—when we should expect biases in *our own* assessments of the efficacy of reductionist methods, with application to various scientific and philosophical problems (see introductions to chapters 10 and 12, and appendixes A and B).

A new wave of attention to heuristic principles of reason may now accelerate Simon's delayed revolution. Gigerenzer, Todd, and the ABC Research Group's (1999) diverse and rich studies complement this work, as has a growing torrent since. They argue convincingly that a small variety of simple "fast and frugal" principles underlie many of our reasoning and decision-making practices. Their heuristics perform remarkably well in realistic decision environments. In environments with partial information, they often match or outperform and obviate the need for LaPlacean demons. But they do not consider the "divide-and-conquer" or "near-decomposability" strategies crucial to reductionist analyses and complex problem-solving. So this book should also provide a philosophical and methodological complement to theirs.

Is reductionism limited to "high science"? Every time we ignore the context-sensitivity of an attribution (e.g., talking about the fitnesses of organisms or genes, or the personality traits of individuals without considering their environments), we are employing a reductionist heuristic. It may sometimes be appropriate to the context, and at others a dangerous oversimplification to do so.

Chapter 6: False Models as Means to Truer Theories

Models with false assumptions are commonly and deliberately used to leverage better ones.[4] This paradoxical and ingenious strategy is discussed in Chapter 6 (based on a 1987 paper). It is central to theoretical investigations everywhere and commonly misunderstood by philosophers, who focus on confirmation and testing and not on the analysis of how models break down. I claimed above that errors can't be idealized out of our normative models of scientists' behavior because they are intrinsic to our methodology. Here that promise is delivered. Standard assumptions about how models are used in science are criticized. The history of genetics illuminates how *deliberate exploitation of known and chosen (i.e., designed!) false assumptions* is central to the variegated tasks accomplished with these well-crafted cognitive tools. The period between 1911 and the early 1920s is an early high point in the history of mechanism and in the history of modern science. Classical genetics emerged through elaboration of theory and experiment in the linear linkage model of genes on chromosomes by members of the "Drosophila group" in Morgan's lab at Columbia. Diverse models used in different ways lead to a confrontation and triumph of a realist-mechanist methodology of science over competing instrumentalist-operationalist approaches. This chapter complements a fuller treatment of the history of linkage mapping (Wimsatt, 1992), which also proposes a new form of "particularistic mechanism" appropriate to the study of complex systems, and further analyzes the heuristic use of law formulations regarded as false.[5]

The chapter benefits from work on idealizations by Nancy Cartwright (1983) and others, but goes further and in different directions.[6] Deliberate use of falsehoods give rich and varied tools to resolve in more detail how we can localize errors in, modify, and extend models to evaluate and eliminate other errors. If we don't always converge asymptotically to the punctate truth, we can usually paint ourselves out of incorrect corners of our conceptual space, getting pragmatic and "realist" guides from our activities for how to live in our world. This methodology is central to new approaches toward the development and refinement of theory. Jeffrey Schank and I (1993) have developed software to teach science students how we build models and analyze them productively and critically.[7]

Chapter 7: Robustness and Entrenchment: How the Contingent Becomes Necessary

This short, new chapter introduces a new concept, *generative entrenchment* (GE). Consider an adapted structure; for examply, say an organism, an organized methodology, a social institution, or a theory. A generatively entrenched feature of such a structure is one that has many other things depending on it because it has played a role in generating them. This crucially impacts processes of change. In an evolving system, with repetitive generational cycles, changing deeply generative features (like the genetic code) will cause major (commonly lethal) malfunctions somewhere in the large number of downstream consequences. Thus GE'd features tend to be stabler across generations on longer (evolutionary) time scales. GE and its consequent conservatism are degree properties, and will tend to change over time. Deeply GE'd features in scientific theories (evolving across intellectual lineages) tend to greater conservatism: entrenchment can give assumptions and procedural commitments a quasi-analytic or normative status. Parallel processes in the social realm explain how practices become fixed and conventionalized, and how divergent contingencies can become constitutively normative individuating features of different cultures. For organisms, GE explains the parallels between ontogeny and phylogeny, the contingent radiations of phylogenetic descent, and, more generally, why history matters in evolutionary explanations, and how culture can be cumulative (without necessarily being progressive!) Other implications are important but less obvious. GE and the processes it describes are as important to the tasks of this book and the nature of science as *robustness*, or the fundamentally *heuristic* character of our adaptations and our means for getting knowledge. Because it is the subject of another book (currently in process), I just sketch here its importance and interactions with robustness, heuristics, and the extension of the larger naturalistic project outlined in the first three chapters.

Chapter 8: Lewontin's Evidence (That There Isn't Any)

Chapter 8 (first published in 1994) exploits the tools described above to reconceptualize the role of evidence in elaborating and testing theories of complex phenomena. It exemplifies how the philosophical landscape could be changed through their use. The nature of evidence and its role in testing theories are philosophical problems with a long history of inconclusive analyses. I show how one can make a (local and limited) realist's response to vagaries in the use of evidence raised by Richard Lewontin (1991). These problems at first seem to torpedo the notion of evidence as a useful concept—or to give aid and comfort to social relativist and deconstructionist critics of scientific objectivity (see also the last part of Chapter 10). Fallibilist assumptions force us to reassess the functions of evidence taken for granted by virtually all players, and to reassess the methods, aims, and accomplishments of science while preserving more realistic and useful conceptions of objectivity. Resistance to counterevidence becomes not arbitrary but comprehensible when the target is a deeply entrenched assumption. Evidence may play a more definitive role in exploratory elaboration of theory than it does in justification. New moves should force significant reassessments of our ideas on objectivity in any case. Daston's (1992) philosophical and methodological history of our changing conceptions of objectivity since the seventeenth century complements Lloyd's broad survey of philosophers' modern uses (and misuses) of that concept, and the debits imposed by their creation myths about it (Lloyd, 1995). Both breathe fresh air into our ideas of objectivity and evidence in ways resonating with themes urged here.

Robustness, Reliability, and Overdetermination

Philosophy ought to imitate the successful sciences in its methods, so far as to proceed only from tangible premises which can be subjected to careful scrutiny, and to trust rather to the multitude and variety of its arguments than to the conclusiveness of any one. Its reasoning should not form a chain which is no stronger than its weakest link, but a cable whose fibers may be so slender, provided they are sufficiently numerous and intimately connected.

—PEIRCE (1936), P. 141

Our truth is the intersection of independent lies.

—LEVINS (1966), P. 423

The use of multiple means of determination to "triangulate" on the existence and character of a common phenomenon, object, or result has a long history in science but has seldom been a matter of primary focus. As with many things, it is traceable to Aristotle, who valued having multiple explanations of a phenomenon, and it may also be involved in his distinction between special objects of sense and common sensibles. It is implicit though not emphasized in the distinction between primary and secondary qualities from Galileo onward. It is arguably one of several conceptions involved in Whewell's method of the "consilience of inductions" (Laudan, 1971) and is to be found in several places in Peirce.

Indeed, it is to be found widely among the writings of various scientists and philosophers but, remarkably, seems almost invariably to be relegated to footnotes, parenthetical remarks, or suggestive paragraphs that appear without warning and vanish without further issue. While I will point to a number of different applications of multiple determination that have surfaced in the literature, Donald Campbell has done far more than anyone else to make multiple determination a central focus of his work and to draw a variety of methodological, ontological, and epistemological conclusions from its use (see Campbell, 1958, 1966, 1969, 1977; Campbell and Fiske, 1959; Cook and Campbell, 1979). This theme is as important a contribution as his work on evolutionary epistemology; indeed, it must be a major unappreciated component of the latter. Multiple determination, because of its implications for increasing reliability, is a fundamental and universal feature of sophisticated organic design and functional organization and can be expected wherever selection processes are to be found.

Multiple determination—or *robustness,* as I will call it—is not limited in its relevance to evolutionary contexts, however. Because of its multiplicity of uses, it is implicit in a variety of criteria, problem-solving procedures, and cognitive heuristics that have been widely used by scientists in different fields, and is rich in still insufficiently studied methodological and philosophical implications. Some of these I will discuss, some I will only mention, but each contains fruitful directions for future research.

Common Features of Concepts of Robustness

The family of criteria and procedures that I seek to describe in their various uses might be called *robustness analysis.* They all involve the following procedures:

1. To analyze a *variety* of *independent* derivation, identification, or measurement processes.
2. To look for and analyze things that are *invariant* over or *identical* in the conclusions or results of these processes.
3. To determine the *scope* of the processes across which they are invariant and the *conditions* on which their invariance depends.
4. To analyze and explain any relevant *failures of invariance.*

I call things that are invariant under this analysis *robust,* extending the usage of Levins (1966, p. 423), who first introduced me to the term and idea and who, after Campbell, has probably contributed most to its analysis (see Levins, 1966, 1968).

Features of robustness are expressed in very general terms, as they must be to cover the wide variety of different practices and procedures to which they apply. Thus, the different processes in clause 1 and the invariances in clause 2 may refer in different cases to any of the following:

a. Using different sensory modalities to detect the same property or entity (in the latter case by the detection of spatiotemporal boundaries that are relatively invariant across different sensory modalities) (Campbell, 1958, 1966).

b. Using different experimental procedures to verify the same empirical relationships or generate the same phenomenon (Campbell and Fiske, 1959).

c. Using different assumptions, models, or axiomatizations to derive the same result or theorem (Feynman, 1967; Levins, 1966; Glymour, 1980).

d. Using the agreement of different tests, scales, or indices for different traits, as measured by different methods, in ordering a set of entities as a criterion for the "validity" (or reality) of the constructed property (or "construct") in terms of which the orderings of entities agree (Cronbach and Meehl, 1955; Campbell and Fiske, 1959).

e. Discovering invariance of a macro-state description, variable, law, or regularity over different sets of microstate conditions, and also determining the microstate conditions under which these invariances may fail to hold (Levins, 1966, 1968; Wimsatt, 1976a, 1976b, 1980b).

f. Using matches and mismatches between theoretical descriptions of the same phenomenon or system at different levels of organization, together with Leibniz's law (basically, that if two things are identical, no mismatches are allowed), to generate new hypotheses and to modify and refine the theories at one or more of the levels (Wimsatt, 1976a, 1976b, 1979).

g. Using failures of invariance or matching in a through f above to calibrate or recalibrate our measuring apparatus (for a, b, or f) or tests (for d), or to establish conditions (and limitations on them) under which the invariance holds or may be expected to fail, and (for all of the above) to use this information to guide the search for explanations as to why the invariances should hold or fail (Campbell, 1966, 1969; Wimsatt, 1976a, 1976b).

h. Using matches or mismatches in different determinations of the value of theoretical parameters to test and confirm or infirm component hypotheses of a complex theory (Glymour, 1980) and, in a formally analogous manner, to test and localize faults in integrated circuits.

One may ask whether any set of such diverse activities, as would fit all these items (and as exemplified in the expanded discussion below), is usefully combined under the umbrella term *robustness analysis.* I believe that the answer must be yes, for two reasons. First, all the variants and uses of robustness have a common theme in the distinguishing of the real from the illusory; the reliable from the unreliable; the objective from the subjective; the object of focus from artifacts of perspective; and, in general, that which is regarded as ontologically and epistemologically trustworthy and valuable from that which is unreliable, ungeneralizable, worthless, and fleeting. The variations of use of these procedures in different applications introduce different variant tools or consequences that issue from this core theme and are explicable in terms of it. Second, all these procedures require at least partial *independence* of the various processes across which invariance is shown. And each of them is subject to a kind of systematic error leading to a kind of *illusory robustness* when we are led, on less than definitive evidence, to presume independence and our presumption turns out to be incorrect. Thus, a broad class of fallacious inferences in science can be understood and analyzed as a kind of failure of robustness.

Nonetheless, the richness and variety of these procedures require that we go beyond this general categorization to understand robustness. To understand fully the variety of its applications and its central importance to scientific methodology, detailed case studies of robustness analysis are required in each of the areas of science and philosophy where it is used.

Robustness and the Structure of Theories

In the second of his popular lectures on the character of physical law, Feynman (1967) distinguished two approaches to the structure of physical theory: the Greek and the Babylonian. The Greek (or Euclidean) approach is the familiar one in which the fundamental principles of a science are taken as axioms, from which the rest are derived as theorems. There is an established order of importance, of ontological or epistemological priority, from the axioms out to the farthest theorems. The "Greek" theorist achieves postulational economy or simplicity by making only a small number of assumptions and deriving the rest—often reducing the assumptions, in the name of simplicity or elegance, to the minimal set necessary to derive the given theorems. The "Babylonian," in contrast, works with an approach that is much less well ordered and sees a theoretical structure that is much more richly connected:

> So the first thing we have to accept is that even in mathematics you can start in different places. If all these various theorems are interconnected by reasoning there is no real way to say, "These are the most fundamental axioms," because if you were told something different instead you could also run the reasoning the other way. It is like a bridge with lots of members, and it is over-connected; if pieces have dropped out you can reconnect it another way. The mathematical tradition of today is to start with some particular ideas that are chosen by some kind of convention to be axioms, and then to build up the structure from there. What I have called the Babylonian idea is to say, "I happen to know this, and I happen to know that, and maybe I know that; and I work everything out from there. Tomorrow I may forget that this is true, but remember that something else is true, so I can reconstruct it all again. I am never quite sure of where I am supposed to begin or where I am supposed to end. I just remember enough all the time so that as the memory fades and some of the pieces fall out I can put the thing back together again every day. (Feynman, 1967, pp. 46–47)

This rich connectivity has several consequences for the theoretical structure and its components. First, as Feynman (1967) observes, most of the fundamental laws turn out to be characterizable and derivable in a variety of different ways from a variety of different assumptions:

> One of the amazing characteristics of nature is the variety of interpretational schemes that is possible. It turns out that it is only possible because the laws are just so, special and delicate. . . . If you modify the laws much you find that you can only write them in fewer ways. I always find that mysterious, and I do not understand the reason why it is that the correct laws of physics seem to be expressible in such a tremendous variety of ways. They seem to be able to get through several wickets at the same time. (pp. 54–55)

Although Feynman nowhere explicitly says so, his own choice of examples and other considerations (which will emerge later) suggest another ordering principle for fundamentality among laws of nature: *The more fundamental laws will be those that are independently derivable in a larger number of ways.* I will return to this suggestion later.

Feynman also observes that this multiple derivability of physical laws has its advantages, for it makes the overall structure much less prone to collapse:

> At present we believe that the laws of physics have to have the local character and also the minimum principle, but we do not really know. If you

> have a structure that is only partly accurate, and something is going to fail, then if you write it with just the right axioms maybe only one axiom fails and the rest remain, you need only change one little thing. But if you write it with another set of axioms they may all collapse, because they all lean on that one thing that fails. We cannot tell ahead of time, without some intuition, which is the best way to write it so that we can find out the new situation. We must always keep all the alternative ways of looking at a thing in our heads; so physicists do Babylonian mathematics, and pay but little attention to the precise reasoning from fixed axioms. (p. 54)

This multiple derivability not only makes the overall structure more reliable but also has an effect on its individual components. Those components of the structure that are most insulated from change (and thus the most probable foci for continuity through scientific revolutions) are those laws that are most robust and, on the above criterion, most fundamental. This criterion of fundamentality would thus make it natural (though by no means inevitable) that the most fundamental laws would be the least likely to change. *Given that different degrees of robustness ought to confer different degrees of stability, robustness ought to be a promising tool for analyzing scientific change.* Alternatively, the analysis of different degrees of change in different parts of a scientific theory may afford a way of detecting or measuring robustness.

I wish to elaborate and illustrate the force of Feynman's remarks arguing for the Babylonian rather than the Greek or Euclidean approach by some simple considerations suggested by the statistical theory of reliability. (For an excellent review of work in reliability theory, see Barlow and Proschan, 1975. No one has, to my knowledge, applied it in this context.)

A major rationale for the traditional axiomatic view of science is to see it as an attempt to make the structure of scientific theory as reliable as possible by starting with, as axioms, the minimal number of assumptions that are as certain as possible and operating on them with rules that are as certain as possible (deductive rules that are truth preserving). In the attempt to secure high reliability, the focus is on total elimination of error, not on recognizing that it will occur and on controlling its effects: it is a structure in which, if no errors are introduced in the assumptions and if no errors are made in choosing or in applying the rules, no errors will occur. No effort is spared in the attempt to prevent these kinds of errors from occurring, but it does not follow that this is the best structure for dealing with errors (for example, by minimizing their effects or making them easier to find) if they do occur. In fact, it is not. To see how well it handles errors that do occur, let us try to model the effect of applying the Greek or Euclidian strategy to a real (error-prone) theory constructed and manipulated by real (fallible) operators.

For simplicity, assume that any operation, be it choosing an assumption or applying a rule, has a small but finite probability of error, p_0. (In this discussion, I will assume that the probability of error is constant across different components and operations. Qualitatively similar results obtain when it is not.) Consider now the deductive derivation of a theorem requiring m operations. If the probabilities of failure in these operations are independent, then the probability of a successful derivation is just the product of the probabilities of success, $1-p_0$, at each operation. Where p_s stands for the probability of failing at this complex task (p_s because this is a serial task), then we have for the probability of success, $1-p_s$:

$$(1-p_s)=(1-p_0)^m$$

No matter how small p_0 is, as long as it is finite, longer serial deductions (with larger values of m) have monotonically decreasing probabilities of successful completion, approaching zero in the limit. *Fallible thinkers should avoid long serial chains of reasoning.* Indeed, we see here that the common metaphor for deductive reasoning as a chain is a poor one for evaluating probability of failure in reasoning. Chains always fail at their weakest links, chains of reasoning only most probably so.

When a chain fails, the release in tension protects other parts of the chain. As a result, failures in such a chain are not independent, since the occurrence of one failure prevents other failures. In this model, however, we are assuming that failures are independent of each other, and we are talking about probability of failure rather than actual failure. These differences result in a serious disanalogy with the metaphor of the argument as a chain. A chain is only as strong as the weakest link, but it is that strong. One often hears this metaphor as a rule given for evaluating the reliability of arguments (see, for example, the quote from C. S. Peirce that begins this chapter), but a chain in which failure could occur at any point is always weaker than (in that it has a higher probability of failure than) its weakest link, except if the probability of failure everywhere else goes to zero. This happens when the weakest link in a chain breaks, but not when one link in an argument fails.

Is there any corrective medicine for this cumulative effect on the probability of error, in which small probabilities of error in even very reliable components cumulatively add up to almost inevitable failure? Happily there is. *With independent alternative ways of deriving a result, the result is always surer than its weakest derivation.* (Indeed, it is always surer than its *strongest* derivation.) This mode of organization—with independent alternative modes of operation and success if any one works—is parallel organization, with its probability of failure, p_p. Since failure can occur if and only if each of the m independent alternatives fails (assume with identical probabilities p_0), then:

$$p_p = p_0^{\ m}$$

But p_0 is presumably always less than 1; thus, for $m > 1$, p_p is always less than p_0. Adding alternatives (or redundancy, as it is often called) always increases reliability, as von Neumann (1956) argued in his classic paper on building reliable automata with unreliable components. Increasing reliability through parallel organization is a fundamental principle of organic design and reliability engineering generally. It works for theories as well as it does for polyploidy, primary metabolism, predator avoidance, microprocessor architecture, Apollo moon shots, test construction, and the structure of juries.

Suppose we start, then, with a Babylonian (or Byzantine?) structure—a multiply connected, poorly ordered scientific theory having no principles singled out as axioms, containing many different ways of getting to a given conclusion and, because of its high degree of redundancy, relatively short paths to it (see Feynman, 1967, p. 47)—and let it be redesigned by a Euclidean. In the name of elegance, the Euclidean will look for a small number of relatively powerful assumptions from which the rest may be derived. In so doing, he will eliminate redundant assumptions. The net effects will be twofold: (1) With a smaller number of assumptions taken as axioms, the mean number of steps in a derivation will increase, and can do so exponentially. This increased length of seriation will decrease reliability along any path to the conclusion. (2) Alternative or parallel ways of getting to a given conclusion will be substantially decreased as redundant assumptions are removed, and this decrease in "parallation" will also decrease the total reliability of the conclusion.

Each of these changes increases the unreliability of the structure, and both of them operating together produce a cumulative effect—if errors are possible, as I have supposed. Not only is the probability of failure of the structure greater after it has been Euclideanized, but the consequences of failure become more severe: with less redundancy, the failure of any given component assumption is likely to *infirm* a larger part of the structure. I will elaborate on this point shortly. It has not been studied before now (but see Glymour, 1980) because of the dominance of the Cartesian Euclidean perspective and because of a key artifact of first-order logic.

Formal models of theoretical structures characteristically start with the assumption that the structures contain no inconsistencies. As a normative ideal, this is fine; but as a description of real scientific theories, it is inadequate. Most or all scientific theories with which I am familiar contain paradoxes and inconsistencies, either between theoretical assumptions or between assumptions and data in some combination. (Usually these could be resolved if one knew which of several eminently plausible assumptions to give up, but each appears to have strong support; so the assumptions—and the inconsistencies—remain.) This feature of scientific theories has not until now (with the development of non-monotonic logic) been modeled, because of the fear of total collapse. In first-order logic, anything whatsoever follows from a contradiction; so systems that contain contradictions are regarded as useless.

But the total collapse suggested by first-order logic (or by highly Euclidean structures with little redundancy) seems not to be characteristic of scientific theories. Inconsistencies are walled off or encapsulated and do not appear to affect the theory other than very locally—for things very close to and strongly dependent on one of the conflicting assumptions. Robustness provides a possible explanation, perhaps the best explanation, for this phenomenon.

When an inconsistency occurs, results that depend on one or more of the contradictory assumptions are compromised. This infection is transitive; it passes to things that depend on these results, and to their logical descendants, like a string of dominoes—until we reach something that has independent support. The independent support of an assumption sustains it, and the collapse propagates no further. If all deductive or inferential paths leading from a contradiction pass through robust results, the collapse is bounded within them, and the inconsistencies are walled off from the rest of the network. For each robust result, one of its modes of support is destroyed; but it has others, and therefore the collapse goes no further. Whether this is the only mechanism by which this isolation of contradictions could be accomplished, I do not know, but it is a possible way, and scientific constructs do appear to have the requisite robustness. (I am not aware if anyone has tried to formalize or to simulate this, though Stuart Kauffman's work on "forcing structures" in binary, Boolean switching networks seems clearly relevant. In Kauffman, 1971, these models are developed and applied to gene control networks.)

Robustness, Testability, and the Nature of Theoretical Terms

Another area in which robustness is involved is Clark Glymour's account of testing and evidential relations in theories. Glymour argues systematically that parts of a theoretical structure can be and are used to test other parts of the theory, and even themselves. (His name for this is "bootstrapping.") This testing requires the determination of values for quantities of the theory in more than one way:

> If the data are consistent with the theory, then these different computations must agree [within a tolerable experimental error] in the value they determine for the computed quantity; but if the data are inconsistent with the theory, then different computations of the same quantity may give different results. Further and more important, what quantities in a theory may be computed from a given set of initial data depends both on the initial data and on the structure of the theory. (Glymour, 1980, p. 113)

Glymour argues later (pp. 139–140) that the different salience of evidence to different hypotheses of the theory requires the use of a variety of types of evidence to test the different component hypotheses of the theory. Commenting on the possibility that one could fail to locate the hypothesis whose incorrectness is producing an erroneous determination of a quantity or, worse, mislocating the cause of the error, he claims:

> The only means available for guarding against such errors is to have a variety of evidence so that as many hypotheses as possible are tested in as many different ways as possible. What makes one way of testing relevantly different from another is that the hypotheses used in one computation are different from the hypotheses used in the other computation. Part of what makes one piece of evidence relevantly different from another piece of evidence is that some test is possible from the first that is not possible from the second, or that, in the two cases, there is some difference in the precision of computed values of theoretical quantities. (p. 140)

A given set of data and the structure of the theory permit a test of a hypothesis (or the conjunction of a group of hypotheses) if and only if they permit determination of all the values in the tested entity in such a way that contradictory determinations of at least one of these values could result (in the sense that it is not analytically ruled out). This requires more than one way of getting at that value (see Glymour, 1980, p. 307). To put it in the language of this book, *only robust hypotheses are testable.* Furthermore, a theory in which most components are multiply connected is a theory whose faults are relatively precisely localizable. Not only do errors not propagate far, but we can find their source quickly and evaluate the damage and what is required for an adequate replacement. If this sounds like a design policy for an automobile, followed to facilitate easy diagnostic service and repair, I can say only that there is no reason why our scientific theories should be less well designed than our other artifacts.

The same issues arise in a different way in Campbell's discussions (Campbell and Fiske, 1959; Campbell, 1969a, 1969b, 1977; Cook and Campbell, 1979) of single or definitional versus multiple operationalism. *Definitional operationalism* is the view that philosophers know as operationalism, that the meaning of theoretical terms is to be defined in terms of the experimental operations used in measuring that theoretical quantity. Multiple means of determining such a quantity represent a paradox for this view—an impossibility, since the means is definitive of the quantity, and multiple means means multiple quantities. *Campbell's multiple-operationalism is not operationalism at all in this sense but a more tolerant and eclectic empiricism,* for he sees the multiple operations as contingently associated with the thing measured. Being contingently associated, they cannot have a definitional relation to it; consequently, there is no barrier to accepting that one (robust) quantity has a number of different operations to get at it, each too imperfect to have a definitional role but together triangulating to give a more accurate and complete picture than would be possible from any one of them alone.

Campbell's attack on definitional operationalism springs naturally from his fallibilism and his critical realism. Both of these forbid a simple definitional connection between theoretical constructs and measurement operations:

> One of the great weaknesses in definitional operationalism as a description of best scientific practice was that it allowed no formal way of expressing the scientist's prepotent awareness of the imperfection of his measuring instruments and his prototypic activity of improving them. (Campbell, 1969a, p. 15)

For a realist, the connection between any measurement and the thing measured involves an often long and indirect causal chain, each link of which is affected and tuned by other theoretical parameters. The aim is to make the result insensitive to or to control these causally relevant but semantically irrelevant intermediate links:

> What the scientist does in practice is to design the instrument so as to minimize and compensate for the stronger of these irrelevant forces. Thus, the galvanometer needle is as light as possible, to minimize inertia. It is set on jeweled bearings to minimize friction. It may be used in a lead-shielded and degaussed room. Remote influences are neglected because they dissipate at the rate of l/d^2, and the weak and strong nuclear forces dissipate even more rapidly. But these are practical minimizations, recognizable on theoretical grounds as incomplete. (pp. 14–15)

The very same indirectness and fallibility of measurement that rule out definitional links make it advantageous to use multiple links:

> We have only *other invalid measures* against which to validate our tests; we have no "criterion" to check them against . . . A theory of the interaction of two theoretical parameters must be tested by imperfect exemplifications of each . . . In this predicament, great inferential strength is added when each theoretical parameter is exemplified in 2 or more ways, each mode being as independent as possible of the other, as far as the theoretically irrelevant components are concerned. This general program can be designated *multiple operationalism.* (p. 15)

Against all this, then, suppose one did have only one means of access to a given quantity. Without another means of access, even if this means of access were not made definitional, statements about the value of that variable would not be independently testable. Effectively, they would be as if defined by that means of access, and since the variable was not connected to the theory in any other way, it would be an unobservable, a fifth wheel: anything it could do would be done more directly by its operational variable. It is, then, in Margenau's apt phrase, a peninsular concept (1950, p. 87), a bridge that leads to nowhere.

Philosophers often misleadingly lump this "peninsularity" and the existence of extra axioms permitting multiple derivations together as redundancy. The implication is that one should be equally disapproving of both. Presumably, the focus on error-free systems leads philosophers to regard partially identical paths (the paths from a peninsular concept and from its "operational variable" to any consequence accessible from either) and alternative independent paths (robustness, bootstrapping, or triangulation) as equivalent—because they are seen as equally dispensable if one is dealing with a system in which errors are impossible. But if errors are possible, the latter kind of redundancy can increase the reliability of the conclusion; the former cannot.

A similar interest in concepts with multiple connections and a disdain for the trivially analytic, singly or poorly connected concept is to be found in Hilary Putnam's (1962) classic paper "The Analytic and the Synthetic." Because theoretical definitions are multiply connected law-cluster concepts, whose meaning is determined by this multiplicity of connections, Putnam rejects the view that such definitions are stipulative or analytic. Though for Putnam it is theoretical connections, rather than operational ones, which are important, he also emphasizes the importance of a multiplicity of them:

> Law-cluster concepts are constituted not by a bundle of properties as are the typical general names [cluster concepts] like "man" and "crow," but by a cluster of laws that, as it were, determine the identity of the concept. The concept "energy" is an excellent sample . . . It enters into a great many laws. It plays a great many roles, and these laws and inference roles constitute its meaning collectively, not individually. I want to suggest that most of the terms in highly developed sciences are law-cluster concepts, and that one should always be suspicious of the claim that a principle whose subject term is a law-cluster concept is analytic. The reason that it is difficult to have an analytic relationship between law-cluster concepts is that . . . any one law can be abandoned without destroying the identity of the law-cluster concept involved. (p. 379)

Statements that are analytic are so for Putnam because they are singly connected, not multiply connected, and thus trivial:

> Thus, it cannot "hurt" if we decide always to preserve the law "All bachelors are unmarried" . . . because bachelors are a kind of synthetic class. They are not a "natural kind" in Mill's sense. They are rather grouped together by ignoring all aspects except a single legal one. One is simply not going to find any . . . [other] laws about such a class. (p. 384)

Thus, the robustness of a concept or law—its multiple connectedness within a theoretical structure and (through experimental procedures) to observational results—has implications for a variety of issues connected with theory testing and change, with the reliability and stability of laws and the component parts of a theory, with the discovery and localization of error when they fail, the analytic-synthetic distinction, and accounts of the meaning of theoretical concepts. But these issues have focused on robustness in existing theoretical structures. It is also important in discovery and in the generation of new theoretical structures.

Robustness, Redundancy, and Discovery

For the complex systems encountered in evolutionary biology and the social sciences, it is often unclear what is fundamental or trustworthy. One is faced with a wealth of partially conflicting, partially complementary models, regularities, constructs, and data sets with no clear set of priorities for which to trust and where to start. In this case particularly, processes of validation often shade into processes of discovery—since both involve a winnowing of the generalizable and the reliable from the special and artifactual. Here too *robustness* can be of use, as Richard Levins suggests in the passage that introduced me to the term:

> Even the most flexible models have artificial assumptions. There is always room for doubt as to whether a result depends on the essentials of a model or on the details of the simplifying assumptions. This problem does not arise in the more familiar models, such as the geographical map, where we all know that contiguity on the map implies contiguity in reality, relative distances on the map correspond to relative distances in reality, but color is arbitrary and a microscopic view of the map would only show the fibers of the paper on which it is printed. But in the mathematical models of population biology, it is not always obvious when we are using too high a magnification.
>
> Therefore, we attempt to treat the same problem with several alternative models, each with different simplifications, but with a common biological assumption. Then, if these models, despite their different assumptions, lead to similar results we have what we can call a robust theorem that is relatively free of the details of the model. Hence, our truth is the intersection of independent lies. (Levins, 1966, p. 423)

Levins is making heuristic use of the philosopher's criterion of logical truth as true in all possible worlds. He views robustness analysis as "sampling from a space of possible models" (Levins, 1968, p. 7). Since one cannot be sure that the sampled models are representative of the space, one gets no guarantee of logical truth but, rather, a heuristic (fallible but effective) tool for discovering empirical truths that are relatively free of the details of the various specific models.

Levins talks about the robustness of theorems or phenomena or consequences of the models rather than about the robustness of the models themselves. This is necessary, given his view that any single model makes a number of artifactual (and therefore non-robust) assumptions.

A theory would presumably be a conceptual structure in which many or most of the fundamental theorems or axioms are relatively robust, as is suggested by Levins' statement, "A theory is a cluster of models, together with their robust consequences" (p. 7).

If a result is robust over a range of parameter values in a given model or over a variety of models making different assumptions, this gives us some independence of knowledge of the exact structure and parameter values of the system under study: a prediction of this result will remain true under a variety of such conditions and parameter values. This is particularly important in scientific areas where it may be difficult to determine the parameter values and conditions exactly.

Robust theorems can thus provide a more trustworthy basis for generalization of the model or theory and, through their independence of many exact details, *a sounder basis for predictions from it.* Theory generalization is an important component of scientific change, and thus of scientific discovery.

Just as robustness is a guide for discovering trustworthy results and generalizations of theory, and distinguishing them from artifacts of particular models, it helps us to distinguish signal from noise in perception generally. Campbell has furnished us with many examples of the role of robustness and pattern matching in visual perception and its analogues, sonar and radar. In an early paper, he described how the pattern and the redundancy in a randomly pulsed radar signal bounced off Venus gave a new and more accurate measurement of the distance to that planet (Campbell, 1966).

The later visual satellite pictures of Mars and its satellite Deimos have provided an even more illuminating example, again described by Campbell (1977) in the unpublished William James Lectures (lecture 4, pp. 89–90). The now standard procedures of image enhancement involve combining a number of images, in which the noise, being random, averages out; but the signal, weak though usually present, adds in intensity until it stands out. The implicit principle is the same one represented explicitly in von Neumann's (1956) use of "majority organs" to filter out error: the combination of parallel or redundant signals with a threshold, in which it is assumed that the signal, being multiply represented, will usually exceed threshold and be counted; and the noise, being random, usually will fall below threshold and be lost. There is an art to designing the redundancy so as to pick up the signal and to setting the threshold so as to lose the noise. It helps, of course, if one knows what he is looking for. In the case of a television camera focused on Mars, Deimos was a moving target and—never being twice in the same place to add appropriately (as were the static features of Mars)—was consequently filtered out as noise. But because the scientists involved knew that Deimos was there, they were able to fix the image enhancement program to find it. By changing the threshold so that Deimos and some noise enters as a signal (probably smeared)—changing the sampling rate or the integration area (stopping Deimos at the effectively same place for two or more times) or introducing the right kind of spatiotemporal correlation function (to track Deimos's periodic moves around Mars)—it was possible to restore Deimos to the pictures again. Different tunings of the noise filters and different redundancies in the signal were exploited to bring static Mars and moving Deimos into clear focus.

We can see exactly analogous phenomena in vision if we look at a moving fan or airplane propeller. We can look through it (filtering it out as noise) to see something behind it. Lowering our threshold, we can attend to the propeller disk as a colored transparent (smeared) object. Cross-specific variation in flicker-fusion frequency indicates different sampling rates, which are keyed to the adaptive requirements of the organism (see Wimsatt, 1980a, pp. 292–297). The various phenomena associated with periodic stroboscopic illumination (apparent freezing and slow rotation of a rapidly spinning object) involve detection of a lagged correlation. Here, too, different tunings pick out different aspects of or entities in the environment. This involves a use of different heuristics, a matter I return to later.

I quoted Glymour (1980) earlier on the importance of getting the same answer for the value of quantities computed in two different ways. What if these computations or determinations do not agree? The result is not always disastrous; indeed, when such mismatches happen in a sufficiently structured situation, they can be very productive.

This situation could show that we were wrong in assuming that we were detecting or determining the same quantity; however (as Campbell, 1966, was the first to point out), if we assume that we *are* determining the same quantity but "through a glass darkly," the mismatch can provide an almost magical opportunity for discovery. Given imperfect observations of a *thing-we-know-not-what,* using experimental apparatus with *biases-we-may-not-understand,* we can achieve both a better understanding of the object (it must be, after all, that one thing whose properties can produce these divergent results in these detectors) and of the experimental apparatus (which are, after all, these pieces that can be affected thus divergently by this one thing).

The constraint producing the information here is the identification of the object of the two or more detectors. If two putatively identical things are indeed identical, then any property of one must be a property of the other. We must resolve any apparent differences either by giving up the identification or locating the differences not in the thing itself but in the interactions of the thing with different measuring instruments. And this is where we learn about the measuring instruments. Having then acquired a better knowledge of the biases of the measuring instruments, we are in a better position not only to explain the differences but also, in the light of them, to give a newly refined estimate of the property of the thing itself. This procedure, a kind of "means-end" analysis (Wimsatt, 1976a; Simon, 1996), has enough structure to work in any given case only because of the enormous amount of background knowledge of the thing and the instruments that we bring to the situation. What we can learn (in terms of localizing the source of the differences) is in direct proportion to what we already know.

This general strategy for using identifications has an important subcase in reductive explanation. I have argued extensively (Wimsatt, 1976a, part II; 1976b, 1979) that the main reason for the productiveness of reductive explanation is that inter-level identifications immediately provide a wealth of new hypotheses: each property of the entity as known at the lower level must be a property of it as known at the upper level, and conversely; usually very few of these properties from the other level have been predicated of the common object. The implications of these predictions usually have fertile consequences at both levels, and even where the match is not exact, there is often enough structure in the situation to point to a revised identification, with the needed refinements. This description characterizes well the history of genetics, both in the period of the localization of the genes on chromosomes (1883 to 1920) and in the final identification of DNA as the genetic material (1927 to 1953). (For the earlier period see, for example, Allen, 1978; Moore, 1972; Darden, 1974; Wimsatt, 1992, and Chapter 6. For the later period see Olby, 1974.) Indeed, the overall effect of these considerations is to suggest that *the use of identities for the detection of error in a structured situation may be one of the most powerful heuristics known and certainly one of the most effective in generating scientific hypotheses.*

Also significant in the connection between robustness and discovery is Campbell's (1977) suggestion that things with greater *entitativity* (things whose boundaries are more robust) ought to be learned earlier. He cites suggestive support from language development for this thesis, which Quine's (1960) views also tend to support. I suspect that robustness could prove to be an important tool in analyzing not only what is discovered but also the order in which things are discovered.

There is some evidence from work with children (Stein and Glenn, 1979; Trabasso and Stein, 1997) that components of narratives that are central to the narrative because they are integrated into its causal and its purposive or intentional structure are most likely to be remembered and abstracted out to include in summaries of the story. This observation is suggestively related both to Feynman's (1967, p. 47) remark quoted above, relating robustness to forgetting relationships in a multiply connected theory, and to Simon's (1996) concept of a blackboard work space, which is maintained between successive attempts to solve a problem and in which the structure of the problem representation and goal tree may be subtly changed through differential forgetting. These suggest other ways in which robustness could affect discovery processes through differential effects on learning and forgetting.

Robustness, Objectification, and Realism

Robustness is widely used as a criterion for the reality or trustworthiness of the thing that is said to be robust. The boundaries of an ordinary object, such as a table, as detected in different sensory modalities (visually, tactually, aurally, orally), roughly coincide, making them robust; this is ultimately the primary reason why we regard perception of the object as veridical rather than illusory (see Campbell, 1958, 1966). It is a rare illusion indeed that could systematically affect all of our senses in this consistent manner. (Drug induced hallucinations and dreams may involve multi-modal experience but fail to be consistent through time for a given subject, or across observers, thus failing at a higher level to show appropriate robustness.)

Our concept of an object is of something that exemplifies a multiplicity of properties within its boundaries, many of which change as we move across its boundary. The idea of a one-dimensional object is a contradiction in terms and usually turns out to be a disguised definition—a legal or theoretical fiction. In appealing to the robustness of boundaries as a criterion for objecthood, we are appealing to this multiplicity of properties (different properties detected in different ways) and thus to a time-honored philosophical notion of objecthood.

Campbell (1958) has proposed the use of the coincidence of boundaries under different means of detection as a methodological criterion for recognizing entities such as social groups. For example, in a study of factors affecting the reproductive cycles of women in college dormitories, McClintock (1971, and personal communication) found that the initially randomly timed and different-length cycles of 135 women became, after several months, synchronized into 17 groups, each oscillating synchronously in phase and with a common period. The members of these groups turned out to be those who spent the most time together, as determined by sociological methods. After the onset of synchrony, group membership of an individual could be determined either from information about her reproductive cycle or from a sociogram representing her frequency of social interaction with other individuals. These groups are thus multiply detectable. This illustrates the point that there is nothing sacred about using perceptual criteria in individuating entities. The products of any scientific detection procedure, including procedures drawn from different sciences, can do as well, as Campbell suggests:

> In the diagnosis of middle-sized physical entities, the boundaries of the entity are multiply confirmed, with many if not all of the diagnostic procedures confirming each other. For the more "real" entities, the number of possible ways of confirming the boundaries is probably unlimited, and the more our knowledge expands, the more diagnostic means we have available. "Illusions" occur when confirmation is attempted and found lacking, when boundaries diagnosed by one means fail to show up by other expected checks. (1958, pp. 23–24)

Illusions can arise in connection with robustness in a variety of ways. Campbell's remark points to one important way: Where expectations are derived from one boundary, or even more, the coincidence of several boundaries leads us to predict, assume, or expect that other relevant individuating boundaries will coincide. Perhaps most common, given the reductionism common today, are situations in which the relevant system boundary is far more inclusive than one is led to expect from the coincidence of a number of boundaries individuating an object at a lower level. Such functional localization fallacies are found in neurophysiology, in genetics, in evolutionary biology (with the hegemony of the selfish gene at the expense of the individual or the group; see Wimsatt, 1980b), in psychology, and with methodological individualism in the social sciences. In all these cases the primary object of analysis—be it a gene, a neuron, a neural tract, or an individual—may well be robust, but its high degree of entitativity leads us to hang too many boundaries and explanations on it. Where this focal entity is at a lower level, reductionism and robustness conspire to lead us to regard the higher-level systems as epiphenomenal. Another kind of illusion—the illusion that an entity is robust—can occur when the various means of detection supposed to be independent are not. (This is discussed further in the final section of this chapter.) Another kind of illusion or paradox arises particularly for functionally organized systems. This illusion occurs when a system has robust boundaries, but the different criteria used to decompose it into parts produce radically different boundaries. When the parts have little entitativity compared to the system, the holist's war cry (that the whole is more than the sum of the parts) will have a greater appeal. In Chapter 8 I explore this kind of case and its consequences for the temptation of antireductionism, holism, or, in extreme cases, vitalisms or ontological dualisms.

Robustness is a criterion for the reality of entities, but it also has played and can play an important role in the analysis of properties. Interestingly, the distinction between primary and secondary qualities, which had a central role in the philosophy of Galileo, Descartes, and Locke, can be made in terms of robustness. Primary qualities—such as shape, figure, and size—are detectable in more than one sensory modality. Secondary qualities—such as color, taste, and sound—are detectable through only one sense. I think it is no accident that seventeenth-century philosophers chose to regard primary qualities as the only things that were "out there"—in objects, their cross-modal detectability seemed to rule out their being products of sensory interaction with the world. By contrast, the limitation of the secondary qualities to a single sensory modality seemed naturally to suggest that they were "in us," or subjective. Whatever the merits of the further seventeenth-century view that the secondary qualities were to be explained in terms of the interaction of a perceiver with a world of objects with primary qualities, this explanation represents an instance of an explanatory principle that is widely found in science (though seldom if ever explicitly recognized): *the explanation of that which is not robust in terms of that which is robust* (for other examples see Wimsatt, 1976a, pp. 243–249; Feynman, 1967).

Paralleling the way in which Levins' use of robustness differs from Feynman's, *robustness, or the lack of it, has also been used in contexts where we are unsure about the status of purported properties, to argue for their veridicality or artifactuality,* and thus to discover the properties in terms of which we should construct our theories. This is the proposal of the now classic and widely used methodological paper of

Campbell and Fiske (1959). Their convergent validity is a form of robustness, and their criterion of discriminant validity can be regarded as an attempt to guarantee that the invariance across test methods and traits is not due to their insensitivity to the variables under study. Thus, method bias, a common cause of failures of discriminant validity, is a kind of failure of the requirement for robustness that the different means of detection used are actually independent, in this case because the method they share is the origin of the correlations among traits.

Campbell and Fiske point out that very few theoretical constructs (proposed theoretical properties or entities) in the social sciences have significant degrees of convergent and discriminant validity, and they argue that this is a major difference between the social and natural or biological sciences—a difference that generates many of the problems of the social sciences. (For a series of papers which in effect claim that personality variables are highly context dependent and thus have very little or no robustness, see Shweder, 1979a, 1979b, 1980a.)

While the natural and biological sciences have many problems where similar complaints could be made (the importance of interaction effects and context dependence is a key indicator of such problems), scientists in these areas have been fortunate in having at least a large number of cases where the systems, objects, and properties they study can be effectively isolated and localized, so that interactions and contexts can be ignored.

Robustness and Levels of Organization

Because of their multiplicity of connections and applicable descriptions, robust properties or entities tend to be *(1) more easily detectable, (2) less subject to illusion or artifact, (3) more explanatorily fruitful, and (4) predictively richer than nonrobust properties or entities.* With this set of properties it should be small wonder that we use robustness as a criterion for reality. It should also not be surprising that—since we view perception (as evolutionary epistemologists do) as an efficient tool for gathering information about the world—robustness should figure centrally in our analysis of perceptual hypotheses and heuristics. Finally, since ready detectability, relative insensitivity to illusion or artifact, and explanatory and predictive fruitfulness are desirable properties for the components of scientific theories, we should not be surprised to discover that robustness is important in the discovery and description of phenomena and in analyzing the structure of scientific theories.

One of the most ubiquitous phenomena of nature is its tendency to come in levels. If the aim of science, to follow Plato, is to cut up nature at its joints, then these levels of organization must be its major vertebrae. They have become so major, indeed, that our theories tend to follow these levels, and the language of our theories comes in strata. This has led many linguistically inclined philosophers to forgo talk of nature at all, and to formulate problems—for example, problems of reduction—in terms of analyzing the relation between theoretical vocabularies at different levels. But our language, as Campbell (1974b) would argue, is just another (albeit very important) tool in our struggle to analyze and to adapt to nature. In Chapter 10 and in Wimsatt (1976a, part III), I applied Campbell's criteria for entification to argue that entities at different levels of organization tend to be multiply connected in terms of their causal relations, primarily with other entities at their own level, and that they, and the levels they comprise, are highly robust. As a result, there are good explanatory reasons for treating different levels of organization as dynamically, ontologically, and epistemologically autonomous. There is no conflict here with the aims of good reductionist science: there is a great deal to be learned about upper-level phenomena at lower levels of organization, but upper-level entities are not "analyzed away" in the process because they remain robustly connected with other upper-level entities, and their behavior is explained by upper-level variables.

To see how this is so, we need another concept—that of the *sufficient parameter,* introduced by Levins (1966):

> It is an essential ingredient in the concept of levels of phenomena that there exists a set of what, by analogy with the sufficient statistic, we can call sufficient parameters defined on a given level . . . which are very much fewer than the number of parameters on the lower level and which among them contain most of the important information about events on that level.
>
> The sufficient parameters may arise from the combination of results of more limited studies. In our robust theorem on niche breadth we found that temporal variation, patchiness of the environment, productivity of the habitat, and mode of hunting could all have similar effects and that they did this by way of their contribution to the uncertainty of the environment. Thus uncertainty emerges as a sufficient parameter.
>
> The sufficient parameter is a many-to-one transformation of lower-level phenomena. Therein lies its power and utility, but also a new source of imprecision. The many-to-one nature of "uncertainty" prevents us from going backwards. If either temporal variation or patchiness or low productivity leads to uncertainty, the consequences of uncertainty alone

cannot tell us whether the environment is variable, or patchy, or unproductive. Therefore, we have lost information. (pp. 428, 429)

A sufficient parameter is a parameter, a variable, or an index that, either for most purposes or merely for the purposes at hand, captures the effect of significant variations in lower-level or less abstract variables (usually only for certain ranges of the values of these variables) and can thus be substituted for them in the attempt to build simpler models of the upper-level phenomena.

Levins claims that this notion is a natural consequence of the concept of levels of phenomena, and this is so, though it may relate to degree of abstraction as well as to degree of aggregation. (The argument I give here applies only to levels generated by aggregation of lower-level entities to form upper-level ones.) Upper-level variables, which give a more "coarse-grained" description of the system, are much smaller in number than the lower-level variables necessary to describe the same system. Thus, there must be, for any given degree of resolution between distinguishable state descriptions, far fewer distinguishable upper-level state descriptions than lower-level ones. The smaller number of distinguishable upper-level states entails that for any given degree of resolution, there must be many-one mappings between at least some lower-level and upper-level state descriptions with many lower-level descriptions corresponding to a single upper-level description. Those upper-level state descriptions with multiple lower-level state descriptions are robust over changes from one of these lower-level descriptions to another in its set.

Furthermore, the stability of (and possibility of continuous change in) upper-level phenomena (remaining in the same macro-state or changing by moving to neighboring states) places constraints on the possible mappings between lower-level and upper-level states. In the vast majority of cases neighboring microstates must map without discontinuity into the same or neighboring macro-states; and, indeed, most local microstate changes will have no detectable macro-level effects. *This fact gives upper-level phenomena and laws a certain insulation from* (through their invariance over: robustness again!) *lower-level changes and generates a kind of explanatory and dynamic (causal) autonomy of the upper-level phenomena and processes* (see also Wimsatt, 1976a, pp. 249–251; 1976b).

If one takes the view that causation is to be characterized in terms of manipulability (see, for example, Gasking, 1955; Cook and Campbell,

1979), the fact that the vast majority of manipulations at the micro-level do not make a difference at the macro-level means that macro-level variables are almost always more causally efficacious in making macro-level changes than micro-level variables. This gives explanatory and dynamic autonomy to the upper-level entities, phenomena, laws, and relations, within a view of explanation that is sensitive to problems of computational complexity and the costs and benefits we face in offering explanations. As a result, it comes much closer than the traditional hypothetico-deductive view to being able to account for whether we explain a phenomenon at one level and when we choose to go instead to a higher or lower level for its explanation (see Chapter 10, particularly sections IV–VI, the appendixes, and Wimsatt, 1976a, part III).

The many-one mappings between lower- and upper-level state descriptions mentioned above are consistent with correspondences between types of entities at lower and upper levels but do not entail them. There may be only token-token mappings (piecemeal mappings between instances of concepts, without any general mappings between concepts), resulting in upper-level properties being supervenient on rather than reducible to lower-level properties (Kim, 1978; Rosenberg, 1978). The main difference between Levins' notion of a sufficient parameter and the notion of supervenience is that the characterization of supervenience is embedded in an assumed apocalyptically complete and correct description of the lower and upper levels. Levins makes no such assumption and defines the sufficient parameter in terms of the imperfect and incomplete knowledge that we actually have of the systems we study. It is a broader and less demanding notion, involving a relation that is inexact, approximate, and admits of both unsystematic exceptions (requiring a *ceteris paribus* qualifier) and systematic ones (which render the relationship conditional).

Supervenience could be important for an omniscient LaPlacean demon but not for real, fallible, and limited scientists. The notion of supervenience could be regarded as a kind of ideal limiting case of a sufficient parameter as we come to know more and more about the system, but it is one that is seldom if ever found in the models of science. The concept of a sufficient parameter, by contrast, has many instances in science. It is central to the analysis of reductive explanation (Wimsatt, 1976a; 1976b, pp. 685–689; 1979) and has other uses as well (Wimsatt, 1980a, section 4).

Heuristics and Robustness

Much or even most of the work in philosophy of science today that is not closely tied to specific historical or current scientific case studies embodies a metaphysical stance that, in effect, assumes that the scientist is an omniscient and computationally omnipotent LaPlacean demon. Thus, for example, discussions of reductionism are full of talk of "in principle analyzability" or "in principle deducibility," where the force of the "in principle" claim is held to be something like "If we knew a total description of the system at the lower level, and all the lower-level laws, a sufficiently complex computer could generate the analysis of all the upper-level terms and laws and predict any upper-level phenomenon." Parallel kinds of assumptions of omniscience and computational omnipotence are found in rational decision theory, discussions of Bayesian epistemology, automata theory and algorithmic procedures in linguistics and the philosophy of mind, and the reductionist and foundationalist views of virtually all the major figures of twentieth-century logical empiricism. It seems almost to be a corollary to a deductivist approach to problems in philosophy of science (see Wimsatt, 1979) and probably derives ultimately from the Cartesian vision criticized earlier in this volume.

I have already written at some length attacking this view and its application to the problem of reduction in science (see Wimsatt, 1974; 1976a, pp. 219–237; 1976b; 1979; 1980b, section 3; and also Boyd, 1972). The gist of this attack is threefold:

1. On the "LaPlacean demon" interpretation of "in principle" claims, we have no way of evaluating their warrant, at least in science. (This is to be distinguished from cases in mathematics or automata theory, where "in principle" claims can be explicated in terms of the notion of an effective procedure.)
2. We are in any case not LaPlacean demons, and a philosophy of science that could have normative force only for LaPlacean demons thus gives those of us who do not meet these demanding specifications only counterfactual guidance; that is, it is of no real use to practicing scientists. More strongly, it suggests methods and viewpoints that are less advantageous than those derived from a more realistic view of the scientist as problem solver (see Wimsatt, 1979).

3. An alternative approach, which assumes more modest capacities of practicing scientists, does provide real guidance, better fits with actual scientific practice, and even (for reductive explanations) provides a plausible and attractive alternative interpretation for the "in principle" talk that so many philosophers and scientists use frequently (see Wimsatt, 1976a, part II; 1976b, pp. 697–701).

An essential and pervasive feature of this more modest alternative view is the replacement of the vision of an ideal scientist as a computationally omnipotent algorithmizer with one in which the scientist as decision maker, while still highly idealized, must consider the size of computations and the cost of data collection, and in other very general ways must be subject to considerations of efficiency, practical efficacy, and cost-benefit constraints. This picture has been elaborated by Herbert Simon and his coworkers, and their ideal is "satisficing man," whose rationality is bounded, in contrast with the unbounded omniscience and computational omnipotence of the "economic man" of rational decision theory (see Simon, 1957 [reprinted as chapter 1 of Simon, 1979]; see also Simon, 1996). Campbell's brand of fallibilism and critical realism from an evolutionary perspective also place him squarely in this tradition.

A key feature of this picture of man as a *boundedly* rational decision maker is the use of heuristic principles where no algorithms exist or where the algorithms that do exist require an excessive amount of information, computational power, or time. I take a heuristic procedure to have three important properties (see also Wimsatt, 1980b, section 3):

1. By contrast with an algorithmic procedure (here ignoring probabilistic automata), *the correct application of a heuristic procedure does not guarantee a solution* and, if it produces a solution, does not guarantee that the solution is correct.
2. *The expected time, effort, and computational complexity of producing a solution with a heuristic procedure is appreciably less* (often by many orders of magnitude for a complex problem) *than that expected with an algorithmic procedure.* This is indeed the reason why heuristics are used. They are a cost-effective way, and often the *only* physically possible way, of producing a solution.
3. *The failures and errors produced when a heuristic is used are not random but systematic.* I conjecture that *any heuristic, once we*

> *understand how it works, can be made to fail.* That is, given this knowledge of the heuristic procedure, we can construct classes of problems for which it will always fail to produce an answer or for which it will always produce the wrong answer. This property of systematic production of wrong answers will be called the *bias* of the heuristic.

This last feature is exceedingly important. Not only can we work forward from an understanding of a heuristic to predict its biases, but we can also work backward from the observation of systematic biases as data to hypothesize the heuristics that produced them; and, if we can get independent evidence (for example, from cognitive psychology) concerning the nature of the heuristics, we can propose a well-founded explanatory and predictive theory of the structure of our reasoning in these areas. This approach was implicitly (and sometimes explicitly) followed by Tversky and Kahneman (1974) in their analysis of fallacies of probabilistic reasoning and of the heuristics that generate them (see also Shweder, 1977, 1979a, 1979b, 1980a, who applies this perspective to judgments about personality; and Mynatt, Doherty, and Tweney, 1977, for a further provocative study of bias in scientific reasoning). The systematic character of these biases also allows for the possibility of modifications in the heuristic or in its use to correct for them (see Wimsatt, 1980b, pp. 52–54).

The notion of a heuristic has far greater implications that are explored in Chapter 5 and the appendixes. In addition to its centrality in human problem solving, the notion of a heuristic is a pivotal concept in evolutionary biology and in evolutionary epistemology. It is a central concept in evolutionary biology because any biological adaptation meets the conditions given for a heuristic procedure. First, it is a commonplace among evolutionary biologists that adaptations, even when functioning properly, do not guarantee survival and production of offspring. Second, adaptations are cost-effective ways of contributing to this end. Finally, any adaptation has systematically specifiable conditions, derivable through an understanding of the adaptation, under which its employment will actually decrease the fitness of the organism employing it by causing the organism to do what is, under those conditions, the wrong thing for its survival and reproduction. (This, of course, seldom happens in the organism's normal environment, or the adaptation would become maladaptive and be selected against.) This fact is indeed systematically exploited in the functional analysis of organic adaptations. It is a truism of functional inference that learning the conditions under which a system malfunctions, and how it malfunctions under those conditions, is a powerful tool for determining how it functions normally and the conditions under which it was designed to function. (For illuminating discussions of the problems, techniques, and fallacies of functional inference under a variety of circumstances, see Gregory, 1962; Lorenz, 1965; Valenstein, 1973; Glassman, 1978.)

The notion of a heuristic is central to evolutionary epistemology because Campbell's (1974a, 1977) notion of a vicarious selector, which is basic to his conception of a hierarchy of adaptive and selective processes spanning subcognitive, cognitive, and social levels, is that of a heuristic procedure. For Campbell, a vicarious selector is a substitute and less costly selection procedure acting to optimize some index that is only contingently connected with the index optimized by the selection process it is substituting for. This contingent connection allows for the possibility—indeed, the inevitability—of systematic error when the conditions for the contingent consilience of the substitute and primary indices are not met. An important ramification of Campbell's idea of a vicarious selector is the possibility that one heuristic may substitute for another (rather than for an algorithmic procedure) under restricted sets of conditions, and that this process may be repeated, producing a nested hierarchy of heuristics. He makes ample use of this hierarchy in analyzing our knowing processes (Campbell, 1974a). I believe that this is an appropriate model for describing the nested or sequential structure of many approximation techniques, limiting operations, and the families of progressively more realistic models found widely in progressive research programs, as exemplified in the development of nineteenth-century kinetic theory, early twentieth-century genetics, and several areas of modern population genetics and evolutionary ecology.

Simon's work, that of Tversky and Kahneman, and, more recently, that of Gigerenzer and colleagues, and many others, have opened up a whole new set of questions and areas of investigation of pragmatic inference (and its informal fallacies) in science, which could revolutionize our discipline in the next decade. (For a partial view of how studies of reduction and reductionism in science could be changed, see Wimsatt, 1979.) This change in perspective would bring philosophy of science much closer to actual scientific practice without surrendering a normative role to an all-embracing descriptivism, and it would reestablish ties with psychology through the study of the character, limits, and biases of processes of empirical reasoning. Inductive procedures in science are heuristics (Shimony, 1970), as are Mill's methods (1843) and other methods for discovering causal relations, building models, and generating and modifying hypotheses.

Heuristics are also important in the present context, because the procedures for determining robustness and for making further application of these determinations for other ends are all heuristic procedures. Robustness analysis covers a class of powerful and important techniques, but they are not immune to failures. There are no magic bullets in science, and these techniques are no exception.

Most striking of the ways of failure of robustness analysis is one that produces illusions of robustness: the failure of the different supposedly independent tests, means of detection, models, or derivations to be truly independent. This is the basis for a powerful criticism of the validity of IQ scales as significant measures of intelligence (McClelland, 1973). Failures of independence are not easy to detect and often require substantial further analysis. Without that, such failures can go undetected by the best investigators for substantial lengths of time. In addition, different heuristics can be mutually reinforcing, each helping to hide the biases of the others (see Wimsatt, 1980b, sections 5 and 8), which can make it much harder to detect errors that would otherwise lead to discovery of failures of independence. The failure of independence in its various modes, and the factors affecting its discovery, emerge as some of the most critical and important problems in the study of robustness analysis, as is indicated by the history of the group selection controversy.

Robustness, Independence, and Pseudo-Robustness: A Case Study

In recent evolutionary biology (since Williams' seminal work in 1966), group selection has been the subject of widespread attack and general suspicion. Most of the major theorists (including W. D. Hamilton, John Maynard Smith, and E. O. Wilson) have argued against its efficacy. A number of mathematical models of this phenomenon have been constructed, and virtually all of them (see Wade, 1978) seem to support this skepticism. The various mathematical models of group selection surveyed by Wade all admit to the possibility of group selection, yet almost all also predict that group selection should only very rarely be a significant evolutionary factor; that is, they predict that group selection should have significant effects only under very special circumstances—for extreme values of parameters of the models—which should seldom be found in nature. Wade (1978) undertook an experimental test of the relative efficacy of individual and group selection—acting in concert or in opposition—in laboratory populations of different species of the flour beetle, *Tribolium*. This work produced surprising results. Group selection appeared to be a significant force in these experiments, one capable of overwhelming individual selection in the opposite direction for a wide range of parameter values. This finding, apparently contradicting the results of all the then extant mathematical models of group selection, led Wade to a closer analysis of these models, with results described here.

All the models Wade surveyed made simplifying assumptions, most of them different. Five assumptions, however, were widely held in common; of the twelve models surveyed, each made at least three of these assumptions, and five of the models made all five assumptions. Crucially, for present purposes, the five assumptions are biologically unrealistic and incorrect, and each independently has a strong negative effect on the possibility or efficacy of group selection. It is important to note that these models were advanced by a variety of different biologists, some sympathetic to and some skeptical of group selection as a significant evolutionary force. Why, then, did all of them make assumptions strongly inimical to it? Such a coincidence, radically improbable at best, cries out for explanation: we have found a systematic bias suggesting the use of a heuristic.

These assumptions are analyzed more fully elsewhere (Wade, 1978; Wimsatt, 1980b; the discussion here merely summarizes part of the results of the latter work). In Chapter 5 I provide a list of eight reductionist research and modeling strategies. Each is a heuristic in that it has systematic biases associated with it, and these biases will lead to the wrong answer if the heuristic is used to analyze certain kinds of systems. It is the use of these heuristics, together with certain "perceptual" biases (derived from thinking of groups as "collections of individuals" rather than as robust entities analogous to organisms), that is responsible for the widespread acceptance of these assumptions and the almost total failure to notice what an unrealistic view they give of group selection. Most of the reductionist heuristics lead to a dangerous oversimplification of the environment being studied and a dangerous underassessment of the effects of these simplifications. In the context of the perceptual bias of regarding groups as collections of individuals (or sometimes even of genes), the models tend systematically to err in the internal and relational structure they posit for the groups and in the character of processes of group reproduction and selection.

The first assumption, that the processes can be analyzed in terms of selection of alternative alleles at a single locus, is shown to be empirically false by Wade's own experiments, which show conclusively that both individual and group selection is proceeding on multi-locus traits (for an analysis of the consequences of treating a multi-locus trait erroneously as a single-locus trait, see Wimsatt, 1980b, section 4). The fifth assumption, that individual and group selections are opposed in their effects, also becomes untenable for a multi-locus trait (see Wimsatt, 1980b, section 7).

The second assumption (that all migrants from each population go into a common pool—the "migrant pool") is equivalent to the time-honored assumption of *panmixia,* or random mating within a population, but in the context of a group selection model it is equivalent to assuming a particularly strong form of blending inheritance for group inheritance processes. This assumption is factually incorrect and, as R. A. Fisher showed in 1930, effectively renders evolution at that level impossible. This assumption is discussed more fully in Chapter 5.

The third assumption is equivalent to assuming that groups differ in their longevity but not in their reproductive rates. But, as all evolutionary biologists since Darwin have been aware, variance in reproductive rate has a far greater affect on the intensity of selection than variance in longevity. So the more significant component was left out in favor of modeling the less significant one. (The second and third assumptions are discussed in Wimsatt, 1980b, section 7; The fourth assumption is further discussed and shown to be incorrect in Wade, 1978.)

The net effect is a set of cumulatively biased and incorrect assumptions, which, not surprisingly, lead to the incorrect conclusion that group selection is not a significant evolutionary force. If I am correct in arguing that these assumptions probably went unnoticed because of the biases of our reductionist research heuristics, a striking analogy emerges: The phenomenon appears to be a paradigmatic example of Levinsian robustness. A wide variety of different models, making different assumptions, appeared to show that group selection could not be efficacious; however, the robustness was illusory because the models were not independent in their assumptions. The commonality of these assumptions appears to be a species of method bias, resulting in a failure of discriminant validity (Campbell and Fiske, 1959). But the method under consideration is not the normal sort of test instrument that social scientists deal with. Instances of the method are reductionist research heuristics, and the method is reductionism. For the purposes of problem solving, our minds can be seen as a collection of methods, and the particularly single-minded are unusually prone to method bias in their thought processes. This conclusion is ultimately just another confirmation at another level of something Campbell has been trying to teach us for years about the importance of multiple independent perspectives.

Author's Note, 2007: I cover other uses of robustness in the development of inter-level explanations in Chapter 10 and for visual representation in Wimsatt (1991). Recent work by Odenbaugh (2001), Weisberg (2006a), Plutinski (2006), and a special issue of *Biology and Philosophy* (Weisberg, 2006b) have added careful and insightful analyses of Levins' work on robustness and other aspects of his philosophy of science.

In the 25 years since this chapter was written, a somewhat different sense of robustness has become important in evolutionary studies, with arguments that robustness is critical to evolvability. Some of this is anticipated in this chapter. These discussions of robustness have moved closer to a focus on invariance of system properties across wide ranges of parameter values or as found in diverse models of the system. It thus comes closer to Levins' search for results that are invariant across different models than to Campbell's emphasis on similar results across different means of measurement or detection, though just as clearly both remain relevant. Wagner (2005) has an extensive and conceptually penetrating review of this evolutionary literature.

CHAPTER FIVE

Heuristics and the Study of Human Behavior

What do we do when the complexity of the systems we are studying exceeds our powers of analysis? This is an old problem in social science methodology, but it does not indicate a cause for despair, since exactly the same thing has happened frequently in the natural and biological sciences. Roughly, the therapy is the same in both cases: introduce idealizations, approximations, or other devices that, perhaps artificially, reduce the complexity of the problem. Here, however, we needn't be looking only to the natural sciences for guidance, for the best developed theory of such devices has arisen in psychology and the social sciences. I have in mind the work of Simon, Lenat, Tversky and Kahneman, and others on problem-solving heuristics. While it might be argued that this is a part of cognitive psychology or artificial intelligence, Simon's interest in heuristics springs from his "satisficing" theory of decision making, which in turn was motivated jointly by his interest in decision making and administrative organizations (1957), and his dissatisfaction with rational decision theory (1955, 1996).

To regard a system as using heuristics is to regard it as a kind of engineering system. This is implicit in Simon's (1996) characterization of the scope of "the sciences of the artificial"—artificial things are products of design processes or, more generally, of selection processes. In this, I also follow Campbell (1974a), who has argued the same point from a somewhat different perspective. Dennett (1979) urges a similar view in his characterization of the "design stance" as a perspective for analyzing functionally organized systems. To the extent that heuristics are important in the analysis of our reasoning processes and action or behavior, the boundary between D'Andrade's (1986) second and third perspectives (or Dennett's analogous ones) is at least blurred; I think it becomes a matter of degree. I will return to this after I say something more about the nature of heuristics.

Heuristics

Heuristic has become one of the most widely used terms in artificial intelligence and cognitive psychology and, as with other such terms, shows wide variance in its use. To my knowledge, it was introduced in its present context by Herbert Simon (Newell, Shaw, and Simon, 1957), who borrowed it from the mathematician, George Polya (1954), who used it to describe "rules of thumb" used in solving mathematical problems. Lenat (1982) provides a rich and constructive discussion of the nature of heuristics in artificial intelligence work. As I understand them, heuristic procedures, or heuristics, have four important properties that explain a number of characteristics of their use:

1. By comparison with truth-preserving algorithms or with other procedures for which they might be substituted, heuristics make no guarantees (or if they are substituted for another procedure, weaker guarantees) that they will produce a solution or the correct solution to a problem. A truth-preserving algorithm correctly applied to true premises must produce a correct conclusion. But one may correctly apply a heuristic to correct input information without getting a correct output.
2. By comparison with the procedures for which they may be substituted, heuristics are very "cost-effective" in terms of demands on memory, computation, or other resources in limited supply (this, of course, is why they are used).
3. The errors produced by using a heuristic are not random but systematically biased. By this I mean two things. First, the heuristic will tend to break down in certain classes of cases and not in others, but not at random. Indeed, with an understanding of how the heuristic works, it should be possible to predict the conditions under which it will fail. Second, where it is meaningful to speak of a direction of error, heuristics will tend to cause errors in a certain direction, which is again a function of the heuristic and of the kinds of problems to which it is applied.

4. The application of a heuristic to a problem yields a transformation of the problem into a nonequivalent but intuitively related problem. Most important this means that answers to the transformed problem may not be answers to the original problem. (This property of heuristics was pointed out to me by Robert McCauley—see his 1986 work.)

Traditional philosophy of science is a philosophy of deductive structures and algorithms for computationally omnipotent computers—LaPlacean demons for which computation has a negligible cost. Theories are assumed to have an axiomatic structure, and they are assumed to be closed under entailment; that is, anything that follows from a set of axioms is a part of that theory. Thus, for example, discussions of reductionism are full of talk of in principle analyzability or in principle deducibility, where the force of the in principle claim is something like "If we knew a total description of the system at the lower level and all of the lower-level laws, a sufficiently complex computer could generate the analysis of all of the upper-level terms and laws and predict any upper-level phenomenon." Of course, we don't have such a complete lower-level description of higher-level systems in any science, we are not even sure that we have all of the relevant lower-level laws, and we have not yet succeeded in producing any such apocalyptic derivations of the total behavior of any higher-level systems—but these are supposed to be merely "technical" difficulties.

I have criticized this unattainable picture of reductive explanation elsewhere (Wimsatt, 1976a, 1976b), and suggested an alternative account that dovetails naturally with the heuristic picture. But the original picture persists widely in philosophical analyses and in "rationalistic" theories in the social sciences, particularly in decision theory, linguistics, and other areas where algorithmic models are found. Its persistence is aided by regarding such theories as "normative"—as specifying what is the optimal behavior usually under impossibly idealized circumstances (when we have no computational limitations and make no errors in our calculations). The search for larger computers, which can calculate faster, and for various technical improvements to increase reliability (such as doing computations in parallel and cross-checking regularly) shows that our best equipment falls short of this ideal. The gap is far larger if we note that all of our models of phenomena involve simplifications and approximations done to increase analytical tractability, so that the problems we are solving are already less complex than the real world they are designed to mimic.

We are not LaPlacean demons, and any image of science that tells us how to behave as if we were still fails to give useful guidance for real scientists in the real world. In fact, it may suggest viewpoints and methods that are less than optimal for the dinky and error-prone equipment we possess. A more realistic model of the scientist as problem solver and decision maker includes the existence of such limitations and is capable of providing real guidance and a better fit with actual practice in all of the sciences. In this model, the scientist must consider the size of computations, the cost of data collection, and must regard both processes as "noisy" or error-prone. A central feature of it is the use of "cost-effective" heuristic procedures for collecting data, simplifying problems, and generating solutions.

Although a growing number of philosophers and cognitive psychologists have become interested in heuristics, the two groups have focused on different properties of heuristics. Philosophers for the most part have focused on their computational efficiency (property 2 above) and have argued that heuristics play an important role in scientific inference and discovery. Note that essentially all "inductive" and discovery procedures in science are heuristic principles, failing as algorithms in part because they do not represent logically valid argument forms. Indeed, psychologists have focused on this fact and gloried in the "irrationality" of our everyday heuristics, by which they mean that we will in the appropriate circumstances draw erroneous conclusions using them (see, e.g., Tversky and Kahneman, 1974; Nisbett and Ross, 1980; Shweder, 1977).

Both of these properties need to be examined together. It is not irrational to use a procedure that may under some circumstances lead you into error if you take pains to avoid those circumstances and if using it saves you a great deal of effort. All instruments in the natural, biological, and social sciences are designed for use in certain contexts and can produce biased or worthless results if they are used in contexts that may fail to meet the conditions for which they were designed. A fair amount of effort in these sciences is devoted to determining the conditions under which instruments can be used without bias or to calibrating them to determine their biases so that they can be corrected for. This is one of the major activities of statistical methodologists—either constructing new instruments or calibrating or criticizing the use of existing ones.

Campbell's notion of a "vicarious selector" (1974) is employed widely by him to explain and characterize a hierarchy of selection processes in perception, learning, and cultural evolution. It follows from his characterization of vicarious selectors that they are heuristic procedures (see Chapter 4). I believe that Campbell's conception of a hierarchy of selection processes acting to produce structures' "fit," whether physical or ideational, with their relevant environments is the most productive form for functionalist theories in the social sciences. It lacks their panfunctionalist tenor and also has a much closer connection with evolutionary ideas in the biological sciences, which have recently begun to move toward productive models for the microevolution of culture (Boyd and Richerson, 1985).

Biological adaptations (and in Campbell's scheme, social and psychological ones as well) all meet the defining characteristics of heuristic procedures. First, it is a commonplace among evolutionary biologists that adaptations, even when functioning properly, do not guarantee survival and production of offspring. Second, they are nevertheless cost-effective ways of contributing to this end. Third, any adaptation has systematically specifiable conditions under which its employment will actually decrease the fitness of the organism. These conditions are, of course, seldom found in the normal environments of the organism, or the adaptation would be maladaptive and selected against. Fourth, these adaptations serve to transform a complex computational problem about the environment into a simpler problem, the answer to which is usually a reliable guide to the answer to the complex problem. Thus shorter day length is a good predictor of oncoming winter and is used by a variety of plants and animals to initiate appropriate seasonal changes in morphology and behavior—even though heavy cloud cover or artificial conditions in the laboratory can fool this adaptation. Similarly, rapid decreases in general illumination in the frog's visual field are taken to indicate the approach of a predator. Though the frog may be fooled frequently by this adaptation (e.g., cows are not predators but may be frequent parts of the frog's environment), the cost of being wrong is sufficiently great that this is a cost-effective solution.

The third property, that the errors produced in using a heuristic are systematic, is widely exploited in the analysis of organic adaptations. It is a truism of functional inference that studying how a system breaks down (and the conditions under which it does) is a powerful tool for determining how it functions normally and the conditions under which it was designed to function. This fact can also be used systematically in the study of our reasoning processes. First, from an analysis of the heuristic, we can determine the conditions under which (and how) it will break down, thus calibrating the heuristic. This, as already pointed out, is the task of the methodologist, but it should be applied to our heuristic reasoning processes no less than to the study of our machines or to organic adaptations. But a more interesting insight, first employed by Tversky and Kahneman (1974), is that the widespread occurrence of systematic errors is to be recognized as the "footprint" of a heuristic procedure or procedures. Different heuristics leave characteristically different footprints, so an analysis of the biases can lead to plausible inferences about the character of the reasoning processes that produced them. Usually, we need some knowledge of these reasoning processes in order to pare down the field of appropriate candidates, but this can be done. For example, I conducted a study of the systematic biases in mathematical models of group selection, a heated controversy in evolutionary biology, and was able to trace the origin of these biases to heuristics for problem simplification characteristic of reductionist problem-solving strategies. I discuss this in the next section because reductionist problem-solving methods are widely used and just as widely criticized in the social sciences (for more details of this specific case, see Wimsatt, 1980b).

Reductionist Research Strategies and Their Biases

If reductionist problem-solving heuristics have a generic bias, it is to ignore, oversimplify, or otherwise underestimate the importance of the context of the system under study. A number of writers (e.g., many of those included in Fiske and Shweder, 1986) have complained about the frequency with which properties of the systems they study are assumed to be independent of context, when in fact they are disguised relational properties. Context dependence is a frequent problem of translation for linguists. I have argued that context dependence of biological fitness components at a lower level is a necessary (but not sufficient) condition of the existence of higher-level units of selection (Wimsatt, 1980b, 1981b). If reductionist problem-solving heuristics lead to illegitimate assumptions of context independence, there is a *prima facie* case for believing that biases of reductionist problem-solving strategies are extremely pervasive in the social sciences.

How these biases arise can be easily discerned from a general characterization of the problem-solving context of a reductionist problem solver. First, assume that a scientist starts by choosing, designating, or constructing a system for analysis. This immediately partitions his world of study into that system and its environment. (See Star, 1983a, for relevant simplification processes here; Griesemer's concept [1983] of the "conceptual map" as fixing the environment in which problem solving takes place is relevant at this stage of analysis.) Second, we must make Simon's "assumption of bounded rationality" that any real-world system is too complex to study in all of its complexity, so we must make simplifications—through selection of properties or objects for study, simplified assumptions about relationships between these properties or objects, assumptions about what variables must be controlled or randomized, and the like. Third, I assume a very general characterization of what it is to be a reductionist; that is, a *reductionist* is interested in understanding the character, properties, and behavior of the studied system in terms of the properties of its parts and their interrelations and interactions. (This is a sufficiently inclusive description that it probably captures any analytic methods in general, even those of many who would not call themselves reductionists. It should, in any case, be acceptable to any reductionist.) This means that the reductionist is primarily interested in the entities and relations internal to the system of study. But this fact, together with the assumption of bounded rationality, has an interesting consequence. While simplifications will in general have to be made everywhere, the focus of the reductionist will lead him to order his list of "economic" priorities so as to simplify first and more severely in his description, observation, control, modeling, and analysis of the environment than in the system he is studying.

Any reductionist who began with the assumption that his system was totally homogeneous in structure and constant through time would have nothing to study: there would be no parts or relations between them. But commonly found in simple models of systems (and even in not-so-simple ones) is the assumption that the system is isolated (in effect, that it has no environment) or that its environment is constant in space and time. This asymmetry in simplifications is indicative of the kinds of biases induced by using reductionist problem-solving strategies.

Below I outline reductionist problem-solving strategies. Each one is used in some circumstance because its adoption transforms the initial problem into one that is easier to analyze and to solve. Each strategy can be seen as an application of the general schema for making simplifications to a specific scientific activity, whether conceptualizing the system for study and analysis, building or modifying models of its behavior, observing its behavior, designing controlled experiments (or looking for natural data sets that meet desired control conditions), or testing the models. This partial list should suggest a variety of relevant cases in various disciplines. I have somewhat arbitrarily divided the heuristics into heuristics of conceptualization, model building and theory construction, and observation and experimental design, though these activities are seldom as separable as this division might suggest.

Heuristics of Conceptualization

Descriptive localization. Describe a relational property as if it were monadic or a lower-order relational property. Note that if a property is a function of system properties and environment properties, keeping the environment constant will make the property look as if it is a function only of system properties. Thus, for example, fitness is a relational property between organism and environment. Keeping the environment constant makes it look as if fitness can be treated as a monadic property of organisms. Many context dependencies are hidden in this fashion.

Meaning reductionism. Assume that new redescriptions of a property at a lower level or an account of that property in terms of the intra-systemic mechanisms that produce it can result in meaning changes (through redefinition) of scientific terms, whereas higher-level re-descriptions (or an account of the property in terms of inter-systemic mechanisms) cannot. Result: Since philosophers regard themselves as concerned with meaning relations, they are inclined to a reductionist bias. Note that this is not a bias when it is applied to properties that are "correctly" regarded as monadic at the level of analysis in question; that is, that are context independent for wide ranges of conditions like those naturally studied. But if the property in question is a disguised relational property or a functional property (both of which impose conditions on properties or entities outside of the system under study), this assumption can lead to serious mistakes.

Heuristics of Model Building and Theory Construction

Modeling localization. Look for an intra-systemic mechanism rather than an inter-systemic one to explain a systematic property, or if both are available, regard the former as "more fundamental." As derivative corollaries, structural properties are regarded as more important than functional ones, and mechanisms as more important than context (see, e.g., discussions of the assumed stability of personality traits in Shweder, 1979a, 1979b, 1980a).

Context simplification. In reductionist model building, simplify the description of the environment before simplifying the description of the system. This strategy often legislates higher-level systems out of existence or leaves no way of describing inter-systemic phenomena appropriately. This is, in effect, a redescription of the account given above of the origin of reductionist biases against the importance of context, but even this general level of description has been exceedingly important in some areas. This is perhaps the most striking bias in mathematical models of group selection (see Wimsatt, 1980b).

Generalization. When setting out to improve a simple model of the behavior of a system in its environment, focus on generalizing or elaborating the internal structure of the system at the cost of ignoring generalizations or elaborations of its external structure. Because a number of simplifications will have been made both internal and external to the system, there will always be room for improving its internal description and analysis. In effect, this strategy involves the following working maxim: If a model fails to work or predict adequately, it must be because of oversimplifications in the description of internal structure, not because of oversimplified descriptions of external structure (see Star, 1983a, 1983b, for a sociological perspective on this).

Heuristics of Observation and Experimental Design

Observation. The reductionist will tend not to model environmental variables and will thus fail to record data necessary to detect interactional or larger-scale patterns. Note that this can apply on a temporal as well as on a spatial scale. Thus one who studies patients without having taken appropriate case histories may be committing this error, as well as one who does not record appropriate contextual variables of the experiment.

Control. The reductionist in experimental design will construct experimental arrangements so as to keep environmental variables constant (or will often merely assume that they are constant!). He then tends to miss dependencies of system variables on them. This heuristic is particularly interesting, since it follows straightforwardly from applying Mill's canon (1843)—to vary the factors one at a time, keeping all others constant—to the context of reductionist problem solving. If the reductionist is interested in determining causal relations among intra-systemic variables, he will try to vary the intra-systemic variables one at a time, keeping all of the other intra- and extra-systemic variables constant. As he studies different intra-systemic variables, he will be keeping the other intra-systemic variables constant in different combinations and will thus tease out the various intra-systemic causal relations. But the extra-systemic variables will also be kept constant in each of these experiments, so he will never have done an experiment appropriately designed to determine effects of any extra-systemic variables on system properties. One can imagine a reductionist replying to the claim of the causal importance of some extra-systemic variable, "But we have been studying these systems for years, and no one has ever reported any effect of that variable!" But of course not—it has always been kept constant! This is another instance of the old maxim that there are no universal control setups. What one must control is a function of what relationships one is studying.

Testing. Make sure that a theory works out locally (or in the laboratory) rather than testing it in appropriate natural environments or doing appropriate robustness analyses to suggest what are important environmental variables and what are relevant ranges of these parameters for study.

An Example of Reductionist Biases: Models of Group Selection

At the end of the last chapter I discussed how Michael Wade (1978) uncovered five false assumptions that were widely shared among models of group selection, with each biasing the results against its efficacy, to massive cumulative effect. I had argued (Wimsatt, 1980b) that they were likely products of diverse reductionistic model-building heuristics discussed here (and further elaborated in appendixes A and B). The problematic character of these assumptions surely went unnoticed in part because the diverse reductionistic problem-solving strategies all tended to generate errors in the same direction. So rather than producing independent confirmation or anomalous mismatches that would have drawn attention to them, they created an apparently robust, or pseudo-robust, conclusion confirming existing theoretical biases. This example with group selection was used in the last chapter to illustrate that the appearance of robustness can be misleading, and robustness can fail through failures of independence among the different models or assumptions. I return to these heuristics now to show more about how they work, why the assumptions were made, and how—despite their cumulative biases—the errors using them can potentially be both systematically detected and corrected.

I discuss only one of them here, the second assumption in Wade's list. (All of the others are discussed in Wimsatt, 1980b, and further elaborated in Wimsatt, 1981b.) The "migrant pool" assumption was first introduced by Richard Levins (1970b), an advocate of group selection. This is the assumption that all of the migrants from any groups (the "offspring" of these groups) go into a common pool from which new groups are drawn at random. It won ready acceptance for two reasons. First, it provided substantial analytic simplifications (the need to keep track only of gene frequencies in a single migrant pool, rather than recording the independent gene frequencies of the migrants from each parent group). Second, it was equivalent to a time-honored simplifying assumption of population genetics, "panmixia," the assumption that all members of a population have an equal probability of mating with any other member, in this case not applied to the groups but to the migrants. But if there is systematic genetic variation across a species' range and the 1-generation mean migration distance is much less than the species range, this assumption must be incorrect—the real chances of mating would vary with location in the species range. This difference has genetic consequences for which genotypes mate and produce offspring (Wimsatt, 1980b, 1981b), and which genotypes are produced in which frequencies, and therefore would generally impact the evolutionary outcome. It is not a benign assumption.

Moreover, when the process of reproduction is re-examined at the group level (rather than the individual level, at which most theorists conceptualize the problem), this assumption is seen to be equivalent to the assumption of a particularly strong form of "blending inheritance." R. A. Fisher showed in 1930 that blending inheritance (with panmixia and bi-parental inheritance) at the individual level results in loss of half of the variance in each successive generation, leaving evolution with little variation to act upon. Wade and I showed (Wimsatt, 1981b) that in group selection with the "migrant pool" assumption (for n parental groups contributing equal numbers of migrants in each group generation), variance is attenuated in each group generation by a factor $1/n$, thus potentially much more severely than in Fisher's case! The striking thing is that while all population geneticists know that "blending inheritance" is a thing to avoid (it is avoided at the individual level by the processes of Mendelian segregation), they were unable to recognize that they had made assumptions equivalent to introducing it *in a stronger form* at the group level in their models of group selection. To put the point in another way, to assume panmixia is the same as assuming that with respect to probability of mating in the composition of new migrant groups, *there are no groups!* It is thus not surprising that reductionists should have found little effect of group selection in their models. (The impact of the assumptions of blending are elaborated in Wimsatt, 2002b.)

A particularly crucial factor in explaining the inability to see the consequences of this assumption is what I have called "perceptual focus" (Wimsatt, 1980b, 248–249). If groups are thought of as merely "collections of individuals" (a hypothesis since confirmed by discussions with some of the protagonists, including John Maynard Smith in 1982), then the description of processes is referred to at the individual level, and one cannot see that apparently benign assumptions at that level may be revealed as dangerous oversimplifications when viewed at a higher level. This perspectival phenomenon appears to explain not only the ready acceptance of the "migrant pool" assumption but also some of the other assumptions Wade discusses. It also suggests a technique for correcting the systematic biases of the reductionist problem-solving strategies. I return to this after considering why the effects of heuristics can be so easily missed.

Heuristics Can Hide Their Tracks

One of the remarkable things about the case just discussed is that the biases in these models had not been discovered in spite of the fact that a variety of models of the processes of group selection had been investigated. One would hope that this "sampling from a space of possible models" (Levins, 1968) would be an unbiased sampling and would turn up models with different conclusions if the conclusions were artifacts of the simplifying assumptions that were made (see Wimsatt, 1981a). But in this case, it clearly was a biased sample. The question is, Why?

The answer resides in the fact that all of the modelers were using a variety of reductionist modeling strategies and assumptions. While they approached their models in somewhat different ways, this commonality of generic approach constrained the results. Each of the heuristics in the list of reductionist modeling strategies independently biases the models against the inclusion or proper consideration of environmental variables. If one then crosschecks these models as a way of validating the results, one will get an apparent robustness of the conclusions, but a spurious one; a case of pseudorobustness. It is tempting to suggest that much of the appearance of success (I do not want to deny that there have been real successes as well) of reductionist methodologies comes from this phenomenon—that different reductionist methods hide their mutual inadequacies by covering each other's tracks.

Another important means through which methodologies (regarded as related bundles of heuristics and practices) can "hide their tracks" (and this applies not just to reductionist methodologies) is through the fourth property of heuristics—that use of a heuristic causes a redefinition or transformation of the problem to which it is applied. If a transformation yields an analytically tractable, and therefore successful, problem, there will be a tendency to act as if the new problem (to which there is now a solution) captures the core issues of the old problem and thus to argue that the old problem is "really" solved by the solution to the new problem. Taken one step further, the new problem may be put forward as defining the proper formulation of the original, and thus it has replaced the old in a manner rendered largely invisible since it is now regarded as a "clarification of the old problem" that preserves the spirit of the research tradition while removing "confusions" that had earlier prevented solution of the problem. This is a "hidden revolution," a more modest paradigm shift masquerading as none at all. The philosopher Ludwig Wittgenstein (1962) argued that this kind of phenomenon characterizes rule-following behavior and the use of concepts in general. We assimilate new kinds of application of a rule to its original domain of interpretation and thus have the anomalous phenomenon that before it is applied to the new situation, it looks as if we have a real choice as to whether to apply the rule or not; but after the fact, it appears as if we had no other choice!

This kind of invisible paradigm shift has many of the properties Kuhn (1970) ascribes to paradigm shifts in general, except for those connected with the explicit recognition of a revolutionary change. It is still true that the new paradigm carries with it evaluative judgments and defines explanatory standards; but instead of arguing that this represents a rejection of the old paradigm, the reductionist will argue that this is the proper interpretation of the old problems and theory—that this was "contained in the old view all along." When this kind of paradigm shift is accompanied by the acceptance of a new formal model, it will result in a bias of oversimplification—often accompanied by the reification of the abstract system to define a kind of (Weberian?) ideal type, which is talked about as if it existed in the real world. Thus thermodynamicists talk about "ideal gases" or "van der Waal's gases," denoting gases whose behavior fits the ideal-gas law and van der Waal's equation of state, respectively. Biologists, similarly, talk about "Mendelian genes," "Mendelizing traits," "Lotka-Volterra communities," and "panmictic populations"—in each case indicating a supposed conformity of a real-world system with the abstract conditions of the model.

While one may decry this kind of reification, the tendency to do it indicates a real and often ignored function of theoretical models. Although we sometimes set out to test our models, far more often we use them as patterns to organize phenomena and to classify results as fitting or not fitting the model. Using the model as a pattern-matching template in this way enormously simplifies the task of finding order in the data, but it also introduces a kind of inertia (or, as Tversky and Kahneman [1974] would say, an "anchoring" bias), which opposes change in the system. When the results don't fit, we do not throw away the model but instead redescribe the system: the population is *not* panmictic, the gene is *not* Mendelian, the community *not* Lotka-Volterra. The only problem with this activity is that if such pattern matching is mistaken for genuine testing, it can lead advocates of the model to believe that the model has been confirmed, and critics to accuse the advocate of ad hoc-ery and of converting the model into a "meaningless schema" or "covert tautology." The difference between this kind of activity and a real test of the model is that in pattern matching, there often is no attempt to determine the values of the parameters of the model independently. Values are chosen so that the model fits the data, producing a kind of sophisticated curve fitting, rather than a legitimate test of the model. (See Tribe's [1972] excellent discussion of the misuse of the assumptions of rational decision theory in attempts to make real-world decisions.)

A parallel kind of situation can happen when the paradigm is an experimental system rather than a theoretical model. If an experimental system is highly successful, it can become a normative paradigm for how a class of studies should be pursued. "Drosophila genetics" and "Tribolium ecology" mark off, not only model organisms, but sets of procedures for studying them and preferred questions for consideration that were extended as paradigms for studying other organisms in other situations because they represented ideal conceptions of how to study genetics and ecology, respectively. When these conceptions are attacked, they elicit spirited defenses of the general methodological approach (see, e.g., Mertz and MacCauley, 1980). The bias introduced here is one of overgeneralization—whether of the limited model or of the experimental protocol and design or of the questions asked.

Both of these kinds of promotion of a theoretical or experimental model to a paradigm can reinforce the biases of the other problem-solving techniques by hiding the need to reconceptualize the system or the formulation of the problem of analyzing it. They can defer for a long time the noticing or analyzing of questions that were far more obvious at the start of this line of investigation. This phenomenon—the increasing entrenchment of a theoretical or experimental paradigm—in part serves to explain why disciples of an approach are often far less flexible and far less methodologically conscious than the originators of that approach. I have attempted to model and to explore the consequences of this kind of entrenchment in a variety of contexts, in biological evolution and development, in cognitive development, and in models of scientific and cultural change (see Glassman and Wimsatt, 1984; interview in Callebaut, 1993, pp. 425–429; Wimsatt, 1986a, 1999a).

Two Strategies for Correcting Reductionist Biases

I know of two general strategies for correcting for the artifacts and biases of problem-solving heuristics. The first is truly general but is subject to specific problems when applied to reductionist problem solving. The second can be regarded as a specific application of the general approach to reductionist problem solving, with an eye to addressing the special problems of a reductionist approach.

The general approach is what Campbell has called "triangulation," which Campbell and Fiske (1959) incorporated in their now classic "multitrait-multimethod" matrix. I have discussed the variety of functions and applications of this method, which I call "robustness analysis," in Chapter 4. By using a variety of different models, approaches, means of detection, or of derivation and comparing the results, we can hope to detect and correct for the biases, special assumptions, and artifacts of any one approach.

But we have already seen that this approach does not guarantee success. It too is a heuristic procedure. The models of group selection display an ingenious variety of assumptions and approaches, but they all share biasing assumptions in common, assumptions whose biasing effects were not apparent before Wade's review (1978). How can we prevent our array of models and approaches from being a biased sample? To this question, there is no general answer, but there is an answer derived from the character of the biases in this specific case. Recall that the effect of each reductionist bias is to ignore or to underestimate the effects of variables in the environment. But is this attached unalterably to the variables in question? No—it occurs merely because of where they are located. This bias would be removed for anything that could be brought within the system, and this can be accomplished merely by changing the boundaries of the system being investigated. Thus the strategy for eliminating biases in the description and analysis of groups as collections of individuals is to build models in which the groups are treated as individuals in a larger system in which they are parts and in which we focus on modeling intergroup relations. (This strategy is deliberately exploited in Wimsatt, 1981b, in the development of models of group inheritance.) The biases of the reductionist heuristics will still apply, but because the system boundaries have changed, they will have different effects. The comparison of intragroup with intergroup models is the right comparison for testing the robustness or artifactuality of lower-level reductionist assumptions for group processes. A comparable strategy should be equally appropriate for analogous problems in the social sciences.

Similarly, going down a level is the right medicine for testing more holistic models, which may ignore microlevel details. Geneticists regularly (and rightly) complain when higher-level optimization models predict optimal states that may not be genetically possible or that may take some time to attain so that the "equilibrium state" cannot be assumed (see, e.g., the last chapter of Oster and Wilson, 1979, and Wimsatt, 2006c, for detailed critiques of optimization modeling in population biology). I have called this approach "multilevel reductionist analysis" (Wimsatt, 1980b), though it is misleading to regard it as reductionist when the move is from a lower to a higher level. It is reductionist only in that the approach uses reductionist problem-solving techniques at any given level. It is not reductionist in the suggestion that one should go to a higher level of analysis to correct for the biases induced at a lower level.

The Importance of Heuristics in the Study of Human Behavior

I have argued that the study of our heuristics—of the nature of our real reasoning processes, their "cost-effective" advantages, and their systematic biases—should be a major topic on the agenda for the human sciences. Let me summarize the reasons why I think this is important. It is important first of all because these heuristics of reasoning are part of our equipment, and as with any equipment, they must be calibrated and used as tools for evaluating hypotheses and experimental and observational studies. This has already been done, for example, in the original investigation of Tversky and Kahneman (1974), as well as their more recent work (Mynatt, Doherty, and Tweney, 1977; Tweney,

Doherty, and Mynatt, 1981; Shweder, 1977, 1979a, 1979b, 1980a, 1980b; Gigerenzer, Todd, and the ABC Research Group, 1999; and Wimsatt, 1980b).

Second, as part of our equipment, these heuristics are part of our subject matter, one of our objects of study. This is true of us not only as individual problem solvers, but also as social beings. I should at this point confess that I have fallen prey to my own reductionist biases with my limited focus on the processes of individuals. We should focus also on the heuristics of group processes and on the biases of groups on the social processes of science. There are, for example, group identification processes that suppress intragroup disagreement, processes of competition that exacerbate disagreements between groups and restrict flow of information and recruits between them, as well as disciplinary and subject-matter biases that lead us to overestimate the importance of and to overextend the subject matter and theories that we know well and to take insufficient account of those we do not. I have formulated these as processes in the sociology of science, but they obviously apply more broadly to other spheres of human action. We need to understand these biases, both for their own sake and also to learn how to correct for their effects.

Early "post-modern" sociology of science (e.g., Latour and Woolgar, 1979) made useful progress in this direction, but from a determinedly "externalist" perspective. Both Campbell and I agree strongly that a more appropriate approach would be to try to integrate internalist and externalist perspectives. Two studies taking important and productive steps in this direction are Star (1983a, 1983b) and Griesemer (1983). Star focused on the development of the localizationist perspective in neurophysiology from 1870 through 1906 and made important new analyses of the individual and social processes for handling anomalies and other processes used for legitimizing data and methodological approaches (Star and Gerson, 1986). Griesemer focused on the macroevolution controversy in evolutionary biology and developed tools for analyzing "conceptual maps" (a scientist's models of the relationships between subject areas in his domain and the resultant structure of the problem to be solved) and for tracking the changes in these maps in the conceptual life history of individual scientists or in the diffusion of ideas and research problems from one research group to another. (Griesemer and Wimsatt later pursue this line in 1989.)

Third, we need to take stock of and to incorporate an understanding of our reasoning heuristics and their biases into our accounts of human action. These act as constraints on our decision processes, and can lead to a variety of unintended suboptimal consequences, as is documented in the theory and case studies of Janis and Mann (1977). On the positive side, things of which we cannot take account are things we can afford to ignore in all explanatory theory of human behavior. It is a truism that it is not the way the world is but how we conceive of and conceptualize it that determines our actions. If we do conceive it in various oversimplified ways, this can lead us into various sorts of error, but if these ways are indeed simpler than the way the world is, our explanatory tasks should be correspondingly simpler as well. We must be careful to correct for these biases in studying human action, but at the same time we must expect to find them in our accounts of it.

For these reasons, I do not see D'Andrade's (1986) third world of meanings and intentions as being quite so different from the engineering and functional systems studied in the biological sciences as he supposes. If I am right in believing that meanings, intentions, plans, strategies, decisions, beliefs, and the like are basically engineered structures, then we should expect them to have the same mix of strengths and weaknesses as any of our artifacts and to be best studied with tools that have at least a family resemblance to the conceptual tools of the other engineering disciplines. They may be less predictable, but complex machines have always been less predictable than simpler machines—and our cognitive and social worlds are nothing if not complex.

If this sounds like facile reductionism, it is not: if our cognitive structures share design principles with evolutionary and engineering artifacts, this does not (by itself) make them biological or hardware entities. It reflects, rather, principles of optimal or "satisficing" design common to all three areas. I fully expect that elucidation of some of our social and cognitive heuristics—as studied by social scientists—will provide insights that will be useful in these other disciplines, just as work in them has provided (for better or worse) metaphors and paradigms for social scientists. Before the rise of Darwinism, genetics, and molecular biology gave biology distinctive well-established theories, biologists borrowed freely from psychology and theories of society, and they are now doing so again. Thus evolutionary biologists have adopted the theory of games and claim that much more is to be mined for application to biological problems from economics and learning theory (Maynard-Smith, 1982), and at least two computer scientists argue that large parallel-processing computers should be developed with an architecture modeled on the structure of the scientific research community (Kornfeld and Hewitt, 1981). I suspect that much more is soon to follow.

Author's Note, 2007: This speculation tracked a line that bore diverse fruits. Danny Hillis' "connection machine" became in the late 1980s the apex of computer power—a mainframe using large numbers of similar processors. Multi-processor architectures now inhabit desktops, and grid computing in which data analysis and simulation processes borrow unused cycles from desktop computers across the Internet stretch the limits, but in ways different than Kornfeld and Hewitt imagined. (Attempts to estimate probabilities of complex events through "auctions" that pool large numbers of diverse expertises and experiences come closer.) Rosenblatt's early work on Perceptrons and brain-like architectures for artificial machines (1962) was reborn in the middle 1980s as "connectionism" and expanded both along research lines and as an applied tool for pattern recognition found in many Internet applications. The biggest innovations in integrated computing have been in the Internet itself, and in the software protocols and architectures that have allowed distributed computing shared among extremely heterogeneous communities of processors. Heuristics permeate these architectures to the core, and the Internet has transformed the scientific research communities they sought to emulate.

CHAPTER SIX

False Models as Means to Truer Theories

Many philosophers of science today argue that scientific realism is false. They often mean different things by this claim, but most would agree in arguing against something like the view that scientific theories give, or at least aim to give, approximate, or are approaching asymptotically to giving, a true description of the world. All theories, even the best, make idealizations or other false assumptions that fail as correct descriptions of the world. The opponents of scientific realism argue that the success or failure of these theories must therefore be independent of or at least not solely a product of how well they describe the world. If theories have this problematic status, models must be even worse, for models are usually assumed to be mere heuristic tools to be used in making predictions or as an aid in the search for explanations, and which only occasionally are promoted to the status of theories when they are found not to be as false as assumed.

While this rough caricature of philosophical opinion may have some truth behind it, these or similar views have led most writers to ignore the role that false models can have in improving our descriptions and explanations of the world. (Nancy Cartwright's [1983] excellent studies of the use and functions of models, though aiming at other conclusions, are a striking exception here.) While normally treated as a handicap, the falsity of scientific models is in fact often essential to this role. I will not discuss the larger issue of scientific realism here: the way in which most philosophers have formulated (or mis-formulated) that problem renders it largely irrelevant to the concerns of most scientists who would call themselves realists. These philosophers attack a realism that is "global" and metaphysical. Most scientists use and would defend a more modest (or local) realism, and would do so on heuristic rather than on metaphysical grounds.

By *local realism,* I mean something like the following: On certain grounds (usually, for example that the existence of an entity or property is known, derivable, or detectable through a variety of independent means—the "robustness" of Chapter 4), scientists would argue that an entity or property is real, and they cannot imagine plausible or possible theoretical changes that could undercut this conclusion. Furthermore, they might argue that their experimental and problem-solving approaches require them to presuppose the existence of that entity, property, or phenomenon—a heuristic argument. I suspect that many philosophical opponents of scientific realism could accept this kind of local and heuristic realism. I think that it is the kind of realism most worth defending, though I think that it may also give handles for defending a more ambitious kind of scientific realism. In any case, there is much more of use to be found in the topic of false models, to which I now turn.

Neutral models are discussed mostly indirectly in this chapter. (See Nitecki and Hoffman, 1987, from which this chapter came.) After an attempt at characterizing what a neutral model is, I show that this idea relates more generally to the use of false models, and discuss a variety of ways in which false models are used to get to what we, at least provisionally, regard as the truth. Many or most of the papers in Nitecki and Hoffman (1987) use their models in one or more of these ways. Thus, the focus of this chapter generalizes beyond the case of neutral models to consider the variety of uses of models, in spite of, or even because of, the fact that they are assumed to be false, to get at the truth or our best current approximations to it.

Even the Best Models Have Biases

The term *neutral model* is a misnomer if it is taken to suggest that a model is free of biases that might be induced by acceptance of a given hypothesis (for example, that the patterns to be found among organisms are products of selection). Any model must make some assumptions and simplifications, many of which are problematic, so the best working hypothesis would be that there are no bias free models in science.

This observation has a parallel in the question, "What variables must be controlled for in an experimental design?" There are no general specifications for what variables should be controlled, since what variables should be controlled or factored out through appropriate attempts to isolate the system, what variables should be measured, what errors are acceptable, and how the experiment should be designed are all functions of the purpose of the experiment. Similarly, what models are acceptable, what data are relevant to them, and what counts as a "sufficiently close fit" between model and data is a function of the purposes for which the models and data are employed. (As one reviewer for this chapter pointed out, many people who would applaud the use of appropriate controls and isolations in an experimental design inconsistently turn on mathematical or causal models of the system and criticize them for doing the same thing! This activity is desirable and necessary in either case.)

Any model implicitly or explicitly makes simplifications, ignores variables, and simplifies or ignores interactions among the variables in the models and among possibly relevant variables not included in the model. These omitted and simplified variables and interactions are sources of bias in cases where they are important. Sometimes certain kinds of variables are systematically ignored. Thus, in reductionist modeling, where one seeks to understand the behavior of a system in terms of the interactions of its parts, a variety of model-building strategies and heuristics lead us to ignore features of the environment of the system being studied (Wimsatt, 1980b, 231–235, and Chapter 5). These environmental variables may be left out of the model completely, or if included, treated as constant in space or in time, or treated as varying in some particularly simple way, such as in a linear or random fashion. In testing the models, with the focus on the interrelations among internal factors, environmental variables may be simply ignored (Wimsatt, 1980b) or treated in some aggregate way to simplify their analysis (Taylor, 1985).

A model may be bias free in one case (where its variables and parameters refer, at least approximately correctly, to causal factors in nature, and where its "accessory conditions" [Taylor, 1985] are satisfied) but biased for other cases it is applied to because it worked well in the first case (Taylor, 1985, discusses examples of this for ecological models in greater detail). Even where it is recognized that a model must be changed, there may be biases (1) in how or where it is changed, or (2) in the criteria accepted for recognizing that it must be changed.

An example of the first type is found in reductionist modeling: whereas both the description and interaction of parts internal to and of variables external to the system are usually oversimplified, there will often be a bias toward increasing the internal realism of the model in cases where the failure of fit of the model with the data is due to unrealistic assumptions about the environment (Wimsatt, 1980b).

A potential example of the second kind of case is provided by Williams' (1966, p. 17) "principle of parsimony," in which he recommends "recognizing adaptation at no higher level than is required" (presumably by the data). If this recommendation is taken as an invitation to find *some* set of parameter values for the simple model for which it fits the data, then one may be engaged in a "curve-fitting" exercise that may hide the need for a higher level or more complex model, unless one takes the pains to determine whether the parameter values for which fit is achieved are actually found in nature. This is seldom done, at least in the kind of optimization modeling frequently found in evolutionary biology.

The Concept of a Neutral Model

In evolutionary biology and ecology, a *neutral model* usually means a model without selection. Raup (1987b), in his work with Gould, Schopf, and Simberloff (1973), considers "random phylogenies"—phylogenetic descent trees in which originations and extinctions are determined by random variables. These artificial phylogenies in many respects resemble those found in nature. Raup et al. (1973) reasoned that the similarities between the artificial and natural phylogenies were not products of selection processes operating at that level. Their model did not, as they pointed out, rule out selectionist explanations for speciations and extinctions at a lower (e.g., intra- or inter-populational) level. Similarly, the work of Kimura (1983), Crow (1987), Crow and Kimura (1970), and others on "neutral mutation" theories modeled and evaluated patterns of molecular variability and change on the assumption that selection forces on these variants could be ignored. Similarities between their predicted patterns and what was found in nature led to various versions of the hypothesis that the evolution of various systems or of various kinds of traits was driven not by selection, but by various forms of genetic drift. Finally, Kauffman's (1985) work on "generic constraints" in development identified features which, because they were near universal properties of his randomly constructed model genetic control networks, are taken to provide "baselines" for the properties of systems on which selection acts. He argues that these generic properties will usually survive *in spite of* selection rather than because of it. Here Kauffman uses a comparison of models with and without selection to argue that selection is not important.

One must not assume that if the data fit the neutral model then the excluded variables are unimportant. The research of Raup, Crow, and Kauffman each suggest that selection may be unimportant in some way in producing the phenomena being modeled, but they do not rule it out entirely. Thus, Raup's "random phylogenies" do not exclude selection as a causal agent in producing individual extinctions (e.g., through overspecialization to a temporally unstable niche) or speciations (e.g., through directional selection in different directions producing isolating mechanisms between two different geographically isolated subpopulations of the same species), because his model simply does not address those questions.

Similarly, the work of Crow and Kimura could be consistent with the (very plausible) hypothesis that the very neutrality of most mutations is a product of selection. This could occur through selection for developmental canalization of macroscopic morphological traits, selection for redundancy of design through duplicated DNA, codon synonymy, or redundancy at the level of parallel synthetic paths in metabolic pathways, or at other higher levels. Neutrality of alleles could also be a product of selection for an architecture for the program of gene expression to preserve relatively high heritability of fitness or other phenotypic traits in the face of random sexual recombination, and with any of the preceding mechanisms or others yet to be discovered to accomplish this.

Finally, the ubiquity of Kauffman's generic properties among genetic control networks of a certain type (which he takes as a constraint on his simulations) does not rule out the possibility that that type may itself be a result of prior selection processes. Thus, Kauffman's genetic control networks have an average of two inputs and outputs per control element. But he chose this value of these parameters in part because his own prior work (Kauffman, 1969) showed that such networks had a shorter mean cycle time than networks containing more or fewer inputs and outputs, a feature that he argues is advantageous.

I have elaborated elsewhere a model of developmental processes (Wimsatt, 1986a, and Chapter 7) that presents an alternative hypothesis for the explanation of developmental constraints that seems likely in many such cases, and in even more cases where the feature in question is taxonomically very widely distributed, but not absolutely universal. Such broad or generic universality of some traits could be produced by strong stabilizing selection due to the dependence of a wide variety of other phenotypic features on them. This view is not new, but no one has attempted to model its consequences, with the exception of a model proposed by Arthur (1982, 1984) to explain features of macroevolution. (Either Arthur's model or mine gives a plausible explanation, for example, for the near-universality of the genetic code. See, e.g., Kauffman, 1985.)

Jeffrey Schank and I, in many papers (e.g., Schank and Wimsatt, 1988; Wimsatt and Schank, 2004), tested my model by doing simulations on networks like those of Kauffman (1985), where connections do not contribute equally to fitness (as in Kauffman's selection simulations) but the contribution to fitness of a connection is a function of the number of nodes that are affected by that connection. (We tried three different fitness functions as measures of the topological properties of the connection and of the number of nodes that are "downstream" of it, with similar results in all cases.) The results are quite striking. Our networks, like Kauffman's, show decay in a number of "good" connections under mutation and selection. (This effect, which increases in strength with the size of the network, is what leads him to doubt that selection can maintain a network structure of any substantial size against mutational degradation.) However, if the number of nodes accessible to a given gene through its various connections is taken as a measure of "generative entrenchment" (Wimsatt, 1986a), then our results show that the genes (or connections of genes) that have been lost in this mutational decay are those with low degrees of generative entrenchment, and that virtually all of the genes (and connections) that are significantly generatively entrenched are still there after 1,000 to 5,000 generations!

Thus, in effect, Kauffman's models confirm my theory, and the two theories need to be regarded as complementary rather than contradictory. Simulations have demonstrated the robustness of this phenomenon for different mutation rates, population sizes, numbers of generations, genome sizes, and connection densities. For the most recent review, see Wimsatt and Schank (2004) and Wimsatt (2001). The general phenomena of generative entrenchment are discussed in Chapter 7.

The models of Raup, Crow, and Kauffman use and rule out selection factors in different circumstances and in different ways, but they have two things in common. In each, the neutral model either does not include selection operating on the postulated variants, or (in Kauffman's case) supports arguments that selection is not efficacious under the circumstances considered. In each, the neutral model is treated as specifying a baseline pattern with which natural phenomena and data are to be compared in order to determine whether selection is required (if the natural phenomena do not fit the neutral pattern) or not (if they do fit). These neutral models are treated as *null hypotheses* (a term frequently found in the literature on ecological models, and of course in statistics—see Stigler's, 1987), which are to be rejected only if the natural phenomena deviate sufficiently from those predicted by the model.

These cases suggest that we characterize a *neutral model as a baseline model that makes assumptions, which are often assumed to be false for the explicit purpose of evaluating the efficacy of variables that are not included in the model.* This leaves out selection, and thus perhaps falls short of a more complete characterization of neutral models in evolutionary biology, but this more general characterization makes more explicit the connection with hypothesis testing in statistics and allows us to bring in features of the use of models from other areas of biology to focus on the heuristic advantages of their use. The characterization of neutral models in this way leads naturally to the more general question of *when, how, and under what conditions models that are known or believed to be false can be used to get new or better information about the processes being modeled.* If neutral models are useful in biology, this has less to do with their neutrality than with more general features of the use of models in science.

The fit of data with a neutral model or null hypothesis usually establishes that omitted variables do not act in a way specific to the models under comparison, not that they do not act at all. This is consonant with my earlier claim that the adequacy of models is highly context-dependent, and that their adequacy for some purposes does not guarantee their adequacy in general. However, the use of these models as templates can focus attention specifically on where the models deviate from reality, leading to estimations of the magnitudes of variables left out, or to the hypothesis of more detailed mechanisms of how and under what conditions these variables act and are important. This is a pattern of inference that is both common and important, and deserves closer scrutiny. The variety of ways in which this is done is the primary focus of the remainder of this chapter.

How Models Can Misrepresent

The simple observation that most models are oversimplified, approximate, incomplete, and in other ways false gives little reason for using them; therefore, their widespread use suggests that there must be other reasons. It is not enough to say (e.g., Simon, 1996) that we cannot deal with the complexities of the real world so simple models are all that we can work with, for unless they could help us do something in the task of investigating natural phenomena, there would be no reason for choosing model building over astrology or mystic revelation as a source of knowledge of the natural world.

Nor does the instrumentalist suggestion that we use them because they are effective tools rather than realistic descriptions of nature give us much help, for it presupposes what we want to understand—namely, *how* false models can be effective tools in making predictions and generating explanations. I want to suggest various ways in which false models can (1) lead to the detection and estimation of other relevant variables; (2) help to answer questions about more realistic models; (3) lead us to consider other models as ways of asking new questions about the models we already have; and, in evolutionary or other historical contexts, (4) determine the efficacy of forces that may not be present in the system under investigation, but may have had a role in producing the form that it has.

Before discussing ways in which false models can help us to find better ones, it is useful to have a classification of the ways in which a model can be false. A model, in this context, means one of two alternative things: (1) a *mathematical model*—an equation or set of equations together with the interpretations necessary to apply them in a given context, or (2) a *causal model* or proposed mechanism through which a phenomenon or set of phenomena is to be explained. Often one will have both, but the following comments apply roughly equally to either.

The following are ways in which a model can be false, and are ordered roughly in terms of increasing seriousness (except for 6 and 7).

1. A model may be of only very *local applicability*. This is a way of being false only if it is more broadly applied.
2. A model may be an *idealization* whose conditions of applicability are never found in nature (e.g., point masses, the uses of continuous variables for population sizes, etc.) but which has a range of cases to which it may be more or less accurately applied as an approximation.
3. A model may be *incomplete*—leaving out one or more causally relevant variables. (Here it is assumed that the included variables are causally relevant, and are so in at least roughly the manner described.)

4. The incompleteness of the model may lead to a *misdescription of the interactions* of the variables that are included, producing apparent interactions where there are none ("spurious" correlations) or apparent independence where there are interactions—as in the spurious "context independence" produced by biases in reductionist research strategies. Taylor (1985) analyzes the first kind of case for mathematical models in ecology, but most of his conclusions are generalizable to other contexts. (In these cases, it is assumed that the variables identified in the models are at least approximately correctly described.)
5. A model may give a *totally wrong-headed* picture of nature. Not only are the interactions wrong, but also a significant number of the entities and/or their properties do not exist.
6. A closely related case is that in which a model is purely *phenomenological.* That is, it is derived solely to give descriptions and/or predictions of phenomena without making any claims as to whether the variables in the model exist. Examples of this include: the virial equation of state (a Taylor series expansion of the ideal gas law in terms of T or V); using automata theory (Turing machines) as a description of neural processing; and linear models as curve-fitting predictors for extrapolating trends.
7. A model may simply *fail to describe or predict the data* correctly. This involves just the basic recognition that it is false, and is consistent with any of the preceding states of affairs. But sometimes this may be all that is known.

Most of the cases discussed in this chapter represent errors of types 2 thru 5, and the productive uses of false models would seem to be limited to cases of types 1 thru 4 and 6. It would seem that the only context in which case 5 could be useful is where case 6 also applies, and often models that are regarded as seriously incorrect are kept as heuristic curve-fitting devices. There is no hard and fast distinction between phenomenological and non-phenomenological models, and the distinction between the two often appears to depend on context (see Cartwright, 1983). Thus, the Haldane mapping function (discussed below) can be derived rigorously from first principles, but as such it makes unrealistic assumptions about the mechanisms of recombination (Haldane, 1919). It is sometimes treated as an idealization (case 2) and sometimes as a phenomenological predictive equation (case 6). Finally, a model may make false or unrealistic assumptions about lower level mechanisms, but still produce good results in predicting phenomena at the upper level. In this case, the upper level relationships may either be robust (see Chapter 4; Levins, 1966; Wimsatt, 1980a) or may be quite fragile results that are secured through obvious or unintended curve-fitting as with many optimization models in evolutionary biology and economics. (The former is a productive use of falsehood [function 10 below] and the latter is an often unproductive instance of case 6 above.)

Twelve Things to Do with False Models

It may seem paradoxical to claim that the falseness of a model may be essential to its role in producing better models. Isn't it always better to have a true model than a false one? Naturally it is, but this is never a choice that we are given, and it is a choice that only philosophers could delight in imagining. Will any false model provide a road to the truth? Here the answer is just as obviously an emphatic "no!" Some models are so wrong, or their flaws so difficult to analyze, that we are better off looking elsewhere. Cases 5 and 7 above represent models with little useful purchase, and curve-fitting phenomenological models (case 6) would be relatively rarely useful for the kinds of error correcting activity I propose. The most productive kinds of falsity for a model are cases 2 or 3 above, though cases of types 1 and 4 should sometimes produce useful insights. *The primary virtue a model must have if we are to learn from its failures is that it, and the experimental and heuristic tools we have available for analyzing it, are structured in such a way that we can localize its errors and attribute them to some parts, aspects, assumptions, or subcomponents of the model.* If we can do this, then "piecemeal engineering" (Simon, 1996) can improve the model by modifying its offending parts.

There is a mythology among philosophers of science (the so-called Quine-Duhem thesis) that this cannot be done, that a theory or model meets its experimental tests wholesale and must be taken or rejected as a whole. Not only science, but also technology and evolution would be impossible if this were true in this and in logically similar cases. That this thesis is false is demonstrated daily by scientists in their labs and studies, who modify experimental designs, models, and theories piecemeal; by electrical engineers, who localize faults in integrated circuits; and by auto mechanics and pathologists, who diagnose what is wrong in specific parts of our artifacts and our natural machines, and correct them. Chapter 4 and Glymour (1980) give general analyses of how this is done, and descriptions of the revised view of our scientific methodology that results. The case for evolutionary processes is exactly analogous. Lewontin (1978) argues that without the quasi-independence of traits (which allows us to select for a given trait without changing a large number of other traits simultaneously), any mutation would be a (not-so-hopeful) monster, and evolution as we know it—a process of small piecemeal modifications—would be impossible. (Implications of this insight are elaborated in Chapter 4.) In all of these cases then, piecemeal engineering is both possible and necessary.

The following is a list of functions served by false models in the search for better ones:

1. An oversimplified model may act as a starting point in a series of models of increasing complexity and realism.
2. A known incorrect but otherwise suggestive model may undercut the too ready acceptance of a preferred hypothesis by suggesting new alternative lines for the explanation of the phenomena.
3. An incorrect model may suggest new predictive tests or new refinements of an established model, or highlight specific features of it as particularly important.
4. An incomplete model may be used as a template, which captures larger or otherwise more obvious effects that can then be "factored out" to detect phenomena that would otherwise be masked or be too small to be seen.
5. A model that is incomplete may be used as a template for estimating the magnitude of parameters that are not included in the model.
6. An oversimplified model may provide a simpler arena for answering questions about properties of more complex models, which also appear in this simpler case, and answers derived here can sometimes be extended to cover the more complex models.
7. An incorrect simpler model can be used as a reference standard to evaluate causal claims about the effects of variables left out of it but included in more complete models, or in different competing models to determine how these models fare if these variables are left out.
8. Two or more false models may be used to define the extremes of a continuum of cases in which the real case is presumed to lie,

but for which the more realistic intermediate models are too complex to analyze or the information available is too incomplete to guide their construction or to determine a choice between them. In defining these extremes, the "limiting" models specify a property of which the real case is supposed to have an intermediate value. (See the discussion of Haldane, 1919 in Wimsatt, 1992.)

9. A false model may suggest the form of a phenomenological relationship between the variables (a specific mathematical functional relationship that gives a "best fit" to the data, but is not derived from an underlying mechanical model). This "phenomenological law" gives a way of describing the data, and (through interpolation or extrapolation) making new predictions, but also, because its form is conditioned by an underlying model, may suggest a related mechanical model capable of explaining it.
10. A family of models of the same phenomenon, each of which makes various false assumptions, has several distinctive uses: (a) One may look for results that are true in all of the models, and therefore presumably independent of different specific assumptions that vary across models. These invariant results (Levins' [1966] "robust theorems") are thus more likely trustworthy or "true." (b) One may similarly determine assumptions that are irrelevant to a given conclusion. (c) Where a result is true in some models and false in others, one may determine which assumptions or conditions a given result depends upon (see Levins, 1966, 1968; and Wimsatt, 1980a, and Chapter 4 for a more detailed discussion).
11. A model that is incorrect by being incomplete may serve as a limiting case to test the adequacy of new, more complex models. (If the model is correct under special conditions, even if these are seldom or never found in nature, it may nonetheless be an adequacy condition or desideratum of newer models that they reduce to it when appropriate limits are taken.)
12. Where optimization or adaptive design arguments are involved, an evaluation of systems or behaviors that are not found in nature, but that are conceivable alternatives to existing systems can provide explanations for the features of those systems that are found.

The core of this chapter, which considers the development of the "chromosomal mechanics" of the Morgan school, is intended to illustrate the first point. Points 2 and 3 are not discussed here, but all of the remaining ones are. I illustrate these functions by reanalyzing a debate between members of the Morgan school and W. E. Castle in the period 1919–1920 over the linearity of the arrangement of the genes in or on the chromosomes.

Background of the Debate over Linkage Mapping in Genetics

In the three short years between 1910 when Morgan (1910b) isolated and first crossed his "white eye" mutant *Drosophila,* and 1913 when Sturtevant published his paper mapping relative locations of mutants on the X chromosome using the frequencies of recombination between them, the Morgan school laid out the major mechanisms of the "linear linkage" model of the location of the genes on the chromosomes. During the next decade Morgan and his colleagues elaborated and defended this model. It won many adherents, and by the early 1920s had become the dominant view, despite several remaining unresolved questions.

This linear linkage model explained many things. Primary among these was the linkage between traits, some of which showed a nonrandom tendency to be inherited together. Two factors or traits linked in this way (as first discovered by Bateson, Saunders, and Punnett in 1906) violated classical Mendelian models of inheritance (Mendel, 1866/1956). They were not always inherited together, as would be expected for pleiotropic factors that produce multiple effects, such as Mendel's flower and seed coat color. Nor did they assort independently, as Mendel's seven factors seemed to do. Rather, they showed a characteristic frequency of association that fell in between these extremes. This and similar cases of partial linkage were first presented as falsifying instances for the Mendelian theory or model of inheritance. *The classical Mendelian model provides a pattern or template of expectations against which the phenomenon of linkage acquires significance, as in items 4 or 5 of the above list. Since cases of partial linkage represent intermediates between the two classical Mendelian types of total linkage and independent assortment, their classification in this way sets up the problem of linkage in a manner that also represents an instance of item 8.* When this phenomenon was first noticed in 1906 there were no theories or models to explain it, though the Boveri-Sutton hypothesis (Boveri, 1902; Sutton, 1903) provided a good starting point, one that Morgan and Sturtevant later developed.

With two alternative alleles at each of two loci, denoted by (**A**, **a**) for the first pair and (**B**, **b**) for the second, and starting with genotypes **AABB** and **aabb** as parents, the cases of total linkage or pleiotropy, independent assortment, and partial linkage would show the following proportions among gametes going to make up F2 offspring:

(Parents: **AABB**, **aabb**):

	Gametic types:			
Type of Linkage:	AB	Ab	aB	ab
Total:	50%	0%	0%	50%
Independent:	25%	25%	25%	25%
Partial:	50 – r%	r%	r%	50 – r%

The proportion **r** (which is bounded between 0 and 25%) was found to be (1) constant for any given pair of mutations or factors, (2) different for different pairs of factors, and, most importantly, (3) **r** was independent of the starting combinations of genes in the parents. (Properties (1) and (3) were given by Haldane, 1919, as a definition of *linkage.*) Thus, from property (3), if we start with the parental types **AAbb** and **aaBB** instead of **AABB** and **aabb**, the proportions of gametic types in the F2 generation are the exact complement of the original pattern:

(Parents: **AAbb**, **aaBB**):

Gametes:	AB	Ab	aB	ab
	r%	50 – r%	50 – r%	r%

Let us denote by **R** the proportion of gametic types not found in the parents, for example, **Ab** and **aB** for parents **AABB** and **aabb** (obviously, **R** = 2**r**). Something is causing reassortment of factors in **R**% of the cases to produce new gametic combinations. Furthermore, this proportion is not a function of what genes are found together in the parents, since the same proportion of **AB** and **ab** gametes are found if we

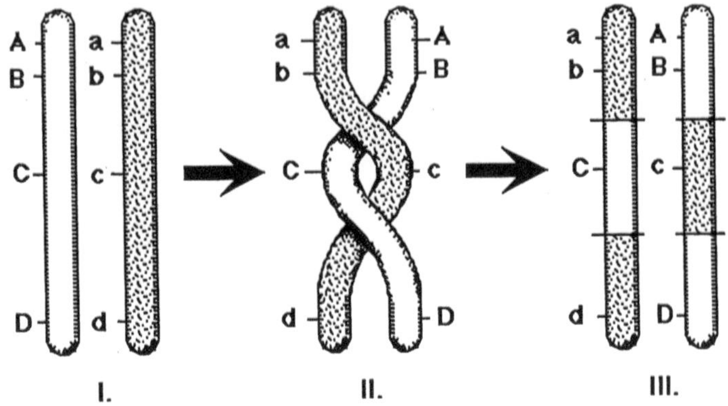

Figure 6.1 Crossing-over and recombination between homologous chromosomes.

start with parental types **AAbb** and **aaBB.** The hypothesis of the Morgan school (Morgan, 1911) was that the homologous chromosomes were winding around one another and then separating, exchanging corresponding segments.

In Figure 6.1, factor **C** is separated from the others (**A, B,** and **D**) through such an intertwining and separation. Morgan suggested that factors that were further apart would be separated more frequently through such a separation. This explains the different values of **r** or **R** for different pairs of factors. Factors with a constant linear location along the chromosome would always separate with the same characteristic frequency. This explains the constant proportion of new types produced by both (**AABB, aabb**) and (**AAbb, aaBB**) parental combinations. Finally, if the factors (and their alleles) kept the same relative locations along the chromosome through such interchanges, **r** or **R** should be constant for a given pair of factors, and independent of what other specific alleles are found at other locations along the chromosome.

By 1913, six mutations had been localized to the X chromosome, and Sturtevant (1913) noted that the recombination frequencies between pairs of factors were approximately additive. Thus, of the frequencies of recombinant types between factors, **R(AB)**, **R(BC)**, and **R(AC)**, one of them was usually equal to or slightly less than the sum of the other two. This suggested to Sturtevant that he could use the basic intuitions of Morgan's 1911 paper to construct a map of their relative locations on the X chromosome from the recombination frequencies between all pairs of the mutations, according to successive applications of the following basic scheme. If three factors, **A, B,** and **C** are arranged in a line, then one of them is between the other two. Which one is in the middle can be determined by seeing which of the following two conditions holds (approximately):

Equation:	Arrangement:
	A B C •-----•--------•
(1a) R(AB)+R(BC)=R(AC)	
	A C B •-----•--------•
(1b) R(AB)−R(BC)=R(AC)	

With this scheme of mapping and Morgan's proposed mechanism for the reassortment of different factors on different chromosomes, Sturtevant could not only give the relative locations of genes on the chromosomes, but he could also explain a new phenomenon that became obvious once the factors were ordered in this way. While one or the other of the above two equations is always met approximately, sometimes the largest recombination frequency is slightly less than the sum of the other two. These deviations from strict additivity in the observed recombination frequencies between different factors are systematic. They occur only for larger recombination frequencies, and increase in magnitude with the size of the recombination frequencies. *This is an instance of the fourth function of false models given above.* It involves, first, recognition of a causal factor (relative location in the chromosome map) that causes behavior like that described in the above equations. Second, it involves recognition of deviations from this behavior that, thirdly, are attributed to the action of a causal factor not taken into account in the simple model.

Suppose that crossovers occur at random with equal probability per unit distance along the length of the chromosome map. Then, for relatively short distances in which we should expect only one crossover event, as between **A/a** and **B/b** or **C/c** in Figure 6.1, the proportion of observed recombination events should increase linearly with the map distance. But if one is looking at a pair of factors that are sufficiently far apart that two or more crossovers have occurred between them, then the observed recombination frequency should be an *underestimate* of the map distance between the two factors. This is because a double crossover between two factors (as with **A/a** and **D/d**) brings the factors

A and **D** back to the same side so that they are included in the same chromosome. At this early stage in the development of their model, they had misidentified the stage in meiosis when crossing over occurs, but the basic argument remains the same today. Under these circumstances, if one is not tracking any factors between the two factors in question, an even number of crossovers will be scored as no crossovers, since the factors will come out on the same side, and an odd number of crossovers will be scored as one crossover, since the factors will come out on opposite sides. Thus an increasing proportion of recombination events will be missed as the distance between the two observed factors increases.

With this bias, factors that are further separated should show increasing deviations between observed recombination frequency and actual map distance. In 1919, J. B. S. Haldane gave a mathematical analysis of the problem and derived from first principles what has come to be known as the "Haldane mapping function," according to which the relation between observed recombination frequency (**R**) and map distance (**D**) is given by the equation: $\mathbf{R} = .5(1 - e^{-2D})$. Haldane characterizes the mechanism underlying this equation as assuming an "infinitely flexible chromosome" (1919, p. 299). He describes the case in which multiple crossing over does not occur (producing an exact proportionality between recombination frequency and map distance) as assuming a "rigid" chromosome (p. 300). He also suggests an equation for behavior between these two extremes as representing the case of a stiff or partially rigid chromosome.

This kind of mechanical argument underlying the derivation of a mathematical model is extremely useful, because it suggests sources of incorrectness in the idealizing assumptions of the models, and serves to point to specific factors that should be relaxed or modified in producing better models. *As such it exemplifies case 8 from the list of functions of false models.* Haldane does not have a preferred choice for how to model the behavior of a partially rigid chromosome, so the best that he can do is to present two largely unmotivated models for the intermediate case, one of which is an analytically tractable but rather ad hoc model and the second of which is explicitly just a curve-fitting attempt for use in prediction.

The Haldane mapping function was either unknown to or at least never used by the members of the Morgan school during the period of debate with Castle over the linear arrangement of the genome in 1919 and 1920 (but see Wimsatt, 1992, for a further account of this mysterious and ultimately crucial and deliberate omission)! This was possibly because their empirically determined "coincidence curves" for the various chromosomes, which gave frequency of double crossovers as a function of map distance, indicated the interaction of more complex and variegated causal factors determining multiple crossing over (see Muller, 1916a–1916d).

Whatever the explanation for their failure to use Haldane's model, the Morgan school adherents clearly had a sound qualitative understanding of the predicted behavior of their proposed mechanism from the first. Such an understanding is displayed already in Sturtevant's 1913 paper, where he pointed out that with the proposed mechanism, there should be a very small frequency of double crossovers even between factors that are quite close together. These double crossovers could be detected by tracking a factor that was between the two factors in question (as **C** is between **A** and **D** in Figure 6.1). But Sturtevant detected no close double crossovers, and the frequency of crossovers between relatively distant factors was substantially less than expected if one crossover had already occurred in that interval. (The relatively distant factors were 34 map units apart, where one map unit is that distance for which 1% recombination is expected.) He therefore hypothesized an "interference effect" acting over a distance around a crossover within which a second crossover was prevented or rendered exceedingly unlikely. *Here is yet another application of the template matching function of models of case 4, where the deviation from the expected performance of the template is used to postulate another causal factor whose action causes the deviation.*

This additional hypothesis of an interference effect was required for the model of the Morgan school to account for the exact linearity found in map relations for close map distances (see Figure 6.4). This was the subject of extensive investigations by Muller in 1915 and 1916, who considered the mechanisms of crossing-over, proposed a number of alternative mechanical hypotheses to explain this effect, and sought to gather data to choose among them (Muller, 1916a–1916d). The mechanical model for recombination was a fruitful source of hypotheses on the cause of this phenomenon. For example, if we assume that chromosomes have a certain rigidity and torsional strength, it is clear that they can be wound up only so tightly (like a double-stranded rubber band) before they break. Thus chromosome rigidity and torsional strength determine a maximum tightness of the loops before breakage would occur. This in turn determines a minimum length of chromo-

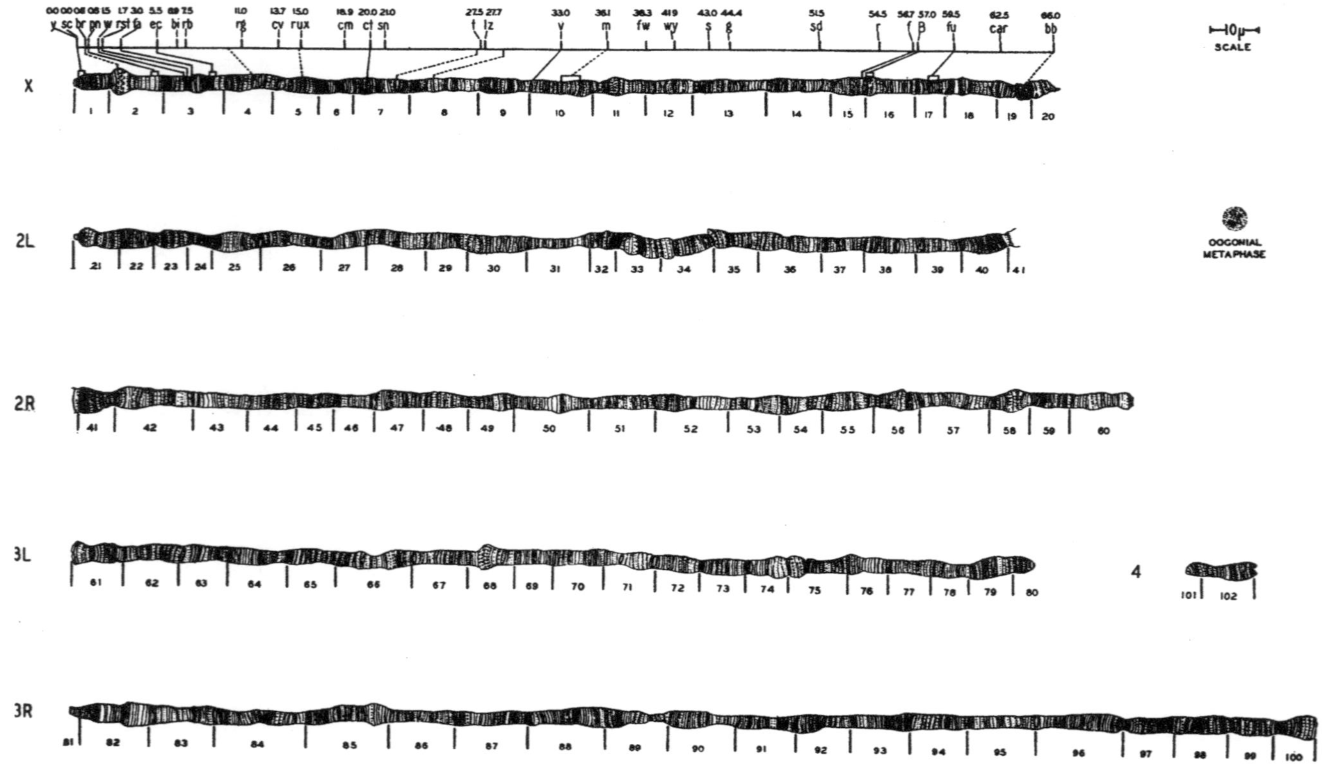

Figure 6.2 Linkage map of factors in the X chromosome and their corresponding bands in the physical (salivary gland) chromosome. From Sturtevant and Beadle (1939), pp. 130–131, fig. 48.

some between crossover events—an "interference distance" (this is just a simple elaboration of the argument of Haldane, 1919).

The occurrence of interference also had a beneficial secondary effect, since it meant that recombination frequencies between closely linked factors could be used as a direct measure of map distance, rather than correcting them for unobserved multiple crossovers. (If recombination events occurred at random along the chromosome map, and without interference, so that they were statistically independent, then the multiplicative law for the co-occurrence of multiple events would apply and for a distance in which there was a probability **p** of one crossover, there would be a probability $\mathbf{p}^2$ of two crossovers, $\mathbf{p}^3$ of three crossovers, $\mathbf{p}^n$ of n crossovers, and so on.)

To appreciate the character of the debate with Castle, it is important to realize the theoretical character of the parameter. One could not infer from a chromosome map exactly where the factors were on the chromosome. As Sturtevant noted in 1913, the chromosome might have differential strengths and probabilities of breakage along its length, leading in general to a non-linear relation between map distance and distance along the chromosome. Nor was it possible to determine which end of the map corresponded to which end of the chromosome without the production of aberrant chromosomes having visibly added or subtracted segments at an end, something that was not done until the 1930s.

In the case of *Drosophila,* the debates continued on a theoretical plane until the (re)discovery by Painter (1934) of the giant salivary gland chromosomes, which had visible banding patterns whose presence or absence and arrangement could be visibly determined. This allowed the location of genes at specific banding patterns, and the ready detection of inversions, reduplications, translocations, and deletions. Some of these were hypothesized earlier solely on the basis of changed linkage relations—a remarkable (and laborious) triumph that makes the earlier work on chromosome mapping one of the most elegant examples of the interaction of theoretical and experimental work in the history of science. The subsequent localization of genes to bands by studying genetic deletions also permitted confirmation of Sturtevant's conjecture that the mapping from genetic map to chromosome, while order-preserving, was non-linear: the frequency of crossing over was not constant from one unit of physical chromosome length to the next (Figure 6.2).

Castle's Attack on the Linear Linkage Model

Not everyone was enamored of the mechanical models of the Morgan school. A number of geneticists (including Bateson and Punnett, 1911, Goldschmidt, 1917, and Castle, 1919a–1919c) attacked it on a variety of grounds between 1913 and 1920, and Goldschmidt was still an outspoken critic in 1940 (see Carlson's [1966] excellent history). At the time they had what seemed like good reasons for these attacks, reasons that led them (and Morgan too—as late as 1909 and 1910; according to Allen, 1978) to attack the Boveri-Sutton hypothesis of 1903 that the genes were located on the chromosomes. The Boveri-Sutton hypothesis was a direct ancestor of the linear linkage models of the Morgan school.

These skeptics of the Boveri-Sutton hypothesis and the later models of the Morgan school were bothered that the models had nothing to say about gene action, and that they explained only correlations in the inheritance of traits. But the tradition from Aristotle down through the beginning of the twentieth century was that the hereditary elements, whatever they were and however they worked, determined both the transmission of traits and their development. It seemed unreasonable that any theory dealing only with the first of these could be correct: they sought a single theory that would explain both (see Morgan, 1909; Allen, 1978; Carlson, 1967). The Morgan school chose to defer consideration of this problem. Given the many successes of their research program, transmission genetics went on apace without having much if anything to say about development until the rise of molecular biology and the development of models of gene action like the operon model of the early 1960s.

Two things characterized the theoretical attempts of the Morgan school opponents: (1) all their models attempted at least to make room for gene action (usually in a variety of different ways), and (2) they all seemed to retain a healthy skepticism, not only about the Morgan model, but about any conclusions that appeared to follow too closely from it. Virtually everyone accepted the data that the linear linkage model explained—that seemed too hard to object to, but this was combined with a healthy skepticism about their mechanical models. A frequent line taken by geneticists that outlived widespread opposition to the model of the Morgan school (found as late as Stadler, 1954) was to be "operationalists" or "instrumentalists" about the gene. Given that they could trust the data from breeding experiments, but not the models produced to explain the data, it seemed more reasonable to avoid commitments to the theoretical models and to accept the genes only as "operationally defined" entities that produced the breeding results, in all cases staying as close to the data as possible.

Castle's operationalist attack on the model of the Morgan school is particularly interesting because the subsequent debate provides a clear example of the superiority of a mechanistic or realist research program over an operationalist or instrumentalist one. Some of the reasons why this is so will become apparent from the following discussion, but many of the most convincing cases and arguments would take us too far afield, and will have to be left for another occasion (see Wimsatt, 1992).

In 1919, Castle published his first attack on the linear linkage model. He thought that the linear model and the associated theoretical concept of map distance were too complicated and too far from the data. He complained that "it is doubtful . . . whether an elaborate organic molecule ever has a simple string-like form" (1919a, p. 26). He was bothered even more by the fact that map distances for three of the four *Drosophila* chromosomes exceeded 50 units, which he saw as inconsistent with the fact that observed recombination frequencies never exceeded 50 percent. (This apparent conflict arose from his operationalist assumption that map distance should be made proportional to recombination frequency and represents a simple misunderstanding on his part of what the Morgan school was claiming—see below.) To construct the map from recombination frequencies one had to invoke the possibility of multiple crossing over and posit "interference effects," and Castle also regarded both of these assumptions as dubious.

Given his operationalist preferences, Castle suggested that the simplest hypothesis would be to assume that the distance between the factors (whatever it signified) was a linear function of recombination frequency. He proceeded to construct a mechanical model of the arrangement of the genes in the chromosome by cutting wire links with lengths proportional to the recombination frequencies and connecting them, producing the phenomenological model of the X chromosome of *Drosophila,* diagrammed in Figure 6.3. (A *phenomenological model* [Cartwright, 1983] describes the data without explaining it.) He thereby claimed to have produced a model that fit the data and did so without making the further hypotheses of double crossing over and interference effects. In spite of his operationalist stance, he went on to suggest and use two alternative (and mutually inconsistent) mechanistic models to interpret the significance of his construction and to attack the linear linkage model. (Like most other operationalist opponents of the

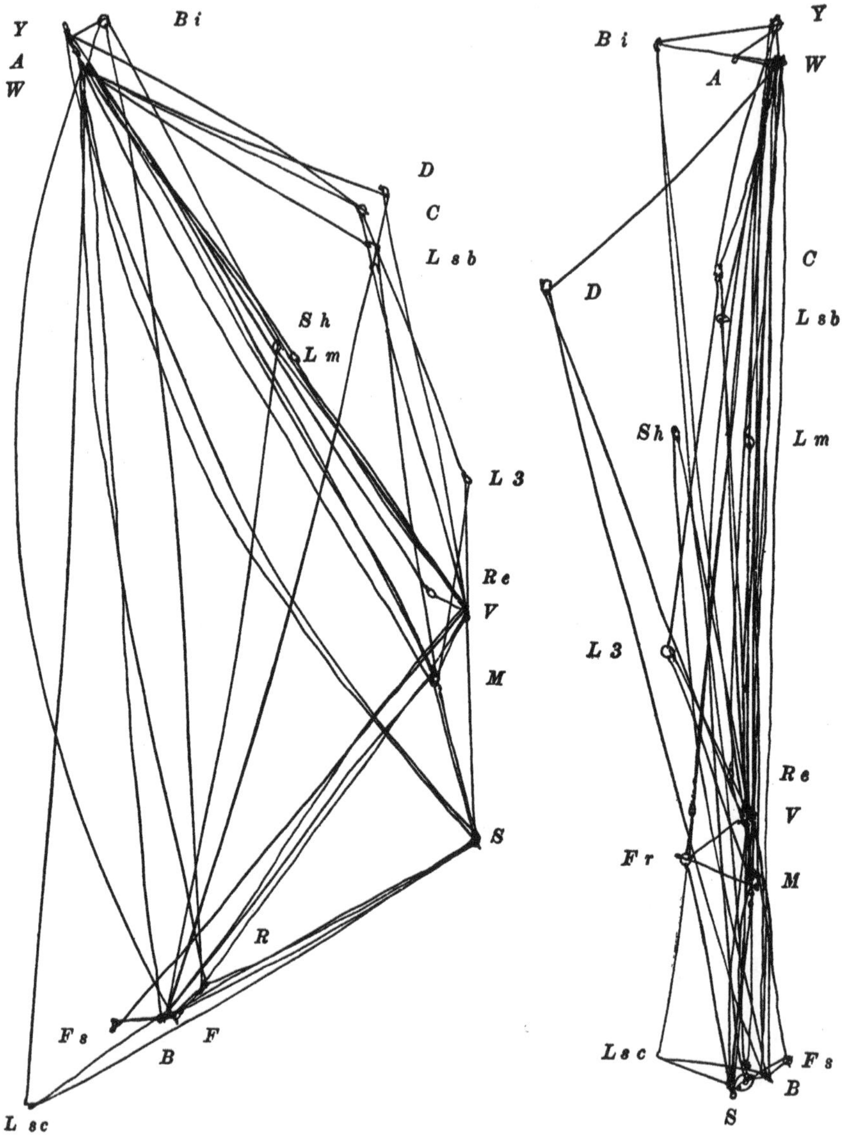

Figure 6.3 Castle's non-linear model derived by making distances proportional to recombination frequencies. From Castle (1919a), p. 29.

Morgan school, Castle seemed to be a frustrated mechanist!) The force of the counterattacks by the Morgan school was to show that (1) the data Castle used did not support his model, (2) even if he used the data acceptably, which he did not, and (3) that his model would fare even worse if more data were included. Furthermore, they showed (4) that neither of the two alternative mechanisms he proposed to account for his phenomenological model would have the desired effects.

Castle's critiques (1919a, 1919c) drew multiple responses, including those by Sturtevant, Bridges, and Morgan (1919), and by Muller (1920). Muller's response covered virtually all the ground of the earlier authors and went much further. His critique is elegant, covering not only the selection of the data and the reasons why it had to be treated differently for map construction and for testing the hypothesis of linear arrangement, but also an explanation of why Castle's model worked as well as it did (in spite of its incorrectness), predictions that it could be expected to break down and where, a defense of the reasonableness (and inevitability) of interference and multiple crossing over, and a complex set of arguments that Castle's claims were not only inconsistent with the data, but were internally inconsistent as well. Muller's critique included a number of additional arguments not discussed here that further strengthen the case for the chromosomal mechanics of the Morgan school.

Muller's Data and the Haldane Mapping Function

In his 1920 paper, Muller used data derived from his own earlier experiments (Muller, 1916a–1916d) to argue for the inadequacy of Castle's model (Wimsatt, 1992, discusses this work extensively). These data are of the observed frequencies of recombination between each possible pair of six of the factors on the X chromosome. In the Morgan school model, the recombination frequencies between nearest neighbor factors are assumed to be the map distances. Deviations from additivity for frequencies of recombination between more distant factors (see discussion of Sturtevant above) are seen as products of multiple crossovers between the observed pair of factors. The greatest distance between nearest neighbors among the factors Muller chose is 15.7 map units, and the lowest distance for which non-additivities are observed is 23.7 map units. Thus the assumption that nearest neighbor distances are real map distances, and not underestimates, seems justified. Another problem emphasized by both Castle and Muller is that the rarity of re-

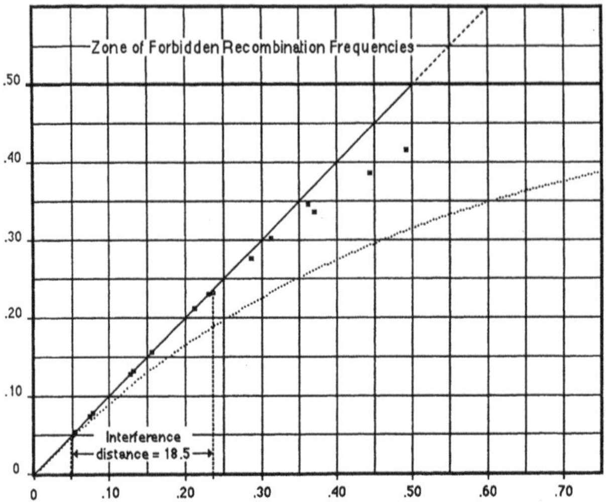

Muller's Data
(from Muller, 1920, p.109)

Map dist.:	Rec. freq.:
primary distances:	
0.055	0.055
0.074	0.074
0.157	0.157
0.080	0.080
0.133	0.133
2nd order distances:	
0.129	0.129
0.231	0.231
0.237	0.234
0.213	0.213
3rd order distances:	
0.286	0.278
0.312	0.303
0.370	0.337
4th order distances:	
0.364	0.347
0.445	0.386
5th order distance:	
0.492	0.416

Figure 6.4 Relation between recombination frequency (vertical axis) and map distance (horizontal axis) according to Castle's linearity assumption (solid line) and the Haldane mapping function (dotted line), with Muller's data (square points on graph) and an empirical estimate of interference distance shown.

combination events between very closely linked factors could make very small distance estimates unreliable. Muller obviated this objection by picking 6 factors (out of 12 for which he had data) that were neither too closely nor too distantly spaced along the chromosome map).

In Figure 6.4 I have graphed the relation between recombination frequency and map distance to be expected in two cases: (1) the linear relation between the two supposed by Castle's model and (2) the curvilinear relationship expected for the Haldane mapping function (or HMF). With the HMF, the relationship for small distances is approximately linear, but deviates from linearity with increasing map distance and asymptotes for very long distances at 50 percent. This suggests that factors very far apart on a long chromosome are essentially uncorrelated in their inheritance. (This is a new prediction of the linkage analysis, since it suggests that two factors can be uncorrelated in their inheritance *either* by being on different chromosomes [as before] *or* by being sufficiently far apart on a chromosome with a long map.)

Both of these curves are to be compared with Muller's data, which suggest a relationship that remains essentially linear for much larger map distances than with the HMF (as a product of interference effects), but that also appears to asymptote at a recombination frequency of 50 percent for very great map distances as with the Haldane mapping function. This actually cannot be seen in Muller's data, and omits significant complexities (Wimsatt, 1992) since the most distant factors he lists do not have very large separations. It can be seen for the data in Haldane's graph (1919, p. 309), which interestingly also shows much greater scatter in the points—largely a product of the fact that Haldane took data from different experiments. (The preceding omits significant complexities discussed in Wimsatt, 1992.) As Muller (1920) argued against Castle, this practice may show confounding effects due to changing linkage relationships (produced by different translocations, inversions, or deletions in the stocks of flies used in the different experiments), different temperatures, and other factors affecting recombination frequency.

Notice several things here:

1. The actual data Muller gives behaves qualitatively like the HMF, and might well be mistaken for it if the HMF were not graphed with it for comparison, pointing to the systematic discrepancies. The HMF is being used here as a template, deviations from which point to the importance of interference, which is a causal factor affecting the data but not taken account of in the mechanisms supposed in deriving the HMF. It is being knowingly used as a model that is false because it leaves out a known causal factor, which produces systematic and increasing deviations from the predictions of the HMF with increasing map distance. *This is an instance of the fourth function of false models.* It was done over and over again by members of the "fly group" (Wimsatt, 1992.)

2. If the data curve did not look qualitatively like that of the HMF, there would be less temptation to use it as a template and more temptation to say that it fundamentally misrepresented the mechanisms involved, rather than being false by being merely incomplete. (In that case the partitioning of the effects of causal factors along the lines suggested by the model could not be trusted.) Given their qualitative similarity (in particular, the linearity of **R** and **D** for small distances and the asymptotic approach of **R** to 50 percent as **D** gets very large), however, this temptation is strong. The tendency to treat the HMF as a baseline model is further increased by the fact that Haldane presents two other models, one for "rigid" chromosomes (in which **R** is supposed to be linear with **D,** at least up to 50 percent), and one for "semirigid" chromosomes in which interference has an effect intermediate between the two extremes. Muller's data also falls in between these two extremes

(though not exactly on the curve of Haldane's intermediate model) and is thus plausibly treated as due to the joint operation of factors found in the HMF model together with the operation of some kind of interference mechanism (see function 8 of false models in the above list).

3. The discrepancies between the HMF and the data can actually be used to get a rough estimate of the interference effect. Thus, if one observes where Muller's data begins to deviate noticeably from the line **R = D** (at 23.4) and estimates where the HMF begins to deviate comparably (at about 5), one can get a rough estimate of interference distance as the difference of these two numbers—about 18 or 19 map units. (I subtract 5 from 23.4 to correct for the fact that double crossovers at close separations may be present in too small a frequency to be detected, but one can use the "measurability lag" for the HMF to estimate that for the real data. *This is another use of a model as template to calibrate a correction factor,* an additional use not on my list since it does not involve the use of false suppositions of a model.) This estimate is a brute empirical estimate from the behavior of the data, not one from any assumptions about the mechanism or mode of action of interference, but illustrates how *a model can be used to estimate a parameter that is not included in it—case 5 from the list of functions of false models.*

Interestingly, the mechanisms of interference have proven since to be quite refractory to analysis in terms of general equations that both apply to a variety of cases and are derived from underlying mechanical models—probably because the mechanisms are both mathematically complex to describe and vary substantially from case to case. (Thus, the location of the two factors relative to the centromere is a crucial determinant of their recombination behavior, a factor not considered in any of these models.) As a result, many more recent treatments, such as those of Kosambi (1944), Bailey (1961), and Felsenstein (1979), have constructed phenomenological mathematical models with no mechanical underpinning, but which have other advantages. These include (1) providing a relatively good fit with known data and (2) a schema for prediction, (3) generating important prior models (such as Haldane's) as special cases, (4) having nice formal properties, and (5) producing nice operational measures that can be applied to more complex breeding experiments. *The third is an instance of the eleventh function of false models.* Felsenstein (1979) explicitly mentions the last four functions. With possibly unintended irony, he summarizes the advantages of his phenomenological model over other more realistic approaches:

> There are a number of papers in the genetic literature in which specific mapping functions are derived which are predicted by particular models of the recombination process. . . . While these models have the advantage of precision, they run the risk of being made irrelevant by advances in our understanding of the recombination process. In this respect the very lack of precision of the present phenomenological approach makes it practically invulnerable to disproof. (p. 774)

Felsenstein's more extended defense illustrates part of the ninth function of model building (predictive adequacy), but he also claims that the lack of underlying mechanical detail may make it more robust in the face of new theory (see also Cartwright, 1983). If his model is more immune to falsification, this "advantage" is purchased at the cost of a descent into the possible abuses of curve fitting, which Felsenstein warns us against by explicitly noting the phenomenological character of his model. As I will show however, being a phenomenological model is no necessary guarantee against falsification.

Muller's Two-Dimensional Arguments against Castle

Castle's model generates a three-dimensional figure because the recombination frequencies between more distant factors are less than the sum of recombination frequencies between nearer factors, a feature earlier referred to as *non-additivity*. This produces triangular figures for each triple of factors which, when connected together, produce complex polyhedral structures in three dimensions, as in Figure 6.4. If we look at three factors at a time, **A**, **B**, and **C**, and arrange them so that they are at distances proportional to their pairwise recombination frequencies, the fact that **R(AC) < R(AB) + R(BC)** implies that **A**, **B**, and **C** will be at the vertices of a triangle (this relationship is the "triangle inequality" of plane geometry). As the deviation gets smaller, the extreme angles in the triangle get smaller, producing for **R(AB) + R(BC) = R(AC)** the "degenerate triangle" of a straight line. Only if **R(AB) + R(BC) < R(AC)** would it be impossible to construct a triangular figure with straight edges.

If we consider another factor, **D**, such that **R(BC) + R(CD) > R(BD)**, we get another triangle, which must be placed on top of the first, since they share side **R(BC)**. The addition of this triangle to the first generates a new distance, **R(AD)**, which provides a prediction of the recombination frequency between **A** and **D**, that can be compared with the data. In this way, with successive applications of this construction, Muller constructed a map of the six factors according to Castle's principles.

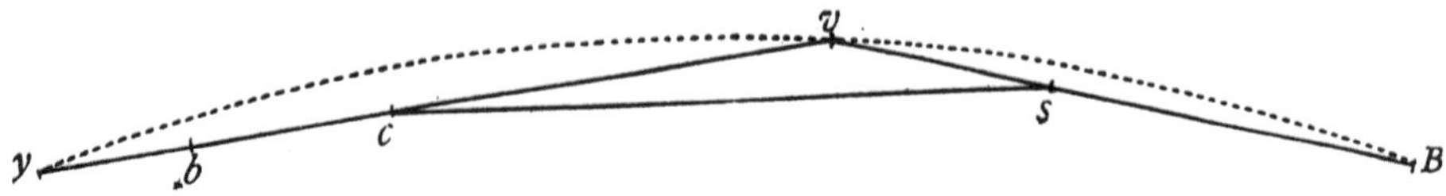

Figure 6.5 Muller's construction of the map of the six factors *y, b, c, v, s,* and *B,* using Castle's method and the closest and next-closest distances. From Muller (1920, figure 3, p. 113). *y b, b c, c v, v s, s B* and *y c, b v, cb s, v B* are each represented by a line of length proportionate to the respective frequency of separation. The dotted curve shows the average angular deviation of the line of factors, according to this system.

Surprisingly, his map is two dimensional, rather than three-dimensional as Castle required! Why was it legitimate to use a figure of lower dimensionality (Figure 6.5), which would allow the use of an additional degree of freedom for fitting the data?

Muller argues (1920, p. 113) that his two-dimensional figure is the only possible one using the "strongest and second strongest" linkages—that is, using the **R**'s for nearest neighbor and next nearest neighbor factors. This is true for a reason that he does not explicitly mention: because he has used relatively tight linkages (sufficiently close to prevent double crossovers in all cases but one), the triangle inequality holds for only one triple of factors—for all others there is an equality. In this special case, we get a "degenerate" figure that is only two-dimensional, because there is a "bend" at one factor and all of the other relations are linear. Two non-collinear straight lines connected at one point determine a plane. The addition of any more data that had non-additivities would have required (at least) three dimensions.

The argument that Muller emphasizes most strongly is that the distance between the most distant factors (with four factors in between) on the model constructed according to Castle's principles is too great—it is 49.3 as opposed to his observed recombination frequency of 41.6. (The predicted distances are also too great for the factors that are three or two factors apart, as he notes.) He claims that if the figure is bent so as to make the longest distance correct, then the shorter distances become too short, and the model still fails to fit the data (Muller, 1920, p. 113, figure 4). Thus, as he argues earlier, "the data . . . could not be represented either in a three-dimensional or in any other geometrical figure" (p. 112).

We can go further than Muller does explicitly, as might be suggested by Muller's references in two places to higher-dimensional figures. Suppose that there were non-linearities or bends at two places, but that they were sufficiently small that the predicted recombination frequency between the two most distant factors was still too large. Then going to three dimensions (i.e., by rotating a part of the figure out of the plane) would not help Castle, since this could serve only to *increase* the distance between the factors, and thus the error in his predictions, if they had already been geometrically arranged in two dimensions so as to minimize the distance and predictive errors. If the predictions had been systematically too small, then going to a three-dimensional figure would help, but if they were already too large, as they were, then nothing would work. Even if the greatest distance were too small, one would still have to fit the shorter distances and, if they are supposed to represent physical distances, one is limited to a maximum of three dimensions. In any case, this procedure would obviously be nothing more than a curve- (or polyhedron!) fitting technique, as Muller's arguments suggest.

Here we see arguments in which simpler models (a two-dimensional model with a reduced data set) are used to draw conclusions about more complex ones (three-dimensional models with larger data sets). In particular, given the data in question, no geometrical figure in any number of dimensions could consistently represent the data as a set of distances without bending wires or other such "cheating." (Note that Castle's model [in Figure 6.4] does contain several bent wires, which represent incorrect predictions, an inconsistency noted by Muller.) *This is an example of the sixth function of false models, in which analysis of a simpler false model is used to argue for the inadequacy of a family of more complex related models.*

Multiply-Counterfactual Uses of False Models

We can take one more step, in this case beyond the grounds of the actual dispute, to illustrate a different kind of use of false models to counter the claim of Castle that he can do without the hypothesis of interference distance. We saw that Muller argued that Castle's triangular models produced predictions for recombination frequencies that were too great at larger distances. If we transform Muller's data (counterfactually) to produce data that would be expected in the absence of interference, the fit of the data with Castle's model is far worse even than be-

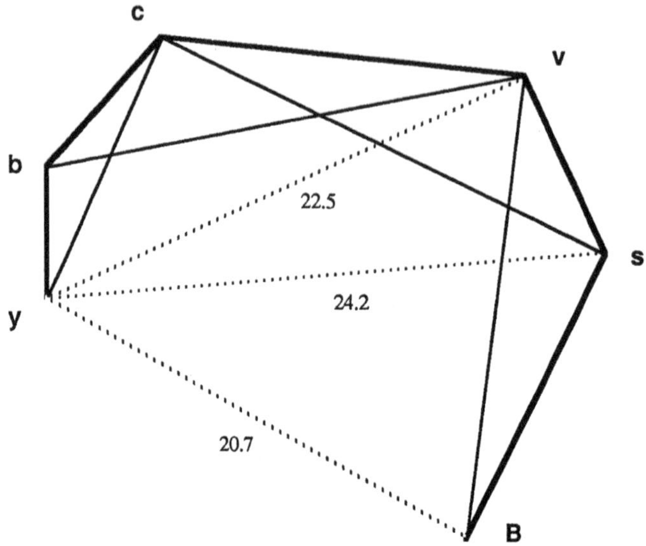

Figure 6.6 Two-dimensional map constructed according to Castle's principles using Muller's data transformed according to the Haldane mapping function to estimate the effect of assuming no interference.

---

fore, with much larger errors in the opposite direction. Thus Castle cannot complain about the Morgan school's use of interference effects, since this additional argument shows that he needs it even more than they do.

I pointed out earlier that interference was very useful to the Morgan school since it allowed them to assume that recombination frequencies between factors that were fairly close together on the chromosome map were true measures of their relative map distances. They would have been biased underestimates if multiple crossovers between the factors were possible. Suppose that we drop this assumption, and assume that close distances are also subject to the systematic bias produced by multiple crossovers. Then the "true" map distances (in a world with "infinitely flexible" chromosomes, and thus no interference) could be calculated from the Haldane mapping function, and these newly, transformed distances could be used to construct a chromosome map according to Castle's model. I have constructed such a map in Figure 6.6 using Muller's data transformed according to the HMF.

In this two-dimensional figure, there is a bend at each factor because the triangle inequality is satisfied for each triple of nearest neighbor factors because of the non-linearities introduced by the (supposed) multiple crossovers. These non-linearities make the map bend so strongly that the predicted recombination frequencies for more distant factors actually *decrease* as the map curves around on itself (compare distances **ys** and **yB**). This data transformation demonstrates that without interference, Castle's model would have been even more strongly inconsistent with the data. Three dimensions might have helped Castle somewhat in this case, but unless the irregular helix that would be produced for a long chromosome by bending in a third dimension was very elongated, the distance function would have oscillated for increasingly distant factors—again inconsistent with data available at the time. Furthermore, a choice of other intermediate factors would have given different relative locations for the two factors at the ends. Castle could have bent successive distances in alternate directions or according to some other still more irregular scheme, but would have had to do so in a way that was consistent with all the other shorter distances, which are not used in this demonstration. At best, the construction would have been very ad hoc, and the addition of any new factors would have produced new problems.

In this argument, the use of false models is very circuitous indeed. True data is transformed according to a false model (the Haldane mapping function) to produce false data, which is then plugged into a false (two-dimensional) but simpler version of a (false) three-dimensional model. This is done in order to argue that the worsened fit of that model with the data undercuts Castle's attack on the Morgan school's use of interference. It shows that he needs interference even more than they do. (This argument is legitimate, because it uses false premises that Castle wishes to accept. It is a kind of *reductio ad absurdum* argument, since it involves accepting a premise that they wish to reject to undercut an argument against a premise that they wish to accept.) *This example illustrates the seventh listed function of false models.* It is contrived, since the Morgan school did not actually make this argument, though it is certainly one that they could have made with the data and models at hand, and it would not have been far from the spirit of either Haldane's analysis or that of Muller to do so. This case is closer to the spirit of many cases of neutral model building, where constructed ("random") data sets are used to test hypotheses about selection.

False Models Can Provide New Predictive Tests Highlighting Features of a Preferred Model

In 1917, Richard Goldschmidt (better known for his later suggestion that macro-evolution took place through the incorporation of macro-mutations or "hopeful monsters") published a paper criticizing the linear linkage model. He proposed an alternative explanation for the factor exchanges in crossing over in which factors were tied to specific locations on the chromosome through biochemical forces. Corresponding allelic factors left the chromosome to accomplish their activities in the cell, an attempt to accommodate gene action. They then returned to their places for the chromosomes to undergo any mitotic or meiotic activities. Each factor had a specific "force" that acted with its binding site, and different alleles had somewhat different forces. The differences between forces for alleles at the same locus were much less than the differences between any one of them and the forces of factors at other loci. When they later reassembled on the chromosome, the similarity of allelic forces with their respective binding sites resulted in occasional interchanges between them. The differences between non-allelic factors were assumed to be too great for non-allelic interchanges. The greater the differences between allelic factors, the stronger would be the (apparent) linkage of those factors with the others on the same chromosome, since they would have less chance of going to the binding site on the homologous chromosome.

Sturtevant (1917) argued that the pattern of multiple crossovers ruled out Goldschmidt's proposed explanation. On Goldschmidt's model, he explained, factor exchanges between alleles at one locus should be uncorrelated with factor exchanges between alleles at other loci, since their occurrence in each case was simply a product of the similarity of alleles at the same locus, and should be independent of events at any other loci. On the model of the Morgan school, the intertwining and linear linkage that produced factor exchange suggested that factors crossed over in blocks, so that if a factor crossed over then other factors near it should also cross over. Thus, if we start with homologous chromosomes:

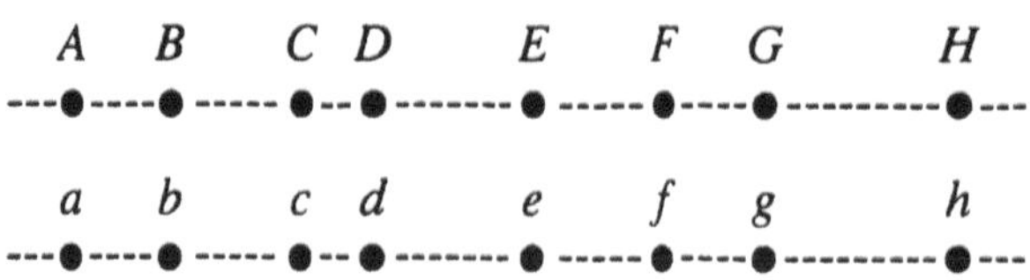

the predicted outcomes of multiple crossovers (looking only at the first chromosome) would be qualitatively as follows:

Goldschmidt model:	*A*	*b*	*C*	*D*	*E*	*f*	*G*	*H*
Morgan school model:	*A*	*b*	*c*	*d*	*e*	*f*	*G*	*H*

The data available to them clearly supported the linear linkage model.

Bridges (1917) made another crucial criticism, that on Goldschmidt's model, factors with a frequency of crossing over of 1 percent in one generation (and whose configuration of forces were stable) should have a 99 percent frequency of "crossing back" to their original location in the next generation. This was at odds with the stability of crossover frequencies for a given pair of factors, and in particular with the independence of these frequencies from the parental types used in the original cross (see the third defining property of linkage in Haldane, 1919).

Of these two predictions (both discussed in Carlson, 1966), the first appears not to have been made before the appearance of Goldschmidt's model. (Muller [1916a–1916d] discusses it with a reference to what could be this model, attributed to "some geneticists." The Morgan school likely had knowledge of the Goldschmidt model prior to publication, since *Genetics* was a house journal for them at that time.) The second made use of an already known property of linkage phenomena, but gave this property a greater importance. *These thus illustrate the third case of the use of false models—to generate new predictive tests of or to give new significance to features of an alternative preferred model.*

Note that Goldschmidt's model required a large array of distinguishable forces that all met the condition that allelic forces were more alike than non-allelic ones. This requirement raised potential problems for mutations that sometimes produced substantial phenotypic effects without changing their linkage relations. Muller used this fact later to argue that the mechanism of gene transmission must be independent of the mechanism of gene expression. Goldschmidt's model would also undercut the advantages for the arrangement of factors in chromosomes first hypothesized by Roux in 1883 and widely accepted since then. In Roux's account, chromosomes were seen basically as devices to increase the reliability of transmission of factors to daughter cells in mitotic division by putting many factors in one package and thus reducing the number of packages that had to be assorted correctly to transmit each factor in mitosis. The linear chromosomal organization of factors gave a simple mechanical solution to the requirement of getting a complete set of genetic factors without having to take care of each and every factor as a special case. The Goldschmidt model would have restored the need for special forces and arrangements for each factor.

False Models and Adaptive Design Arguments

One other common use of false models would seem to be special to evolutionary biology and to other areas like economics and decision theory where optimization or adaptive design arguments are common. Where various forms are possible, and one seeks an explanation for why one form is found rather than any of the others, one may build models of the alternatives and specify fitness functions of their structural properties such that their fitnesses can be estimated and compared. *When the fitnesses of the other alternatives are significantly less than that of the type found in nature, we have a prima facie explanation for why they are not found—at least if they are assumed to arise only as variants of the given type, or it from them.* Technically, we do not have an explanation for the occurrence of the given type: we have shown only that it would be adaptive if it *did* arise, not that it would do so. Nor do we necessarily explain the absence of the other types (if they can arise other than in selective competition with the given type); other factors could independently prevent their evolution or render these variants maladaptive. Nonetheless, such thought experiments can illuminate.

Consider a simple thought experiment that provides an explanation for why there are no species with three (or more) sexes. (This topic was the subject of a paper in the early 1970s by A. D. J. Robertson, which I read at that time but that I have been unable to find since and remember at best unclearly. Thus, some, all, or none of this argument may be due to Robertson.) We must first distinguish two things that could be meant by more than two sexes. It might mean several sexes of which any two could have fertile offspring, but mating was bi-parental. This system could be called "disjunctive," since there are a variety of possible types of successful mating. Alternatively, it might mean that three (or **n**) different types of individuals are all required to produce fertile offspring. This system could correspondingly be called "conjunctive," since only one mating type is successful, and this requires a conjunction of **n** individuals. The first type of system is uncommon, but is found occasionally. The wood rotting fungus, *Schizophyllum com-*

munes has a variety of bi-parental mating types, and almost any two types will do. There are two genetic complexes, **A** and **B**, with 40–50 variants at each "locus." To be compatible types, two individuals must have different variants at each of the two loci (see King, 1974). The second type of system does not exist in nature. Various considerations could explain why this is so. I consider only one.

I will assume the simplest kind of interaction among members of a species, in which matings are modeled on collisions (or on linear reaction kinetics). Suppose that an individual has a probability p of encountering another reproductively potent individual during the time when it can mate successfully. This probability is a partial measure of fitness, which is presumably a product of it with several other factors. With a two-sex system, if the sexes are equally frequent, there is a probability of $p/2$ that an individual can have a pairing that will produce offspring—one with the opposite sex. With a disjunctive n-sex system (where any two can play, if they are of different sexes), the probability of finding a mate becomes $((n-1)/n)p$, since an individual can mate with any other sex, and all other sexes constitute a proportion $(n-1)/n$ of the individuals it could meet. This quantity is greater than $p/2$ for $n>2$, which shows that systems with more sexes have an advantage. This could be one factor explaining the multiplicity of types in *Schizophyllum communes,* and raises the interesting question of why it is not more common. Presumably there are other constraints in addition to those considered in this model.

A conjunctive multi-sex system represents an even more interesting kind of thought experiment yielding an explanation of why such systems have *not* been found in nature. If one sex of each type must get together during their reproductive period, the probability of this occurring is $(p/n)^{n-1}$, since an individual would be expected to meet individuals of each other sex in the relevant period with a probability of (p/n) per sex, and there are $(n-1)$ other sexes. (I assume that meetings are statistically independent. Note also that if the sexes are not equally frequent, the probability of the right conjunction occurring is even smaller, so that in this respect, this is a best case analysis.) $(p/n)^{n-1}$ is always less than $p/2$ for $n>2$, and very much smaller if p is much less than 1. If $p=.5$, an individual of a two-sex species has a chance of 1/4 of finding a mate, an individual of a conjunctive three-sex species has a chance of 1/36 of finding its two mates of the right sexes, and an individual of a conjunctive four-sex species has a chance of only 1/512 of finding its necessary three mates of all the right sexes. Note that slight extensions of these arguments could be deployed against more complex mixed "disjunctive-conjunctive" systems, such as one in which any k

out of **n** sex types could reproduce conjunctively, since whatever the value of **n**, **k** = **2** is selectively superior to any larger value of **k**. A classic puzzle of sexuality is also preserved in this model, since an asexual species obviously does better in this respect than any of these more complex systems, including the familiar two-sex system.

This brutally simple model thus serves to explain why there are no conjunctive three-or-more sex mating systems. Of course, there are other reasons: the classic work of Boveri (1902) on the catastrophic consequences of simultaneous fertilizations of a single egg by two sperm shows that other constraints point in the same direction. Without substantial re-engineering of the mitotic and meiotic cycles, it is hard to imagine how a conjunctive three or more sex system could ever evolve from a two-sex system. The mitotic and meiotic cycles are deeply "generatively entrenched"—they are effectively impossible to modify because so many other things depend upon their current forms. This is a paradigm of the kind of developmental constraint discussed earlier (for more detail see Wimsatt, 1986b; see also Arthur, 1984, Rasmussen, 1987, Schank and Wimsatt, 1988, and the next chapter).

This almost playful example illustrates a point that is of broad importance in evolutionary biology, and equally applicable in feasibility or design studies in engineering. In the latter case, much effort is expended to model the behavior of possible systems to determine whether they should become actual. If they are found wanting, they never get beyond design or modeling stages because the proposals to build them are rejected. The models describing their behavior are "false" because the conditions of their true applicability are never found in nature, not because they make approximations or are idealizations that abstract away from nature (though they are surely that!), but because their idealizations represent *bad designs* that are maladaptive.

Obviously false models have an important role here. It is partially for this reason that much good work in theoretical evolutionary biology looks more than a little like science fiction. This is nothing to be ashamed of: thought experiments have a time-honored tradition throughout the physical sciences. The main advantage that evolutionary biology has is that the greater complexity of the subject matter makes a potentially much broader range of conditions and structures the proper subject matter of biological fictions.

Such biological fictions are interesting because one cannot always tell if nothing like the model is found in nature, whether this is because of the idealizations required to get models of manageable complexity, or because of selection for superior alternatives. An example is provided (in Wimsatt, 1980a) by discussions of the importance of chaotic behavior in the analysis of ecological systems. In that article I argued that the absence or rarity of chaotic behavior in ecological systems did *not* show that avoidance of chaos was not a significant evolutionary factor. It seemed likely that there were evolutionarily possible responses that could act to control or avoid chaotic behavior. If so, we would expect selection to incorporate these changes because chaotic behavior is generally maladaptive. There is not only the possibility of at least local extinction, but also the fact that major fluctuations in population size, sex-ratio, and effective neighborhood can easily generate much smaller effective population sizes and lead to substantially reduced genetic variation (Wimsatt and Farber, unpublished analysis, 1979). Reduced genetic variation could lead chaos-prone populations either to increased probability of extinction in changing environments, or to their more rapid evolution away from that state via mechanisms suggested in Wright's "shifting balance" theory (Wright, 1977).

After many years in which the common wisdom has seemed to be that chaotic behavior was rare in nature, more recent work (e.g., Schaffer, 1984) suggests that chaotic behavior may be much more common than we suspect, and can be seen with the right tools. The kinds of biases discussed in Wimsatt (1980a) would tend to hide chaotic behavior, even in cases where we have the good data over an extended period of time necessary to apply Schaffer's analysis. Thus it may be more common than even he suggests.

Another use of false models not discussed here is Richard Levins' suggestion (1966, 1968) that constructing families of idealized and otherwise false models could allow the search for and evaluation of "robust theorems"—results that were true in all the models and thus independent of the various false assumptions made in any of the models. This was the subject of Chapter 5, but in Wimsatt (1980a) I discuss more specifically Levins' views and apply robustness analysis to chaotic behavior in ecology. Taylor (1985) provides further in-depth study of its use in ecological modeling, and a more sophisticated analysis of the strengths and limitations of different alternative approaches.

Neutral models in biology represent baseline models or null hypotheses for testing the importance or efficacy of selection processes by trying to estimate what would happen in their absence. As such they often represent the deliberate use of false models as tools to better assess the true state of nature. False models are often used as tools to get at the truth. In this chapter I have analyzed and illustrated the many ways in which this can be done, using cases from the history of genetics, drawing on the development of the linear linkage model of the Morgan school, and discussing the different ways in which this model was used in countering contemporary attacks and competing theories. It is often complained by philosophers of science that most models used in science are literally false. Some of them go on to argue that "realist" views of science, in which arriving at a true description of nature is an important aim of science, are fundamentally mistaken. I have tried to illustrate the variety of ways in which such deliberate falsehoods can be productive tools in getting to the truth, and thus to argue that these philosophers have despaired too quickly.

Robustness and Entrenchment

How the Contingent Becomes Necessary

Generative Entrenchment and the Architecture of Adaptive Design

Robustness and heuristics are central themes of this book. Robustness (Chapter 4) is a property of many of the pivotal features of our natural world and of the objects and consequences of our better theories. It is also a fundamental principle of adaptive design: a way of increasing the reliability of structures built with unreliable components.[1] By improving reliability, robustness accomplishes some of the aims of the search for deductive certainty by the logical empiricists, and indeed, by philosophers since Aristotle and Euclid. Robustness is widely employed in our more reliable cognitive, symbolic, and technological structures. Chapter 5 covered important properties of heuristics: fallible, efficient, context-sensitive, but checkable and improvable tools for problem solving. Adaptations also have similar characteristics and share six important properties with heuristics (Appendix A). This is no accident: heuristics, human problem solving and design, and adaptation together reflect key common features of the architecture of a deeper selectionist functional organization (Wimsatt, 1972, 1997a, 2002a). Adaptive design is a layered organization of kluged adaptations acquired sequentially and assembled on the fly: it is heuristic "all the way down."

Generative entrenchment (GE) is a third deep principle of adaptive design. A deeply generatively entrenched feature of a structure is one that has many other things depending on it because it has played a role in generating them. It is an inevitable characteristic of evolved systems of all kinds—biological, cognitive, or cultural—that different elements of the system show differential entrenchment. Systems shaped by selection processes show significant degrees of adaptation, so things that become entrenched in them are commonly elements or parts of functional designs. If the system were not at least moderately well adapted, messing around with deeply GE'd parts of it might not be so strongly selected against. So assuming an adapted system is pivotal to the GE argument. Heuristic architectures built through a series of opportunistically implemented kluges are also inevitably rich in historically contingent entrenched processes and structures. They also commonly have to be robust to work reliably in diverse circumstances. So entrenchment intersects and complements robustness, heuristic structure, and other fundamental architectural features like modularity or near-decomposability.[2] GE is important for its impact on the stability of parts of structures or processes, their resistance to change or being changed, and inevitably makes the assembly of more complex structures roughly cumulative. Like robustness, heuristics, and modularity, generative entrenchment is a crucial part of a broader "engineering" epistemology. This epistemology recognizes and utilizes the characteristics of evolved adapted systems, and how they gain information from their imbedding in their co-evolved environment.[3]

Something may resist change or remain relatively invariant for longer periods of time in two importantly different ways. Both are relevant to epistemological concerns about parts of knowledge-structures, and to engineering concerns about parts of material systems. Something may just be very reliable, very resistant to failure. Philosophers have looked for adamantine certainty, but that is unattainable. Robustness is probably the most important way that high reliability is attained. If differential success or failure is a driver of change, then differential robustness among alternatives should lead to selection for increased robustness and relative stability. Alternatively, an element in an evolutionary lineage may resist change in the exactly opposite way: it might be quite fragile but any attempt to change it leads to collapse of existing structures that depend on it, so modifications don't survive. GE'd things don't need to be fragile, but they will be relatively stable whether fragile or not. Robust or fragile, if many other things *depend* upon it, it is then "essential" or at least conditionally "necessary" for what it does, so structures that survive leave it unchanged or only very slightly modified.

Generative entrenchment as a state of having many downstream elements can be thought of as a structural feature or, through time, in an ensemble of similar but variable structures with differing GE for some of their parts, as a mechanism or trait promoting evolutionary stability. The structure may be material, abstract, social, or mental. With different degrees of entrenchment, this becomes a mechanism producing differential stability. And because GE can change over time, it provides ways in which initially changeable or dynamical processes can become "frozen" through use, or freed through disentrenchment. With accumulating dependencies, seemingly arbitrary contingencies can become profoundly necessary, acting as generative structural elements for other contingencies added later. Contingencies layering upon one another can thus generate an indefinitely reticulate "fractal" structure for the evolving systems they characterize. This fractal hierarchy of contingent details on all scales of size, age, and generality is obvious for organisms and their phylogenies. It is also a property of cultures and their lineages. Gaddis (2002) spends a book celebrating the fractal structure of historical process. Generative entrenchment is the only process we know that is not only capable of generating this kind of structure, but that *inevitably* or *robustly* does so (Wimsatt, 2001). For this reason, it is likely to play a central role in any deep characterization of the nature of historical processes and, most particularly and strongly, for any evolutionary process.

Generative Systems Come to Dominate in Evolutionary Processes

New systems that facilitate mechanisms by which some elements can come to play a generative or foundational role relative to others are always pivotal innovations in the history of evolution, as well as—much more recently—in the history of ideas. Mathematics, foundational theories, generative grammars, and computer programs attract attention as particularly powerful ways of organizing and producing complex knowledge structures and systems of behavior. They not only produce or accumulate downstream products, but they do so systematically and relatively easily. *Once they appear, generative systems become pivotal in any world where evolution is possible:* biological, psychological, scientific, technological, or cultural. Generative systems come to dominate in evolution, and are rapidly retuned and refined for increasing efficiency, replication rate, and fidelity, as soon as they are invented. We must suppose that even modest improvements in them spread like wildfire.

The emergence and spread of the informational macromolecules RNA and DNA were two of the most irreversible reactions in the history of life. Eucaryotic cells, multicellularity, and sociality (in different ways in different lineages) each opened new domains of possibilities, and became similarly irreversible in their effects on the lineages that adopted them. Culture, spoken language, the advent of written and alphabetic languages, printing and broader literacy, and other adaptations that increased the reliability of transmission (teaching) and accumulation of increasingly complex information spread rapidly, generating an efflorescence of structure, processes, and practices building upon their presence. Diamond (1997) attempts global and integrated explanations of the rise and character of civilizations essentially from this perspective, providing a mix of contingencies and autocatalytic and hierarchically dependent processes showing rich signs of generative entrenchment. Adaptations (beginning with agriculture) causing and supporting increasing population densities, cities, role differentiation and interdependencies, and governments make GE inevitable.

It is important to see that there is not *a* cultural level above *the* psychological above *the* biological, but many interpenetrating ones (Chapter 10). New generative foundations have yielded many levels and cross-cutting perspectival systems of adaptation, mind, and culture, and conversely. Campbell (1974a) distinguishes ten levels of "vicarious selectors" that are plausible candidates:[4] varieties of perceptual, cognitive, and cultural widely adaptable lower-cost selection processes acting as extensions of and substitutes for biological selection (thus "vicarious"). These processes range from blind trial-and-error learning as an ontogenetic adaptation to culturally communicated and supported language and science as adaptive tools for generating and selecting among alternatives. For Campbell, the selectors are all initially midwifed by lower processes reaching back to biological selection, but they often can become dynamically decoupled from it (as perceptually mediated sexual selection can produce "runaway" sexual displays apparently reducing survival chances) by processes operating "upstream" or on faster timescales.

The ease of this decoupling (the "dynamical autonomy" of chapters 4 and 10) is why no devotee of an aspect of human mentality or culture need fear that a proper naturalism, even using reductionist tools, could devalue or eliminate our objects of attention. Once each dynamical structure is established, it becomes part of the supporting repertoire or "scaffolding" for further innovations, and as such becomes reinforced for increased reliability and productivity (Wimsatt and Griesemer, 2007). For each, a dominance and irreversibility, the product of runaway positive feedback processes, results in a contingent—but once started, cumulatively unavoidable—freezing in of things essential to their production. Arbitrary features, like the particular association between anti-codons and amino acids, acquire an essential character as organic systems increase in complexity. Everything would be scrambled if these associations changed, making modifications in them impossible. Loss of a generatively entrenched trait characteristically entails loss of any order that depends on it, and organizing forces have to start again from this new floor. Generativity is an extremely efficient way of building complex adaptive structures, while at the same time locking in its generators. *Since these are two sides of the same coin, their association is a deep fact of nature.*

Resistance to Foundational Revisions

But this is common sense. Rebuilding foundations after we have already erected an edifice on them is demanding and dangerous work. It is demanding: unless we do it just right, we'll bring the house down, and not be able to restore it on the new foundations. It is dangerous, and there are rarely guarantees that we *are* doing it right. We'd rather just "make the best of it," doing what we can to patch problems at less fundamental levels. This is not just backsliding. The systematic difficulty of the task makes it well-founded advice that deserves a rule-like status. These—the situation, and the relative difficulty of different tasks—are the phenomena to be explained and exploited. They are very general and have many and diverse consequences for the nature and possibility of change in complex systems.

This is as true for theories or for any complex functional structures—biological, mechanical, conceptual, or normative—as it is for houses. *This is why evolution proceeds mostly via a sequence of layered kluges.* Scientists rarely do foundational work, save when their house threatens to come down about their ears. (Philosophers like to mess around with foundations, but preferably in someone else's discipline!) Neurath's image of this activity as rebuilding the boat while we're in it is heroic (not an activity one recommends lightly!). If we must, actual revisions are preceded by all sorts of vicarious activity. We'd prefer to redesign a *plan,* rebuild only after we're satisfied with the revisions, keep in touch during the reconstruction for problems that inevitably come up, and move in only after the rebuilding is done—complete with local patches and in-course corrections. Jokes, usually painful ones, about construction projects executed by theoreticians or architects who don't visit the site are legion—rich folk-wisdom about the difficulties of foundational revisions, or in getting from new foundations to the finished product.[5]

Cornell's student union, the lovely gothic Willard Straight Hall, was finished in 1923. It has church-like multi-story leaded glass windows facing East, giving a vista uphill for half a block. To the west, the hill drops away 500 feet and largely blank walls face a view of the city below, the flanking hills behind, to the south, and north up the lake reaching 20 miles in the other three directions. Needless to say, when the New York architect finally visited the site late in the construction period, provisions were hurriedly made (that is, *kluged*) for additional windows—not many, and not so well placed—on the downhill side. It was too late to turn the building around. But we adjust. I grew up in Ithaca, so I spent decades in and out of "the Straight." As a kid, I loved the building, and didn't even wonder about the window placement. I resisted telling this story on my alma mater, but as one of the paradigmatic generators of this architectural metaphor for me, I must. Each campus I know has a similar story, though not commonly so dramatic, so its very ubiquity obliges me to tell it as saliently as I can. Others will recognize it.[6]

This bias against doing foundational work is a very general phenomenon. Big scientific revolutions are relatively rare for just that reason. *The more fundamental a proposed scientific change (or change of any other sort!), the broader its effects would be, and consequently, the less chance that it would work. So if adopted, the more work it would make for other practitioners, changing things that they have been able to take for granted, so then the more strongly and widely it would be resisted.* The last two facts are social, institutional, or social psychological, but the first two are not. They are broad, robust, and deeply rooted logical, structural, and causal features of our world—unavoidable features of both material and abstract generative structures.

Major changes in perspective are rarer later in life.[7] We get more conservative as we age. Behavioral and mental habits "build up," and "increase in strength," as we add commitments and increasingly "channel" possibilities and choices. These metaphors reflect deep truths about the characteristic architecture of our behavior. Biological evolution shows related features even more deeply than cognitive or cultural realms (Wimsatt, 2001). Von Baer's "law" that *earlier developmental stages of diverse organisms look more alike than later ones* is broadly true (with some revealing exceptions) and reflects this deep structural truth

(Gould, 1977; Wimsatt, 1986a). It is, most fundamentally, an expression of the evolutionary conservatism of earlier developmental features. More usually depends on these features (they are more likely to be deeply "generatively entrenched"). Mutations affecting them will more likely be strongly deleterious or lethal. So such features persist relatively unchanged or change more slowly than features expressed later in development.

Bootstrapping Feedbacks: Differential Dependencies and Stable Generators

Why is this pattern of developmental conservatism so general? Differential dependencies of components in structures, causal or inferential, are inevitable in nature. Machines depend differentially on the performance of their differentiated parts.[8] Some failures just are more destructive than others, and any machine has some parts whose slight modification will have better or broader effects than others. Things that are generatively fruitful in this way will tend to elaborate products that depend (at least partially) on them. It is overwhelmingly likely that you will have differential dependencies to begin with, but if not, it is overwhelmingly likely that differential dependencies will be spontaneously generated (Wimsatt and Schank, 1988), so differential entrenchment is doubly guaranteed. This natural elaboration of relationships confers upon the generators a relatively foundational character.

But we can go the other way—from stability to generative role—as well. Stable structures, processes, or traits provide opportunities to accumulate constructions presupposing them. Unstable things and their dependent structures "wash away" before they have much chance to grow. We are much more likely to use stable elements than unstable ones in building new structures. This is a bias with consequences. This differential use affects the magnitude of the expected generative roles of these elements. Their usefulness and cumulating history of application then feeds back to make us less willing to change them or to allow them to be changed, further increasing their stability. If nothing else changes, this is a self-amplifying, or symmetry-breaking process (Wimsatt and Schank, 1988; Wimsatt, 2001). This impacts their "foundational" status (viewed as a degree property, rather than as an absolute distinction)—things that are both generative and fixed will tend to become and be treated as increasingly foundational, both in nature and in the tools with which we approach nature. This provides a perspective bridging intentional and natural realms—not collapsing them into one, but richly connecting them.

Implications of Generative Entrenchment

What are the consequences of generative entrenchment? There are correlative impacts for biological organization in development, for the emergence of innate traits, for the formulation and spread of conventions, for the fixation of technologies and practices, and for scientific change. Consider networks of propositions connected by inferential relations. These may be heuristic, abductive, or other forms of informal inference, and need not be entailments. The following are some of the implications of GE:

1. Something that is deeply generatively entrenched is in effect a foundational element, principle, or assumption. A great deal depends upon it, and must be given up, or generated, or justified in other ways if it fails. Changing principles with greater generative entrenchment is revolutionary in effect if successful, and disastrous if not.

2. Propositions downstream provide more constraints on allowable modifications for deeply entrenched things, and fewer possible modifications will meet all of these constraints. Capturing (at least approximately) older, well-confirmed theoretical relationships is an adequacy condition for new theory—as predicted by this account. An important special case is the search for "reductions" of older theories as limiting cases of newer ones, which thereby preserves the older theoretical structures as heuristic approximations while validating the new ones (see successional reduction in Chapter 11).

3. Rather than change deeply entrenched things, we prefer changes elsewhere in the structure—parts that are less entrenched—to meet problems that arise. This is endemic to scientific practice at all levels—to make "fixes" at as low a level of generality as possible, so that successively fewer changes will be made at deeper levels, unless you have more robust reasons to believe that the fault lies deeper (robustness analysis in Chapter 4 gives heuristics for detecting such cases).

4. We may reflect this different attitude toward deeply GE'd things by refusing to allow them to be compromised or falsified. This is a pragmatic change in their status from empirically or contingently true to "covert tautology." These are not static categories and propositions can move through them. With some other help, it points the way for passage of a given claim from contingent through a priori or necessary

truth, to stimulate changes in language that make them conventional, semantic, or analytic relations.[9]

5. The resulting stability of these elements thus makes them natural to build upon, further increasing their entrenchment in a positive feedback cycle.

6. The application of models, or theoretical or semantic structures, to diverse kinds of cases forces their abstraction, which further increases their potential breadth of application, again yielding a positive feedback cycle. But abstraction also reduces their empirical content. If they are now treated as a priori or analytic, their role is now primarily inferential or generative as they become increasingly foundational.

7. Elements that have been around longer (and have become more deeply entrenched) will be things we are less willing to change—predicting and explaining conceptual or genetic "inertia" as kinds of developmental phenomena.

8. Generative elements of smaller structures should be less entrenched and thus easier to change. This helps to explain such diverse phenomena as greater rates of evolution for bacteria, and the fact that models are regarded as more tentative and modifiable than theories (see Chapter 6). Indeed, in modern science, models are effectively the cutting edge of and units of change for theories.

9. Nonetheless, some of the deepest principles in science are among the youngest. In such cases they are commonly "added" at the bottom in ways that generate or regenerate most of the other (once deepest) principles as limiting cases, preserving most of their downstream consequences. But that this is possible at all points to adaptations for cultural evolution; heuristics making the production of adaptive deep modifications easier—though still extremely difficult (Wimsatt and Griesemer, 2007). These heuristics are crucially important; finding "near-decomposability" in problem-structure (Chapter 9) is one of them (Bechtel and Richardson, 1993).

Qualitatively new kinds of phenomena emerge from considering these implications in connection with other design principles, and specific structural forms, but they are enough to anchor relevant connections here and to suggest why GE deserves careful attention.

Generative Entrenchment and Robustness

In Chapter 4, I characterized a robust proposition, assumption, or entity as one that was accessible (measurable, derivable, inferable, detectable . . .) in a variety of different (at least partially) independent ways. A robust node has multiple inferential paths leading *to* it and resists failure because of its multiple sources of support. But one can also ask how big the inferential shadow *from* that node is. If it had to be given up, how many other things must be given up, changed, or weakened? If it is wiggled, how many things jump? In a directed graph of inferences, this would be the number of nodes reachable from that node.[10] Different measures of generative entrenchment are appropriate to different structures, but for directed graphs (as in some kinds of models of causal interaction) this will do and can be thought of as the generative entrenchment of that node.[11]

Imagine a network of propositions ordered by inferential connections and represented in a directed graph, with nodes for propositions[12] and arrows indicating the direction of inferential movement from one proposition to another (Figure 7.1). A node near the bottom is robustly anchored by three others (dotted arrows) and is itself entrenched by many downstream arrows (open arrows). Another entrenchment network feeds in from top left. This kind of picture of theory (or one with more robust nodes in it) is suggested by Feynman's (1967) remarks quoted in Chapter 4, or Levins' (1966) characterization of the structure of theory in ecology. Here I am deliberately vague about whether these arrows are entailment relations, evidential relations, or some other kinds (for example, representing part of a conjunct whose elements jointly entail the proposition). These relations and more are found mixed in the architecture of complex inferences involving theory, experiment, calibration, measurement, application, prediction, and testing.

Generative entrenchment is distinct from but complementary to robustness. Like robustness it bridges nature and our study of it: it is crucial to the architecture of nature, and also to the architecture of our methods, and more, to the architecture of any naturalistic metaphysics, epistemology, or methodology. Like robustness, it affects the stability of its bearer, though for different reasons. Also, like robustness, it is intrinsically relational. To be either robust or entrenched is to be connected to other things. Both are also clearly degree properties. Finally, like robustness, it resonates naturally with a pragmatic, fallibilist, and heuristic methodology, but one in which these stand not against realism, but in part are constitutive of it.

There is one very big difference between entrenchment and robustness. Robustness provides presumptive evidence for truth, but we have no intrinsic reason for arguing that something is true because it is deeply generatively entrenched. Remember that scientific revolutions

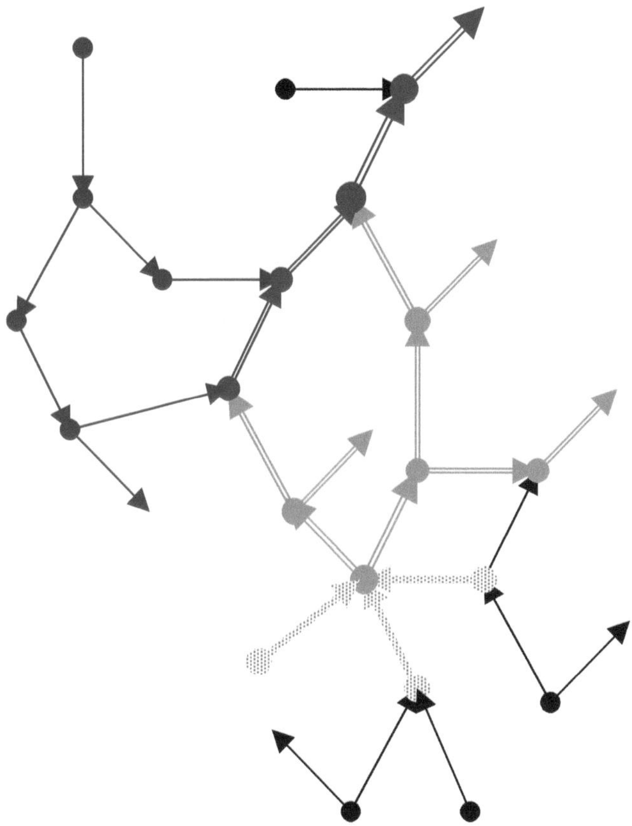

Figure 7.1 Network with robust node (indicated by multiple dotted arrows) and entrenched by multiple downstream consequences (open arrows) that are thus more reliable than GE network intersecting with it from the left.

issue when we overturn entrenched principles. An entrenched assumption is not necessarily true, and might even be empirically false, though we are in big trouble if it is not at least approximately true in most circumstances where it can be applied. We *need* it to be true. Or, if not it, then something sufficiently like it to regenerate most of what follows from it also from the new variant principle. It would obviously be sound methodology in constructing theory to choose as foundational principles things that are as robust as possible, entrenching only things that are robust, so in this way there may be a secondary association between entrenchment and robustness.

One final complexity: looking at Figure 7.1, with arrows and pathways to and from nodes, it may appear straightforward as to whether a given node is robust, entrenched, or (more commonly) some of each. But inferences in theoretical networks (and causes in evolutionary ecology) can run in different directions in different situations (Feynman, 1967, as quoted in Chapter 5). A deductive inference may in different contexts be run the other way as an inductive inference (thus the oft-noted claim that "inductive inference deliberately commits the fallacy of affirming the consequent"). This should alert us to look carefully at context to determine what is going on when evaluating claims to entrenchment or robustness. Can a claim of inferential robustness (seeing a conclusion reflected in many different consequences) just be a mis-assessment of pervasive entrenchment? A mitigating consideration blocking too easy a move to this conclusion is the role of independence in robust inference. Robustness is increased by the independence of the different means of access. But independence or modularity in organized networks has just the opposite effect on entrenchment, decreasing it and allowing independent modifiability of components without generating as many broader consequences. Perhaps the importance of contexts—current ecological, broader theoretical, and above all historical—is the deepest biological principle of all, one for which practitioners of a reductionist methodology should cultivate a particular respect.

Do things have to be propositional to become deeply generatively entrenched? Based on our previous discussion, obviously not. DNA will, and so will TCP/IP (a communication protocol crucial to cross-platform compatibility on the Internet). Any software that becomes sufficiently important that new releases guarantee "backwards compatibility" has reached this status—at least temporarily. Chip architectures matter too: Motorola and Intel played out a competitive game for years with different competing operating systems written for the different chip architectures. Better chip architecture (Motorola/Apple) gradually lost ground to the older Intel architecture, largely because IBM, unlike Apple, adopted an "open architecture" policy. This let developers look into their hardware boxes, eased software and hardware development, and pushed them ahead for the "Wintel" machines by allowing them to utilize chip-dependent and operating system-dependent strategies more effectively. This rapidly GE'd the chip architecture in a more rapidly growing software and hardware environment that presupposed it, generated more money for further development at hardware and software levels and support infrastructure, and imposed a cost (particularly in communications) on the less frequent type. As the Intel chip became significantly more common, economies of scale further pushed developers to develop it for the larger market. Coordination games (of which this is one) provide a kind of entrenchment—what economist Brian Arthur (1994) has called "lock in." A large fraction of the costs and possibilities in this commercial battle play out in terms of generative entrenchment—even to more frequent viruses and more rapidly spreading and widespread infections for the Wintel environment.

In the realm of animal behavior, things that are deeply generatively entrenched meet most criteria (more than on any other analysis) for innateness: species universality, characteristic earliness in development, deep problems if its development is prevented, and about 20 others (Wimsatt, 1986a, 1999b, 2003). This reflects a deeper resonance between the generating features of innateness and "necessary truth" once common in early modern philosophy (Cowie, 1999). This resonance is explained by a generative entrenchment analysis of both. And GE explains a residual paradox: even for learned behavior, if it becomes habitual, smooth, and widely used, it is easy to speak of it as something that has "become instinctive"—a natural extension (whether taken literally or metaphorically) on the generative analysis, but nonsense on traditional accounts.

Entrenchment can reach a lot further, far enough, in fact, to provide alternate (but complementary) foundations for evolutionary theory to genetic approaches—an approach more naturally integrating evolutionary and developmental concerns (Wimsatt, 2001). And for cultural evolution? Without an underlying genetics and no tractable memetics, generative entrenchment may be the most productive way to produce a rich explanatory and predictive structure for culture, science, and technology (Wimsatt, 1999a; Wimsatt and Griesemer, 2007). Indeed, it may be the only way to go. But these are themes for another occasion.

Lewontin's Evidence (That There Isn't Any)

Let me say at the start, that this title is guilty of rhetorical overstatement. Or, since I am a philosopher, and by definition, never use rhetoric, it would be more honest of me to admit to a deeper failing. I saw Dick Lewontin's (1991) article and its many embarrassments for the canonical concept of evidence that philosophers of science used to like so well. Because we still suffer from the stultifying—indeed mortifying—rigors of the belief that only that which is general is worth knowing, on first reaction, I *took* this as doing away with our concept of evidence. (While the defenders of evidence look in vain for a decontextualized, general, and preferably logical relation, the attackers also accept the same ground rules: they seem to expect that a single, or a small number of counter-examples should bring down the house. The first is unattainable, and the second foolish.) But Lewontin is onto something much richer. Nancy Cartwright (1983) has said that the Laws of Physics lie. It would appear that the most general laws of biology (that is, if you thought there were any) don't show any consistent respect for the proper evidential authorities either:

> At least in biology, there may be general statements, but there are no universals, and . . . actual events are the nexus of multiple causal pathways and chance perturbations. As a consequence, in an ironic reversal of the Popperian claim, the least general and most specific statements of science are the least protected against contrary evidence, while the most general can survive numerous apparent factual disagreements. (Lewontin, 1991, p. 144)

Lewontin follows with a list of scientists *in flagranti delicti* providing compromising copy for critics of the leveling voice of scientific evidence. Similar examples are also found in fields other than biology, but there's no need (or time, here) to go further afield. Within biology we have plenty of embarrassments for our traditional beliefs about the power and objectivity of evidence. We have inconclusive arguments (e.g., the neutralist-selectionist debate over which forces if any power evolution), in which both sides bash each other with what appears to be virtually the same evidence. In his earlier discussion of this case (Lewontin, 1974), Lewontin falsifies roughly $Popper_1$ through $Popper_{13}$[1] on his way to a much richer account of the dialectical dynamics of scientific disputes than found in any traditional falsificationist accounts or their descendants. Elsewhere, we have apparently shameless curve fitting and ad hoc-ery (in the service of the adaptationist program and sociobiology—in defense of central place foragers who are supposedly maximizers, but who often appear to settle for less than they might). Here the failure of the generalizations to face the facts seems to entail (to their defenders, at least!) not that they are false, but the existence of yet undiscovered facts yet to come to their rescue. Does *anyone* give up their theories in the face of contrary evidence? No one surrenders gracefully—if at all—in today's philosophy of science.

Perhaps most embarrassing are baseless variations in the standard of evidence used by different disciplines working on the same kinds of problems. Is this perhaps a reflection of a caste system in science, with the standards of rigor matching one's position in the hierarchy of the sciences—with the good guys from the higher sciences still trying to educate the teeming unsophisticates, and the basic scientists educating their applied brethren? But not so fast, look who comes out on top: the genetic speculations of some "pure scientists"—human ethologists and IQ researchers—are compared with the quantitative genetics of animal and plant breeders and found wanting. This last is a cruel slap: those who have most slavishly imitated the positivists in their demands to quantify and operationalize, who have promised us the largest array of "value free" facts, and who have touted themselves as practitioners of "high science" are shown to be grossly inferior in their methodological practice to a bunch of applied agricultural geneticists. (Of course it helps if the "applied boys" learned their stuff from Sewall Wright!) *Generalizing from these cases, the problem seems to be not that evidence is not to be had, but that it's to be had too easily by anyone who wants it, and he who supports everything sustains nothing.*

I can't manage much motivation for defending classical concepts of evidence—or for defending every foible of various scientists' (and sciences') use of evidence—nor am I ready for the anything goes flavor of much of the post-modernist relativism of many current sociologists of science. We need to be much less absolutist, and much more contingent, contextual, and historicist in our analyses of science. But we must do this by recognizing the real complexities we are increasingly able to study in natural systems whose simplicity we have been taking for granted for decades or centuries. A major fraction of these complexities are not a function of our conceptual schemes, language, or interests, but products of the way the world is. Realism lives! But any wise realist must recognize that the social, cultural, and ideational entities of the "social relativists" are real, too, and have causal effects in this world. They must be imbedded in the appropriate (pan-realist) world picture along with the entities of the natural sciences, via the idealized models one constructs, and the carpentered "natural" entities, tools, practices, procedures, and phenomena one detects and experiments with and on—guided, regulated, and maintained by the social structures, languages, and values of science. All of these are real in our world, and we mold and are molded by and in their presence and effects.

Is Evidence Impotent, or Just Inconstant?

Lewontin's (1991) article must be taken as exemplifying his own thesis. Philosophy of science makes happy methodological bedfellows with history and evolutionary biology in this, as in some other respects. All are messy, complex, richly textured subjects that cover an enormous diversity of things with deceptively simple and unitary labels. Lewontin has given "evidence" for a generalization (not a universal) that is immune to counterattack by simply counting counterexamples because the "actual events [of theory construction, discovery, and the marshalling of facts and counter-examples] are the nexus of multiple causal [and inferential] pathways and chance perturbations" (p. 144). In this context, *in principle* arguments now work side-by-side with empirical evidence—sometimes merely bounding possible solutions, sometimes finding accurate pivot points around which the whole argument must turn—all in the service of simply trying to tie the phenomena down. In this underdetermined context (and we should not assume, as philosophers commonly do, that all theoretical contexts are significantly underdetermined!), what used to be thought of as refuting counterevidence often becomes just a stimulus for new elaborative investigation.

Just? But it is an effective stimulus, and that's not at all bad. At least it is a response—showing that the offering of the evidence was not totally without effect. The response may sometimes be just curve fitting, but more often, I think, it is not. There can be responsible denials of counterexamples, and even responsible adoption of them without jumping ship. If the counterexamples are good and deep ones, they should be treasured, and closely studied, even if we don't know how to deal with them. But nor should we be naive falsificationists. If they are good ones, our generalizations should not be as fragile as all that, and even when they are false, they may be very useful tools nonetheless—a topic I return to later in this chapter.

The culprit, if there is one, is the assumption that we must remove any contradictions or tensions in the theory before we proceed any further. (This is traditional advice for beginning philosophers—justifying close conceptual analyses and careful definitions of terms, as if these had to be temporally as well as logically prior to any theory construction or revision. But it is poor counsel for any practicing scientist, and arguably, for philosophers as well.) We need to have some idea of *how* to remove the contradictions. Remove the contradictions, we must, and it is important to identify them, if possible to localize them, and to keep them in mind when revising the theory, but often no way of removing them immediately suggests itself as desirable on other grounds. The further development of the theory in directions guided by other constraints and desiderata will often suggest new and natural ways of resolving the problem if we ignore it for the time being. (Indeed, the faith that this will happen represents an often-unappreciated commitment to scientific realism—nature, after all, does not tolerate contradictions![2]) Contradictions should be treated like holes in the ice that we skate around gingerly, and look for ways to bridge or fix, rather than acting like we have just had a total meltdown, and everything has to stop until we fix it.

The thing we worry about most is that ad hoc responses to counterexamples could enervate our generalizations, rendering their empirical force so unclear that they are deprived of useful consequences we would stand by, thus making the theory so formless that it is unclear what could *either* confirm *or* compromise it. With this worry, it is useful to contemplate some fact-generated paradigm-shifts, just to remind ourselves that it can happen. *If evidence were impotent for forcing theory change, how could we account for the occasional Keystone facts—apparently isolated facts whose acceptance has far-reaching consequences?* This can happen even in messy complex areas, where one would expect more opportunities to "take up the slack" and resist change. Such a case—the "iridium anomaly" that led to the so-called Alvarez hypothesis that the extinction of the dinosaurs was caused by an asteroid impact—has in the 25 years since its initial publication changed almost everyone's mind except perhaps for some of those most strongly affected by it; the vertebrate paleontologists, keepers of the life histories of the dinosaurs. I think it is now only a matter of time until the last of them cave in.

A delightful irony of this case is that Walter Alvarez started out to do an interesting but conservative application of existing science. The amount of iridium was measured on the assumption that its source was a steady rain of meteoric dust at a known slow rate, which could then be used as a "clock" to tell how rapidly the various sedimentary layers were deposited. But the amount was so high that sediment would have had to be deposited at an impossibly slow rate—hundreds of times slower than normal, so the assumption of a constant deposition rate for iridium had to be given up—in an inferential path that led to the hypothesis of a large body impact.

Did the great Cretaceous extinction (65 million years ago) issue from collision with a comet, asteroid, or large meteor? The resulting debates across and within disciplines reoriented awareness to a variety of other facts and led to conceptual and strategic readjustments producing a variety of new foci for research. A very partial list of disciplinary reorientations (for more see Raup, 1987a) would include:[3]

a. the search for the smoking gun(s), with robust evidence of contemporaneous collisions found: craters of the appropriate age off Yucatan and in Iowa;

b. searches with increasing success for iridium (and for carbon layers—indicating collisions and massive resultant forest fires) now abound for similar causes of earlier mass extinctions;

c. revised theories and evidence of collision activities on other bodies, particularly the moon and Mars, and new ideas in planetary geology;

d. relations to other theories of mass extinction (volcanic eruptions and ecological collapse, which aren't necessarily competitors) are more appreciated now than before;

e. periodic extinction theories (and the question of what kinds of causes could be found for very long period fluctuations in extinction rates);

f. new correlative questions about the meaning of long-range stability as applied to terrestrial processes, leading back to Poincaire and chaotic dynamics in the solar system, and to intermediate range orbital periodicities (Milankovitch cycles of 20,000, 40,000, 100,000, and 400,000 years) whose biotic effects had not been seriously considered before;

g. a new interest in theories of disturbance, their causes and effects: fire ecology, nuclear winter, greenhouse effects, ice ages, etc.;

h. relation to punctuated equilibrium theories of evolution (a facilitating theoretical change that made it easier for some biologists to accept Alvarez);

i. correlative new attention to the duration of mass extinction events;

j. more general consideration of extraterrestrial causes of biotic events (and recognition that our planetary and galactic surroundings are not in steady-state equilibrium on a time scale appropriate to even intermediate range evolutionary processes);

k. development and application of stochastic theories regularizing rare events (consider frequency vs. size distributions of collisions, as in Raup, 1991);

l. a changing assessment of the importance of catastrophic/uniformitarian arguments; in which the latter—holder of the high ground for the last 130 years—has recently suffered serious inroads, as illustrated by the following set of hypotheses taken from a research news article in *Science* magazine in the fall of 1991: the suggestion that the earth was kept hot by large body collisions with declining frequency until more recently (4.0–3.8 billion years ago) than had heretofore been supposed; that life may have evolved (and been wiped out) several times by collisions with asteroids having diameters in the 200–400mi range; and that in consequence, life may evolve much more rapidly, and originate much more easily, than we have supposed.

The problem posed by Lewontin's (1991) examples and this additional case isn't that facts are sometimes efficacious in producing theoretical (and factual!) change, and sometimes not. It is that we have no general theory of when they will or when they won't produce change, and furthermore, that they sometimes appear to do so when they shouldn't or fail to do so when they should—epistemologically, morally, and sometimes both. But perhaps this means that we have just been seeking too simple a theory of evidence.

Some of Lewontin's other examples, such as the inconclusive neutralist-selectionist debates that he has participated in and sometimes crucially redirected, show a similar pattern of territorial contests resulting in often productively elaborated arguments. These debates (heightened in 1966 by paired papers by Hubby and Lewontin with their adaptation of gel electrophoresis to discover unanticipated amounts of genetic variation in natural populations and his crucial overview of 1974) led to new understandings of macroscopic constraints on the evolution of complex genotypes and their rates of evolution; the limitations of genetic load arguments; and new tools, ranging from hardware and assay procedures through new statistical techniques and new conceptual insights and models to characterize the new forms of variation. Even if we don't have final answers, the debate has forced clearer formulation of the questions, significant shifts in the grounds of debate, and new attention to the causal factors and mechanisms affecting phenotypic evolution. And that's important—more so, I submit—than exactly how the investigation comes out in this case.[4]

False Models as Means to Truer Theories

There is another practice (one probably affecting some of the cases Lewontin discusses) that can produce apparent waffling over the evidence. We do many more things in science than test theories.[5] For some of them it is not important that the model or theory be right—only that we have some good idea of the place or places where it is wrong. *If so, then we may misinterpret what is going on when a theory or model is not discarded in the face of apparently damaging counterevidence.* In some of these cases it is not the evidence (or theory) that is to blame, but our view that the only relations between facts and models or theories are evidential. We need to look at another important class of ways of using models or theories in science, which could be called *template matching*.

We have to recognize that our best theories and models are idealizations—deliberate simplifications, usually made with knowledge that and where they are false (not to say that we therefore know what is true!). We can use these falsehoods as heuristic tools—as baselines to organize and restructure our perceptions of the data in productive ways.[6] These idealizations can be used in many ways:

1. To study the effects of a subset of the causal interactions without the further complications added by including the rest. This can serve several functions—the effects may not be at all obvious, they may not be experimentally isolable in real systems, they may have been experimentally isolated but studied piecemeal under different conditions and never put together before, or this may just be an initial stage that provides a benchmark to understand the effects of later added complexities.
2. Diverse things can be accomplished by what might be called *residual analysis*—where the aim is not to test the model against the data, but to detect and analyze the deviations between the model and the real world. This could serve:
    a. to estimate and remove main effects captured by the model that are much larger than and submerge the effects you want to study;
    b. to be able to figure out how to model the causes of those deviations, and incorporate them in this or another related model (supposing that you have the factors already included in the model roughly right);
    c. to be able to evaluate the magnitude of a deviation, either as a calibration of the model for accuracy under different circumstances, or as part of a strategy to figure out how detailed a model is needed to predict with reasonable accuracy.
3. To generate limiting results: these are used sometimes for predictive simplicity, to bound or to bracket a range of outcomes, or to provide conservative estimates on the magnitude of effects.[7]
    a. There may be one-way limiting results, which provide upper or lower bounds for the real case—and which can thereby rule out possible causal factors, mechanisms, or theories that would require conditions or values in the "forbidden" region. These one-way bounds may also be important in risk-benefit analysis.
    b. We may construct two-way (or multi-way) limiting results that "bracket" the real case. It is often true that one can model and solve for idealized limiting cases much more easily, and then, with these "spanning" cases, get a better idea of what is going on in the real case. Qualitatively, this was important in conceptualizing the nature of partial linkage in ge-

netics: bounding its nature between pleiotropy or absolute linkage for factors in unbreakable chromosomes and independent assortment for factors in different chromosomes; and suggesting its cause as intra-chromosomal recombination in chromosomes that break sometimes and whose factors separate with a frequency roughly proportional to their distance. Later in the same series of successively better models of linkage, this bracketing strategy was used even more elegantly by Haldane, who constructed three "limiting" models and then abstracted from them a parameterized meta-model to characterize the nature of interference.[8]

4. Idealizations and isolations in experimental design have basically the same character as constructed simplifications in mathematical models. Both the contrast between the conditions in nature and those found in the laboratory, and the differences between treatment and control conditions[9] (as well as deciding when an outlier is a rogue that should be treated differently—and even perhaps not analyzed with the rest of the data)[10] have features very much like those found in modeling.

Narrative Accounts and Theory as Montage

"It is the purpose of all of these [historical sciences] to provide a correct narrative of the sequence of past events, and an account of the causal forces and antecedent conditions that led to that sequence . . . [they] assume the existence of several forces simultaneously operating, and include the importance of chance. . . . The actual event is seen as the nexus of these forces and their chance perturbations" (Lewontin, 1991, pp. 143–144). Lewontin's account seems right here not only for the obviously historical sciences, but more broadly, for any of those that study complex mechanisms.[11] A mechanism is after all a distributed causal structure designed so that its parts articulate through time to produce desired effects under a possible diversity of controlling inputs—in its operation, it undergoes historical trajectories. And, by extension, the things that we call mechanisms undergo sequences of causal interactions. But aren't causal mechanisms designed to behave regularly and reliably; that is, to minimize the effects of chance perturbations? Perhaps so, but then the designer of such a mechanism must consider the range of possible perturbations if his mechanism is to behave reliably. (Indeed, the design of a mechanism to [reliably] behave randomly may be an even harder task—as most computer modelers know.[12]) We need to evaluate how the causes work or would work in a *range* of actual and possible circumstances in part because we often don't know what the actual circumstances were, or will be, and would thus prefer an explanation (or a mechanism) that is not too sensitive to these details. This is also in part to understand how the general-purpose tools work in a range of circumstances so we can tell where else they may be applicable.

But surely in historical explanations, one might urge, our only concern is with the actual. True, but misleading! Even where the search is not for any level of broader generality, but only for historical accuracy in this particular case, the search for how the proposed mechanism would work under various possible circumstances also serves to determine the *robustness* (and thereby the probable correctness) of the proposed explanations. Explanations, which are too sensitive to a variety of detailed conditions, are more demanding, and therefore more likely false. Therefore, we tend not to be satisfied with such an account. We need a robustness analysis to evaluate the appropriate level of the causes in question so as not to suffer unduly from the "for want of a nail" syndrome. "For want of a nail" may have been part of the explanation for Napoleon's loss at Waterloo—and such explanations surely increase the sense of drama—but it was far from the whole cause. In fact, a more plausible characterization of the cause was as a structural condition: it was part of Napoleon's failure that he was unable to secure the conditions necessary to make *this* kind of failure improbable. But for a proper account of generalizations in the messy sciences, we do need to see them as tools as well as ends. *Universals* may be commonly viewed as ends in science, but both unqualified and exceptionless *idealizations* and gappy and sloppy *generalizations* are often happier as tools—admittedly general purpose tools for which we have an aesthetics of design and admiration, and which we critique to improve their function.

We have a reason to expect gaps in our generalizations if we want to have modularity and near-decomposability in our theoretical structures. A "Theory of Everything" (as referred to in the frontispiece) may seem okay in cosmology, where it seems to unify the very large and slow with the very small and quick. (A theory of everything is in effect a fractal theory of the organization of things on the cosmologically intermediate scale driven by a topological theory of 10-dimensional strings that have been unkinking in ways determined by accidents of the very small, fast, and compact in the first micro-instants of the big bang.) But in between these two extreme size ranges, which it covers tolerably well, the details—even major details—of the two dozen or so orders of magnitude of size and time that interest us the most are left almost untouched.

But we don't really want a theory of everything. The ideal Piercean community will not asymptote to one big theory that explains everything in all of its details. Such a theory would be unmanageable, untestable, and incomprehensible, and have to mimic the detailed structure of the world itself—as one big historical entity. The theories of the very large and the very small can have simple structures because it is the middle range theories that have to deal in detail with the largest range of types of entities. What we want is a number of partial theories of selected aspects of classes of things, which we can pick up and carry around as portable templates to organize and explain a variety of systems, and which can be used together to make a multi-perspectival montage of each event in most of its complexity, and guide the articulation of these different pictures (see Chapter 9), together with those uncommon, peculiar, and particular causal linkages that are easier to account for piecemeal. Explain everything, we can, but to do so we customize each explanation out of a bunch of adaptable standard pieces, tied together with selections from our junk-box collection of tinkerable widgets. In doing so, we try to avoid in our theories the uninformative abstract sterility of mass production, the epicyclic decadence of too much unconstrained curve fitting, and the disorganized urban sprawl of totally particularistic conceptual tinkertoys gone wild.

In this activity, we've learned about the danger of "single cause" explanations, but we still need to resist the temptation of "single discipline" explanations. (Witness the claims of all we're supposed to learn from the human genome project—as if the production of an untranslated Book of Nature, but one volume of a library-sized encyclopedia, would solve all of our problems!) Dave Raup has claimed that perhaps the biggest problem with the emerging investigations surrounding the astrophysical, geophysical, and biotic dimensions of large-body collisions (descendants of the Alvarez hypothesis) is that no one is well-placed to evaluate the goodness of work in the myriad of other disciplines that they must draw upon. If the probative force of evidence requires seeing how it fits into a network of other theory and data offered by a variety of other agents, we have to find ways both of bounding that network and of evaluating the heterogeneous mix of characters within those bounds. The same kind of problems affect the analysis of development; the relations between ontogeny and phylogeny; the causes of macro-evolution; the analysis of culture; and that classic Gordian knot—the "mind-body" problem.

In these tasks, our concept of evidence would predictably be much more contingent, contextual, and historicist—complex, relational, and sensitive to context, but that is not to say relativistic. Because we can't place clear bounds in general and in advance on what could be relevant considerations in weighing evidence in a particular case does not mean that there aren't any—that anything goes. There are some things that no one would defend, and some defenses you do not have to accept just because they are offered by someone with good stature and good intentions. So we don't have a general theory of evidence. So what? Who thought the world was that simple? All the cases Lewontin discusses have this character, and this seems likely as a pivotal source for many of their problems. But if so, so describing them is not yet solving them. There are many problems of evidence yet to come.

PART THREE

Reductionism(s) in Practice

This section is the core of the book. Chapters 9–12 take a single pair of connected themes—emergence and reductionism—and show how a once simple topic (*theory reduction*) reveals new dimensions as we turn from justification to discovery, and abandon *in principle* claims to study the heuristics of real-world practice. Here the gap between philosophers' spare deductivist picture of theory reduction and the enormous proliferation of reductionist practices in real science is most striking. It is not just that there are problems with the classical analyses: whole new problem-areas are revealed. Levels of organization are the focus of Chapter 10. These pivotally important macroscopic compound entities are probably *the* major determinants of the geography of science. They are presupposed by but not examined in dozens of philosophical accounts given of inter-level reduction. Related approaches yield strategies for naturalizing perspectives—the "what it's like to be" of bats, philosophers, and other beasts. Perspectives have been focal points for modern anti-reductionist philosophers of mind, who have not viewed them broadly enough or sought to see them in naturalistic perspective. Disciplines also provide perspectives. *Levels* and *perspectives* provide crucial context for evaluating accounts of reduction and subjectivity, as well as the reductionist problem-solving techniques of Chapter 5 and the modeling tactics of Chapter 6. We must calibrate these tools, identifying and analyzing their strengths and limitations. Chapter 12 introduces a new analysis of an old concept—aggregativity—that abolishes traditional conflicts between reduction and emergence, and identifies a new (or recovers an old) one. The heuristic uses of criteria for aggregativity give new handles for detecting functional localization fallacies and abuses of "nothing but-isms" and "greedy reductionism."

A closer look reveals three notions of reduction—not one. Successional and inter-level reductions are distinguished in Chapter 11. Chapter 12 yields two divergent varieties of the latter: mechanistic and aggregative reductions.

Reductionist strategies act differently for each. Some tools are the same (e.g., approximations have important uses in all three) but some are clearly divergent. Functional localization fallacies or their generalized analogues are critically important for inter-level reduction or issues of aggregativity, but irrelevant to successional reduction. Issues of elimination can arise for aggregativity and for successional reductions, but not for mechanistic explanations. This expanded chapter relates emergence and its opposite, aggregativity, to heuristics for finding good decompositions of systems into parts, and to natural kinds. Criteria for aggregativity or non-emergence play roles in discovering parts that compose levels of organization and are articulated to provide mechanistic explanations for upper-level phenomena. Philosophy intersects differently with work on problem solving in cognitive science for each of these, with benefits for both. I have spent more time on these topics than any other.[1]

In chapters 9 and 10, I examine the architecture of complex systems. This is not a topic to be addressed a priori, as Nelson Goodman (1966) tried to do. The structure of our theories and conceptual frameworks are influenced by very general physical and evolutionary constraints, and different kinds of conceptual structures work under different circumstances. These chapters have a special status, being about the theory we use in defining the pieces in our "piecewise approximations." They are broadly realist and naturalistic, using empirical claims about natural systems as starting points to characterize our investigative modes, but also taking heuristic, problem-solving, and cognitivist perspectives on how we structure our pictures of the world. I ask what are easy and what are hard problems in the analysis of these systems, how they arise, and how to recognize them. This provides a new kind of framework for considering problems of reductionist methodologies. Chapters 11 and 12 address more philosophically familiar topics—reduction and emergence—posed in this new framework. Both take a fundamentally problem-solving and heuristic approach, with strikingly different results. Extensive introductions to these four chapters provide meta-commentary relating the different perspectives on reduction and emergence.

Chapter 9: Complexity and Organization

Chapter 9 (originally published in 1974) considers a task usually taken for granted—description. For compositional systems, philosophers characteristically talk about theory reduction; however, we too easily assume that we already have the relevant descriptions of behavior, and need only to relate them. For complex systems, we often don't have theories to relate, so the task of relating them is undefined. We need to know more generally how we order and relate different descriptions of the behavior of a system, particularly partial descriptions, to construct explanatory mechanistic accounts of its performance. Biological systems are difficult to describe and analyze in a *systematically* reductionist manner—even for one with reductionist sympathies. I consider why this is so.

I offer first a phenomenological descriptive account of our situation in dealing with complexity, but it is explanatory too. Even with a thoroughgoing materialism, we can still encounter situations that strain ordinary intuitions derived from looking at simple systems. In biology (and elsewhere!), we often have multiple cross-cutting theoretical perspectives on a system (e.g., the anatomy, physiology, development, and genetics of an organism). With these perspectives we decompose the system into parts in different ways. Each can claim to have some but not all of the necessary information for a total description or causal account of the system (or organism). These multiple boundaries and their relations can be a rich source of ways for detecting order.[2] If the parts from these perspectives are composite entities, they should be in-principle describable in terms of a common set of still lower-level parts (insert here your favorite list of atoms, and organic, biochemical, and macromolecular tinker toys for all higher level biological things). But the objects of these perspectives are not compositionally orderable relative to each other, so we can't order the perspectives in the way we order compositional levels of organization. We rarely have the kind of systematic and complete knowledge needed to reconstruct most larger parts, and often can't tell how much detail is necessary in the reconstruction. At best, we have to study different instances to determine the "don't care" conditions allowing multiple realizability for entities of that type. It is not desirable to put in everything, even if it were possible! Thus we are forced to deal with higher-level, heterogeneously characterized parts *from this multiplicity of different perspectives,* and to characterize their relations to one another to solve most of our problems. 1. (The recent book by Wagner [2005] nicely exemplifies the multiple levels and perspectives brought to bear in the analysis of complex biological systems.)

An anatomist and a physiologist might each have charts of their study organism on their wall: The anatomist's is a familiar arrangement of skeletal, muscular, and organ systems—the modern descendant of Vesalius's engravings in *De Fabrica* extended to other size scales. The physiologist's looks more like a program flow chart for how different parts of various physiological systems interact (connecting respiratory, metabolic, circulatory, nervous, and other systems). The physiologist's chart will usually seem far less complete.[3] At more microscopic levels, the anatomist switches to a diagram of a typical cell (usually an "ideal type," not an actual cell, with cellular ultra-structural components represented—nucleus, membrane, mitochondrion, endoplasmic reticulum, and so forth), or perhaps to electron micrographs of these components in real cells. The physiologist will by now be looking at a flow chart of biochemical pathways (such as the Krebs cycle in primary metabolism). Both are spatial decompositions of the same organism (or part of it) into parts, but parts in anatomical and physiological charts won't look anything alike. Worse, no information for how to map one class of parts to the other is provided. (The biochemist's space is a topological "reaction space," with a reaction's location in a network indicating its causal antecedents and descendants, *not* the place in the cell where it occurs.) More recent awareness that biochemical reactions commonly take place in specific locations supported by elements of cellular ultra-structure is now forcing the first systematic articulation of these perspectives.[4] But tying them together at the level of biochemical reactions and cellular ultra-structure does not, contrary to philosophical expectations, automatically zip anatomy and physiology together all of the way up. Instead, each pair of anatomical and physiological hooks at ascending levels must be separately and laboriously engaged, often in the conjoint context of neighboring hooks of the same or different sorts above, below, and on all sides. 2. (If the deductivist picture of reduction really worked, this would be unnecessary.)

Higher-level parts (and their parts) usually have well delineated *intra*-perspectival relations to one another, but many problems demand that they be related to other parts *across* perspectives in a systematic fashion, either compositionally or causally. Thus the dynamics of interaction of biochemical elements—some at hand, some bound to membranes, some needing biochemical synthesis from a precursor two stages up the pathway, and some needing genetic synthesis from a gene not yet turned on—demands analyses of interactions between components in diverse topological and spatial locations in the two decompositions. Components must be both spatially local (so they can interact physically) and topologically local in biochemical reaction space (reactions connect them). This requires joint use of both decompositions. Such compositional and causal problems reflect two kinds of complexity—*descriptive* and *interactional*—that bedevil the articulation of our different partial theories or perspectives of the organism to produce an integrated account of what it is doing and how. This *problem of conceptual coordination* is the norm for anything other than extremely local and discipline-specific questions in biology. Problems of functional localization represent a special case of this more general phenomenon.

This incommensurability is not between competing theories, but between complementary theories that we must use together. We don't solve the problem by choosing one or another of the theories—we must bring all of them to bear! *Philosophical theories of incommensurability do not address this kind of case, but it is far and away the most common kind of incommensurability we deal with in science.* Such cases are encouraging: heuristics for dealing with them suggest useful tools for articulating relations between different paradigms across time as well as across disciplines. *Real* problems of incommensurability are both systematically underestimated (in frequency) by philosophers who have missed this kind of case, and systematically overestimated (in intractability) because we have more experience (and success) in dealing with them than they suppose. Such integration and coordination issues are major foci of Schank's (1991) work. He argues that computer simulations (of cross-perspectival problems) are essential tools for their analysis.

This situation provides useful diagnostic criteria for such complexities in biology and the human sciences. Nothing illustrates this better—though not his primary intent—than the complexities and qualifications Schaffner is forced to deal with in his magisterial review, *Discovery and Explanation in Biology and Medicine* (1993). Each system and problem—all of the real cases he brings up—could be exemplified simultaneously (though in different ways) in analyzing organisms of any one of diverse biological species. I consider when these kinds of problems of conceptual coordination can be expected to happen, and suggest (enter the systematic errors!) a range of fallacious inferences we would be more likely to make, and with this new understanding, have a basis for avoiding. (This is the premier domain of functional localization fallacies!) I provide concepts for understanding the nature of interperspectival disputes in complex sciences; particularly for seeing how theorists with different paradigms can all be partially right, but fundamentally "out of register." Bringing such views into focus simultaneously to compare and integrate them would be useful throughout the human sciences.

Traditional relativists have tried to recognize truths in different perspectives while in effect assuming they are each complete and all competing. They then seem inexorably pushed into saying that if these can *each* be right, then (in a sense) *nothing*'s right, and (therefore) *anything* goes. But if we see this as the problem of how to integrate complementary partial-truths, then the demands on reality are strong and multifaceted, but not impossible. Accepting a common referent becomes an essential tool in seeing how to reconcile and integrate the different views. If the contributions of different perspectives are partial, there is *room* for them to complement one another. Seeing them as complementary rather than contradictory suggests how to focus and resolve disagreements, or at least how to localize and understand their sources—a precondition for any such resolution. I urge this therapy for relativist disputes almost everywhere.

Chapter 10: The Ontology of Complex Systems: Levels of Organization, Perspectives, and Causal Thickets

The use of multiple complementary views or perspectives on an object is continued in Chapter 10 (based on Wimsatt, 1994), elaborating an account of levels of organization first proposed in Wimsatt (1976a). Reductive explanation in complex sciences presupposes such an account, but no one has tried to characterize them. We tend to think of levels as collections of objects, but they have a more molar order as well (not any collection of objects will do!). They are as much proper subjects of ontological investigation as the more atomistic abstract things normally embraced by philosophical ontologists—objects, events, causes, properties, and the like. *Levels* are pivotal parts of our hierarchical view of nature. Size and time scales of particular characteristic causal processes have a central role in individuating them. These might seem merely contingent features, but they are sufficiently central to the genesis and properties of levels to make a conceptual difference. Levels have interesting properties bearing directly on problems of explanation, reduction, emergence, evolution, and the nature of living and thinking beings. They give us important handles on the nature of macroscopic theoretical structures and on many of the more molar questions in philosophy of science.

Some properties are features of *any* transition between levels. Philosophers of psychology treat many features as if they were special to the mental realm and a challenge to mechanism: emergence, "supervenience," the conceptual and dynamical autonomy of special sciences, multiple realizability, and the "anomalousness" of the mental. But these also characterize relations of chemistry to physics, molecular biology to chemistry, membrane biophysics to molecular biology, the cellular biology of nerve cells to membrane biophysics, neurophysiology to the cellular biology of nerve cells, and so on all of the way up, and (in many cases) for important intermediate levels in between.[5] Understanding how they emerge in terms of relations between levels puts them in a different light. They are not alternatives to mechanism: unpacking intermediate levels of organization—and these kinds of relations between levels—is an integral part of articulating both specific mechanisms and more general mechanistic explanations of higher level phenomena.

Each new level of organization has characteristic emergent properties (Chapter 12). Some systematic trends emerge with increasing size: higher-level interactions almost always take place more slowly, and levels become less well defined as their increasing complexity leads to greater interpenetration of levels (at least up to that of the biosphere).[6] As boundaries between levels break down, we have the emergence of new kinds of macro-entities in and ranging through the biological, psychological, and social realms. One centrally important new type is a *mechanistically explicable* family of things I call *perspectives,* a special case of which we see in Chapter 9.[7] A perspective—something much like a point of view, which we may think of as subjective or as part of *our* theory of a system—may actually be intrinsic to the system if it has characteristic relational properties. This is surprising, because perspectives have been considered as paradigmatically unanalyzable and antiobjectivist since Nagel (1974). (Some—the theoretical perspectives of Chapter 8, or the ecological niche—are relatively objective. Others are paradigmatically subjective. Yet they both have many features in common.) Understanding what perspectives are and how they relate to levels of organization gives useful purchase on naturalistic anchors for some of the mental dimensions of the human sciences.

Causal thickets arise in turn as boundaries between perspectives begin to break down with still further increases in complexity. They do much to explain the character of methodological disputes in psychology and the social sciences. We should expect methodological and conceptual disputes exactly when boundaries among divergent perspectives break down, especially when they lay claim to the same territory. More recent work on embodied consciousness and social cognition (Thompson, Rosch, and Varela, 1991; McClamrock, 1995; Hutchins, 1995) point to ambiguous boundaries between perspectives, as well as issues with how to distinguish perspectives from causal thickets, and how to recognize new emergent perspectives at higher levels. The productive fusion of economics, history, cognitive anthropology, psychology, and cultural evolution (Boyd and Richerson, 1985) required to make sense of what economic historian Douglass North (1990) calls "institutions" raises similar issues. Or consider the increasing confluence of paleontology, macroevolution, morphology, and evolutionary genetics with developmental biology and genetics (Raff, 1996). Here we have whole disciplines as nodes providing increasingly rich contexts for one another. These would have to fuse with developmental psychology and psycholinguistics to engender a new evolutionary theory of development, in which both the internal process-architecture and the rich informational structure of the environment will be seen anew in terms of their intrinsically relational structure (Wimsatt and Griesemer, 2007). These three mega-disciplinary aggregates will themselves be linked—providing connections to and boundary conditions for each others' problems.[8] These confluences of disciplines for particular problems do not unify them generally, and thus do not unambiguously locate the problems. Most people are still trained in particular disciplines, so in solving these complex problems we need to be able to address questions like: Even though language is a complex multi-disciplinary entity—thicket-like if anything is—under what circumstances, and for what purposes, can we get away with studying or approximating it from the perspective of just one of its attendant disciplines? We can do it sometimes, and just as surely, we can't at others. Are there any generalizable ways to recognize when a different approach is called for? Can people trained within disciplines be expected to recognize it?

These three new kinds of ontologically macroscopic entities—levels, perspectives, and causal thickets—increase the coherence and plausibility of a generalized mechanistic view of the world, and without the eliminativism that has characterized so many past attempts. Paradoxically, *our usual conditions for a mechanistically acceptable explanation are seen to break down systematically in some cases in a mechanistically explicable fashion, and—therefore—without fundamentally compromising mechanism.* Chapter 10 uses concepts of robustness, drawing on material in Chapter 4, to delineate an alternative philosophical methodology. Recognizing when logical structure, crisp definitions, and universal generalizations are inappropriate to some problems, I propose other ways of proceeding. This chapter is methodologically self-referential: it uses the new methods it recommends. It is thus the most philosophically radical paper here—with deeper implications for the methods and practice of philosophy, and for strategies in the analysis of key concepts and problems of the human sciences.

Chapter 11: Reductive Explanation: A Functional Account

The reductive methodological perspective provided in Chapter 11 gives a new fundamentally and deliberately more approximate framework in which to understand the use of exact and inexact tools alike—including those of chapters 9 and 10. This chapter comes the closest to traditional accounts of reduction, and spends more time addressing philosophical literature on the topic. In it I argue for the importance of mechanistic explanation (originally published in 1976, this was well before it became the common subject of philosophical attention it is today), and propose several theses uncommon or unknown in 1976, but since widely accepted.[9] Paralleling Nickles (1973), I argue that the "classical" search for *the* logical form of reductions conflates two fundamentally different activities, though our accounts differ somewhat.

In theory succession, a newer and better theory replaces or reduces to an older one. Both either deal with phenomena at the *same* compositional level of organization, or (for some physical theories), lay claim to phenomena at all levels. If structures of older and newer theories are too dissimilar to construct reductive transformations (serving various problem-solving functions elaborated here), there is no reason to keep the ontology of the older theory, and there can be wholesale eliminations. (By contrast, if there *is* a reductive transformation, the ontology of the older theory will tend to be preserved as limiting case approximations of the ontology of the newer one. And the limiting cases will usually be, or include, the ontology of our world.[10])

The other activity called *reduction* need not relate theories at all, but refers instead to the explanation of upper level *phenomena* and regularities in terms of lower-level *mechanisms.*[11] *It is never eliminative.* The idea that there is a kind of *theory reduction* (eliminative reduction—favored by some philosophers of mind) in which higher-level objects and relations are systematically eliminated in favor of lower-level ones, rather than extended, modified, or transformed, arises from a conflation of these two distinct activities. *No such beast as eliminative reduction is to be found anywhere in the history of science, and there is no reason, in terms of the scientific functions served, to expect it in the future. It, and its aims, are largely misconceived philosophical inventions. Robust higher-level entities, relations, and regularities do not disappear wholesale in lower-level scientific revolutions—our conceptions of them transmute and add new dimensions in interesting ways, yes, but disappear, no.*

An Interpolated Excursion on Eliminative Reduction

Though eliminative reduction is a serious error, it has become so widely endorsed that I feel the need to address it here directly. A related position responds to its temptations, but with a much more sensible outcome.

As always, the story is more complicated than it appears, but I will try to provide a good first approximation.[12] *First,* an eliminative view in inter-level cases would be more plausible if all system properties were simple aggregates of parts properties—this would motivate the "nothing but"–style eliminativist talk. But, as explained in Chapter 12, these conditions are essentially never met. Confusions between consequences of total aggregativity and what to expect of inter-level reductions generates illegitimate apparent support for eliminativism. (This confusion is abetted for philosophers by a tradition with views like Russell's [1918/1985] logical atomism, in which derived terms, including theoretical terms and material objects, were mere logical fictions and *in principle* dispensable.)

Seciond, the phlogiston/oxygen case often quoted by eliminativists to support their views was *not* a case of theories or phenomena at different levels of organization as commonly supposed, but one of successional reduction at the same level, where ontological elimination *can* occur. (This dispute took place before Dalton's atomic theory, so during their competition, phlogiston and oxidation theories were accounts offered at the same macroscopic level: oxygen was an extract of or species of air.) So elimination of entities in this case does not bear on the issue of eliminations for inter-level accounts.

Third, (here it gets more complex) most real cases of scientific progress in understanding an inter-level mechanism or coordinated cluster of phenomena involve *both* successional and inter-level change and reduction. So, although elimination through inter-level scientific progress is possible, it would appear as a displacement of one set of conceptions of macro-level things by another—not by *any* conceptions of *micro*-level things. More controversially, most such replacements will preserve so much of the older phenomenology that changes will not be obvious in most operations at the macro level. Wimsatt (1986a, 1999b, 2003) provides a plausible case: the fundamental reconstruction of the innate-acquired distinction necessary for consistency with new theory and data in evolutionary and developmental biology and ethology. If this reconstruction is a replacement, it is unconvincing as a paradigm for the eliminative replacement of folk psychology. The replacement for the innate-acquired distinction preserves the vast majority of its traditional consequences. If anything, it *increases* their number, and for the first time explains their relationships in an integral fashion in such a way as to make the distinction more robust than before—although it is drawn along radically different lines, and thus could be described as a replacement. To be sure, there are crucial differences, but the net appearance is not of eliminating a traditional discipline or perspective. New macro-level conceptions will both have to capture any robust phenomena of the older set and be appropriately related to or anchored at the lower level.

A *fourth* alternative provides intriguing possibilities for agreement.[13] Conceptual change can attenuate or expand what we attribute to a level. This is because functional localization fallacies can lead us to misassign levels for phenomena. Conceptual change can make us aware of this and lead to reassignments. I'm quite happy to explain Mach bands (the heightened contrast at color change boundaries in the visual field) in terms of the activity of lateral inhibition networks—giving them a lower-level anchor—though we still have to deal adequately with the phenomenal colors that *show* the increased contrast at the boundaries. However, some mental phenomena will also be correctly reassigned to higher levels—see Hutchins (1995) and others on social cognition. Indeed, biases of reductionist research strategies (see Chapter 5 and Appendix A) *predict that in periods dominated by reductionist views, more properties should be mis-localized at too low a level than at too high a level!* It is thus plausible that many properties attributed to individual psyches are in fact psychologically reified projections of social practices—a tendency exacerbated by our methodological individualism. This seems especially likely for claims that large fractions of our internal thought processes are deductive inferences. We construct valid inferences, but usually with a great deal of external help—ranging from the development of our language, logic, and mathematics, to our use of external tools (ranging from written language to computers) to check and elaborate the results. The current "computational" worldview suffers from an exactly parallel malady—projecting downwards (or inwards) many features of a high level abstraction.

So could a level, such as the traditional mental realm of individual psychology, evaporate totally? Not likely; but our theories of mentality or cognition may diffuse sideways, as well as downward or upward, as people like Frank (1988) argue the inadequacies of simplistic faculty psychology that put cognition, conation, and affection into separate black boxes, all distanced from the motoric (which I predict will also come to be seen as central). In our current framework, cognition has grabbed too much credit that is better shared with other mental faculties (see the bias-producing phenomenon of *perceptual focus* in Wimsatt, 1980b). I believe that the *apparent* personal level will get thinner: thinned to the stuff that's really there—robustly accessible relatively directly at that level.

This introduces more focus to the dispute: Now we can ask whether there *are* any robust phenomena at the upper or intentional level. Indeed, there are lots of them, including phenomenal colors (not *qualia,* if those are defined as things that can't be explicated in any other way) and a variety of phenomena loosely characterized as intentional. If we do away with beliefs, intentions, and propositions, it could only be in favor of things or assemblages of them that behave an awful lot like them in a wide variety of contexts. I don't feel the same way about propositional attitudes—which in their elaborations of the 1960s and 1970s seem more the constructions of philosophers than the beliefs of any "folk." (On phenomenal colors, see Evan Thompson's "Novel Colors" [1992]. For intentionality and the self, I think that there are non-eliminative ways of reading most of what Dennett argues in his *Consciousness Explained* [1991].) I won't justify these claims here.

They are made to locate the natural extensions of my position in a broader conceptual geography.

Discussions of eliminativism have grabbed the stage, but the distinctions between types of reduction help to clarify inconclusive discussions. The different functions of successional versus inter-level reduction reveal and explain unnoticed differences in their logical form and assumptions, and point to many detailed and powerful heuristics for theory construction for inter-level reductive explanation. The analysis of Chapter 11 thus delivers on a promissory note of increased relevance and usefulness to scientists. It suggests that they should *not* look to construct eliminative theories but look for resources at both upper and lower levels to constrain, and more positively, to use in connecting with and refining the other level, or with new phenomena and entities at levels in between (Wimsatt, 2006a). Purely functionalist approaches favored by advocates of the "special sciences" are also compromised: knowledge of lower-level mechanisms is a crucial source for the refinement of upper-level accounts, and conversely. This is nicely supported in Bechtel and Mundale's (1999) discussion of "multiple realizability." The view argued here makes specific suggestions for how to use these resources in constructing and refining identificatory or localizationist hypotheses.[14] It is also *symmetric:* it does not privilege lower-level accounts over higher-level ones in the development of reductionist explanations. Normally, each has things to contribute to the development of accounts at the other level, and usually each needs revision to fit—a phenomenon suggested by Schaffner in his model of reduction as early as 1969, and elaborated here and in Wimsatt (1976a) as a co-evolutionary process. These views have attracted growing support (and significant elaboration) ever since (see Darden and Maull, 1977; Churchland, 1986; Bechtel and Richardson, 1993; McCauley, 1996).

Wimsatt (1976a) anticipated other trends in other respects: I advocated a non-eliminativist causal interpretation of Salmon's (1971) "statistical relevance" account of explanation nearly a decade before he did. His account (and attack on the deductive nomological model of explanation) dovetails more naturally with a search for causal factors that we articulate into mechanisms than with a search for laws. But to make sense of how scientists deal with mechanistic explanations, we must also modify his scheme (as he did not) to bring our search for causal mechanisms in line with our bounded abilities. In giving mechanisms, scientists stop short of exhaustively complete accounts when the yield from additional causal factors gets too small, too rare, or both (this is a satisficing version of a cost-benefit rule). *Mechanisms are just truncated, partially and roughly characterized, manipulable, and relatively modular and generalizable patches of the causal fabric of the world.* They are the creatures of a backwoods mechanic or an engineer—not of a mathematician or foundational physicist—though they are used by model-building theoretical physicists in all other areas. Mechanisms are portable and applicable in different contexts, and subject to *ceteris paribus* qualifiers in an explicable way (see Glennan, 1992, 1996, on mechanism and causation; and Mikkelson, 1997, on the role of *ceteris paribus* in the analysis of counterfactuals).

I also concurred with Nancy Cartwright's (1983) later rejection (for different but convergent reasons) of laws as central to causal explanation.[15] In 1992, I extended this account: to understand the Morgan school's differences with Haldane over the role of general models of linkage, one has to assume that "laws" played a distinctly derivative role for them—as templates that they expected to be falsified, and used primarily as tools for detecting anomalies, which pointed to deeper mechanisms. We are thus led to a picture of theory structure in genetics in which mechanisms are primary, and are revealed not directly through laws but through a tissue of successive anomalies to tentatively advanced models and generalizations, a view I call (with a bow to Cartwright), *particularistic mechanism*. It fits mechanistic investigations and explanations more generally.

Chapter 11 also converges partially with Kitcher's (1981) explanatory unification account, though the metaphysics is quite different. The flavor here is more strongly realist and mechanist than Kitcher's, and gives explanatory unification with a strictly local flavor as a *consequence* of our activity rather than as its aim. We unify in terms of mechanisms, looking for the same or similar ones in diverse places, rather than in terms of laws. We also seek to integrate our accounts of phenomena by connecting them as richly (as robustly) as possible into the causal network, but do so without use of exceptionless generalizations, laws, and inference rules as Kitcher does.

Finally, to reflect the explanatory and causal autonomy of levels, I invoked cost-benefit considerations a second time to revise Salmon's (1971) "screening off" relation (my "effective screening off"). Salmon's original account is problematic for inter-level phenomena. When the same macro-state is realized by diverse micro-states, his account introduces an unrealistic reductionist bias favoring lower-level redescriptions, like most other philosophical accounts of that period. When lower-level redescriptions are much more expensive, and yield little or no predictive benefits, the causes are properly located at the upper level. A cost-benefit account naturally leads one from multiple realizability (via sufficient parameters and dynamical autonomy—see chapters 4 and 10) to recognize higher-level variables as *causally potent under some but not all conditions*. We can understand when we explain at an upper level, and when it is necessary or profitable to go to a lower level—for phenomena or regularities that are anomalous at the upper level. Levins' (1966) "sufficient parameters" are more realistic heuristic analogues to multiple realizability and supervenience. They make higher-level variables both theoretically cost-effective (thus things upper-level organisms should want to detect and use) and pragmatically causally effective.

This view also leads us away from the unrealistic expectation that causal relationships require or entail generalizations that are both usable and exceptionless. In the real world (of folk psychology, and of science as applied to real-world systems), we usually can have one or the other, but not both. We will also almost always choose usability over exceptionless universality. Blind attempts to construct exceptionless generalizations at any cost would produce cumbersome and useless structures blunting this necessary multi-level dialectic. Only this account correctly captures how and when working scientists choose to work at one level or another.[16] Cartwright (1983, 1989) argues similar points, but embraces probabilistic theories of causality. This is an error that would undercut all the techniques we use for "localizing faults" and to figure out what happened at the macro-level (see also Glennan, 1996).

With accounts of the dynamical autonomy and robustness of upper-level variables (chapters 9, 4; Wimsatt, 1976a), Chapter 11 provides the only analysis consistent with a reductionist methodology while justifying upper-level talk in terms stronger than pragmatic convenience. Neither extreme is acceptable: both fail to capture the integrative articulation in a single mechanistic explanation of entities, phenomena, and causes from different levels—the basic form of explanations that are so common in modern scientific and engineering practice.[17]

Chapter 12: Emergence as Non-Aggregativity and the Biases of Reductionisms

Chapter 12 (see also Wimsatt, 1997b) takes up the idea of emergence as a relationship between a property of a system and properties of its parts. (A system commonly has some properties that are emergent and others that are not.) Philosophers often suppose that emergence implies irreducibility—or if they are reductionists, treat claims of emergence as counsels of ignorance. Scientists would reject this: many properties they claim as emergent are not only reducible, but straightforwardly so. Mechanism is at its most convincing when it can give an intuitive explanation of a system property regarded as emergent in terms of properties of the parts of that system. But having that explanation doesn't deny emergence. Extending my earlier work (Wimsatt, 1986b), I argue that emergence indicates dependence of a system property upon the mode of organization of parts of that system. But how should we characterize this?

Given the many possible modes of organizational dependence, it is better to flip the problem: if the mode of organization of the parts *didn't* matter at all, the system property would seem to collapse as nothing more than an aggregate of parts' properties. What does this entail? Four plausible conditions are required for the system property to be *invariant* over a (specific) class of operations on the parts and their properties. Meeting these conditions rules out any organization.[18] Emergence is thus a failure to meet one or more of these conditions—blocking *aggregativity*. But organization may be internal to the parts as well as embodied in their relationships, thereby requiring the conditions to be met *for all possible decompositions of the system into parts* if the relationship is to be aggregative. There are diverse ways they can fail to be met—providing an interesting variety of kinds of emergence. Various examples are classified according to how they meet or fail these conditions. I'll return to the conditions shortly.

Aggregativity suggests a third (very extreme) sense of reduction that has been influential in a backhanded sort of way. Although rarely satisfied, our misconceptions about it—conflating it with inter-level reduction—probably motivate negative attitudes toward reductionist accounts on the one hand, and the mistaken association of reductionism with eliminativism on the other. These mistakes are endemic on both sides of almost every debate over reductionism.

Mathematical biologist Jack Cowan (personal communication) loves to describe the difference between biophysicists and theoretical biologists. A university president once said to him: "You both use a lot of math and physics to do biology—you must be doing the same thing. Why shouldn't I merge your departments?"

"I'll tell you the difference," Cowan said, "take an organism and homogenize it in a Waring blender. The biophysicist is interested in those properties that are invariant under that transformation." One couldn't get a more graphic image of the difference between aggregativity and emergence.[19]

What if some properties of the parts and system were invariant no matter how you cut it up, aggregated, or rearranged its parts? For such properties, organization wouldn't matter. There are such properties—those picked out by the great conservation laws of physics: mass, energy, charge, and so forth. As far as we know, that's all. These meet very restrictive conditions: for any decompositions of the system into parts, these properties are invariant over appropriate rearrangements, substitutions, and re-aggregations, and their values scale appropriately under additions or subtractions to the system. Meeting these conditions makes them very important properties—properties that became the source of the great unifications of nineteenth-century physics. For these aggregative properties, we are willing to say: "The mass of that steer I gave you was nothing more than the mass of its parts." And we blame the butcher—not vanished emergent interactions—for any shortfalls. *If these four conditions* (informally stated above) *are met for all possible decompositions of the system into parts, aggregativity must be an extremely demanding relationship, one seldom found in nature.* (Even Waring blenders only disrupt organisms down to the macromolecular level, preserving organization at lower levels, which is why they—used with centrifuges to separate out fractions of varying densities from the homogenate—have such a large role in isolating specific macromolecules.)

Emergence in this sense is thus extremely common—much more so than normally supposed. Is this then too weak a notion of emergence? It fits the intuitions of most scientists I know, who want both their reductionism and their emergence, and who agree with its classification of particular cases. Even with stronger notions of emergence to explore, the specificity and power of these four conditions make them extremely useful tools of analysis. Were a stronger notion needed (not yet demonstrated!), this notion could "clear the brush" of the many cases it captures to focus discussion on those that remain. Aggregativity also gives a powerful handle for diagnosing certain important kinds of errors, but doing so first requires a short discussion of approximations.

Since these conditions are expressed in terms of invariance of the system property under operations on the parts, one can easily produce a family of quantitative criteria for approximate or local aggregativity, where variations of the system property are constrained within an "error tolerance" of $\pm \epsilon$ for various values of ϵ. Chapter 9 uses a similar move for different degrees of near-decomposability or modularity in systems. Adding such "tolerances" to an analysis is a useful strategy for qualitative concepts in a messy, inexact, and approximate world having many regularities and stable patterns, but few exceptionless generalizations. (We need "sloppy, gappy generalizations." This is quite consistent with determinism. See Chapter 10 and the discussion of deterministic chaos [with pictures!] in Wimsatt, 1991. Chapter 12 also supports this view.)

Tolerances are essential because we often use quantitative and formal qualitative frameworks as templates for pattern detection and matching that nature may meet in varying degrees. With a particularly adaptable framework to fit onto nature in diverse places in different possible ways, we may try a variety of such mappings, looking for "best fits."[20] The complex ontological objects—levels and perspectives (Chapter 10)—result from natural selection and stabilization processes involving best fits of such frameworks in nature. Our modeling and theory construction also make fundamental uses of such processes. The conditions of aggregativity are just such an approximate and adjustable framework, since one can construct orderings for how well each condition is met, and partial orderings for how well all four of the conditions are met across different decompositions of a system into parts.[21] Indeed, to the extent that modularity exhibits some of the features of aggregativity, the importance of modularity in evolution (and in neural architecture) implicates these conditions causally (through selection) in the architecture of many natural objects (see Wagner and Altenberg, 1996).

Situations where the conditions are met for some decompositions, but not for most, are particularly interesting. These properties look aggregative for those decompositions, but reveal themselves as emergent or organization-dependent for others. *The better a decomposition meets these conditions (meeting more of them, more exactly, and over a broader range of conditions), the more nearly it factors the system into modular parts that can be characterized from that perspective in terms of monadic, intrinsic, or context-independent properties. Such perspectives thus provide particularly simple and theoretically productive decompositions of the system into parts. We are more likely to see these properties as natural, and these parts as instances of natural kinds, as robust, and to regard the system as nothing more than the collection of these parts.* Note the use of aggregativity here: an apparent foundational distinction between kinds of properties is transformed into a search heuristic for finding preferred, simple, "maximally reductionistic" analyses of systems. Given the biases of reductionist problem-solving strategies (see Chapter 5 and Appendix A), such decompositions lead readily to excesses of "nothing but" talk and disciplinary imperialism.

Aggregativity aids the search for good parts decompositions (other criteria are also used!). What else does it have to do with inter-level reductive explanations (as mechanistically explicable) in chapters 9, 10, and 11? The answer is—virtually nothing! (Consider that total aggregativity has almost no instances, and the latter has many.)

But there is a curious operational paradox: with total knowledge of a system, the two senses are clearly distinguishable, but when we know very little about a system, our strategies in approaching them look very similar. This spawns many confusions! The simplest theories we can construct for the interactions of parts tend to model them in a very homogeneous and aggregative fashion. These are "first-order approximation" models we construct (because they are very simple) for systems we don't yet know much about (Wimsatt, 1979, 1980b). We thus use similar starting points in modeling diverse systems: some quite aggregative; others non-linear but still symmetric and homogeneous (tolerating rearrangements and substitutions of parts); and others that are mechanistic, highly differentiated, and whose key properties are very organization-dependent. With partial knowledge but predominant ignorance, it is not surprising that the properties of mechanisms and of aggregates should be easily confused—especially if the former are modular or partially aggregative.

This explains temptations to make unsupported and excessive claims of the sort common in early stages of a reductionist analysis, and then to quietly take them back later. Aggregativity *would* compromise claims for autonomous higher-level properties and systems; *but it never happens.* Mechanistic explanations of phenomena commonly involve highly differentiated parts and behavior that depend on their mode of organization. Some people see such explanations as very threatening. But they misidentify the enemy: only aggregativity, not mechanism, would justify the claims they fear. And aggregative claims are false. Analyzing these excessive claims and their biases (commonly associated with eliminativist positions) is especially important for fields and explanatory tasks whose major questions are still in process. They are just the kinds of evaluations and calibrations of conceptual tools we should seek for limited, fallible, and error-prone scientists. To assess these claims fairly we *must* recognize the limitations and incompleteness of our knowledge, the heuristic character of our tools, and the specific biases that likely issue from their application. This is then another point at which the LaPlacean image of the ideal scientist who already knows everything serves us poorly.

But then why should reductionist methodologies have appeared to be so successful? In part, because they often *are* genuinely successful. But there is more. Heuristic principles characteristically transform hard problems into different but related ones that are easier to solve (Chapter 5). If we then solve them effectively, we will likely identify the new problem with the old one—saying, "Now that we've *clarified* the problem so that it can be solved . . ."—without noticing that it's been changed! *In this way, quite substantial changes in a paradigm can be hidden—particularly a cumulative string of such changes, each too small to be regarded as fundamental.* Without denying the power of reductionist approaches, this kind of ex post facto reification helps to generate excessively high opinions of them, and—in another context—the exaggerated belief that work elaborating a paradigm or applying a theory is merely playing out options that are already given. By looking at such achievements only after the fact, when justification is the focus, they look not only easy but unavoidable. Other things also contribute: the traditional focus on justification rather than discovery—especially if one never looks at cases of the latter—makes it easy to undervalue the difficulty and creativity of applying a theory in a new area or in a new way.[22] Positivists commonly underestimated the importance and difficulty of applying a theory. This is an important lesson one can learn by studying engineering. We will return to this in the final section of the book.

CHAPTER NINE

Complexity and Organization

In his now classic paper, "The Architecture of Complexity," Herbert Simon observed that "In the face of complexity, an in-principle reductionist may be at the same time a pragmatic holist" (Simon, 1962, p. 86). Writers in philosophy and in the sciences then and now could agree on this statement but draw quite different lessons from it. When he wrote this, pragmatic difficulties were commonly things to be admitted and then shrugged off as inessential distractions from the way to the *in principle* conclusions. Now, even among those who would have agreed with the in principle conclusions of the last generation's reductionists, more and more people are beginning to feel that perhaps their ready assumption that the pragmatic issues were not interesting or important must be reinspected. This chapter suggest for the concept of complexity how an in principle reductionist can come to understand his or her behavior as a pragmatic holist.

Reductionism and the Analysis of Complex Systems

A number of features of the reductionist orientation contribute to a point of view that is ill-suited to an adequate treatment of the concept of complexity. First, there is a bias toward theoretical monism. In biology and the social sciences, there is an obvious plurality of large-, small-, and middle-range theories and models that overlap in unclear ways and usually partially supplement and partially contradict one another in explanations of interactions of phenomena at a number of levels of description and organization.[1] Despite this plurality, all of the models phenomena and theories in a given area (however that be defined)[2] tend to be treated as ultimately *derivative* from one primary theory. So questions concerning their relationships to *one another* tend to be ignored on the supposition that all will be made clear when their relationships to the perhaps as yet unknown reducing theory are determined.

But scientists must work with this plurality of incompletely articulated and partially contradictory, partially supplementary theories and models. Philosophers have ignored the requirements of this situation, though this kind of theoretical pluralism has played an important role in the analyses of some biologists,[3] and it is central to the analysis of our intuitive judgments of complexity.

Second, given the difficulty of relating this plurality of partial theories and models to one another, they tend to be analyzed in isolation with unrealistic assumptions of system-closure (to outside "disturbing" forces) and completeness (that the postulated set of variables includes all relevant ones).[4] These incomplete theories and models have, *individually,* impoverished views of their objects. Within each, the objects of the theory are just logical receptacles for the few predicates the theory can handle with manageable degrees of theoretical simplicity, accuracy, closure, and completeness. Nobody attempts to put these views together to see the resultant objects. Suppose that the five blind men of the legend, perceiving different aspects of the elephant, nonetheless recognize their common referent. "Good for them!" you might say. But not so fast: given the tremendous difficulties of reconciling their views of it, they nonetheless decide to treat their views as if they were of *different* objects! The net result is often not to talk about objects at all, but to emphasize predicates, or the systems of predicates grouped together as theories or models.[5]

Thus, although biologists, social scientists, and others who work in areas where *complexity* is a frequent term talk almost invariably of the complexity of *systems* (thereby meaning the objects, in the full-blooded sense, which they study), most analyses of complexity in the philosophical literature have been concerned with the simplicity or complexity of sets of predicates or of theories involving those predicates in a manner jumping off from the pioneering analyses of Nelson Goodman (1966, pp. 66–123).[6] But Goodman's theory of complexity, even if acceptable, is a poor measure of the complexity of the objects of that theory unless the theory gives a relatively complete view of those objects. Absent the ultimate all encompassing reduction to an all-embracing theory, one can only talk about the internal complexity of our different theoretical perspectives or views of an object. Nor could one avoid this conclusion by taking the complexity of the object as some aggregate of the complexities of the different views of the object, since part of its complexity would be located at the interfaces of these views—in those laws, correlations, and conceptual changes necessary to relate them—and not in the views themselves.

Third, we must consider how we tend to relate our different views or theoretical perspectives of objects and in particular of complicated objects. Doing this for just two views, where these views are theories or theoretical perspectives, would be daunting if we take that task to be relating those theories conceptually, unifying them into a single theory. But we can relate the different views through their common referents or objects—if we are willing to assume, contra Berkeley (1709) and modern conceptual relativists (for different reasons in each case), that these different views *do* have common referents. Psychologist Donald Campbell has done very interesting work on the identification, reification (or, as he says, "entification"), delineation, and localization of objects and entities.[7] Most interesting in this context is the emphasis on boundaries of objects. Not long ago, this work would have seemed irrelevant to philosophy of science and appeals to it would have been regarded as "psychologism," but it has an important bearing on the ways in which we decompose a system into subsystems and on how we conceive the results.

Complexity

There are a number of factors relevant to our judgments of the complexity of a system, though I here discuss only two, which I will call *descriptive* and *interactional* complexity, respectively.[8]

Kauffman (1971) advanced the idea that a system can be viewed from a number of different perspectives, and that these perspectives may severally yield *different non-isomorphic* decomposition's of the system into parts. A similar point is useful in the analysis of complexity: systems for which these different perspectives yield decompositions of the system into parts whose boundaries are not *spatially coincident* are properly regarded as more descriptively complex than systems whose decompositions under a set of perspectives are spatially coincident.[9]

Assume we can individuate the different theoretical perspectives, T_i, applicable to a system. Each of these T_i perspectives implies or suggests criteria for identification and individuation of parts, and thus generates a decomposition of the system into parts. These decompositions, $K(T)_i$, I will call *K-decompositions*. The different K-decompositions may or may not give spatially coincident boundaries for some or for all of the parts of the system. The boundaries of two parts are spatially coincident if and only if for any two points in a part under $K(T)_j$ these points are in a single part under $K(T)_k$, and conversely. This is spatial coincidence defined relative to $K(T)_j$ and $K(T)_k$, but it generalizes in an obvious manner. If all K-decompositions of a system produce coincident boundaries for all parts of the system, the system will be called *descriptively simple* relative to those $K(T)_i$.[10]

If two parts from different K-decompositions are not coincident, but have a common point that is an interior point of at least one of them, then various different mapping relations can hold between their boundaries, each of which contributes to its *descriptive complexity*. Specifying these mapping relations for all parts of the system under both decompositions gives a complete description of this complexity from a set theoretic point of view.[11]

Different level theories of the same system (e.g., classical versus statistical thermodynamics) generally exhibit many-one mappings from the micro level to the macro level. Far more interesting, however, is the relation between different K-decompositions that apply at roughly the same spatial order of magnitude. Thus, decomposition of a piece of granite into subregions of roughly constant chemical composition and crystalline form, $K(T)_1$; density, $K(T)_2$; tensile strength (for specified orientations relative to the crystal axes), $K(T)_3$; electrical conductivity, $K(T)_4$; and thermal conductivity, $K(T)_5$, will produce at least roughly coincident boundaries. The granite is thus descriptively simple relative to these decompositions (see Figure 9.1).

By contrast, decomposition of a differentiated multi-cellular organism into parts or systems along criteria of being parts of the same anatomical, physiological, biochemical, or evolutionary functional system into cells having common developmental fates or potentialities, or into phenotypic features determined by common sets of genes, will, almost part by part and decomposition by decomposition, result in mappings that are not 1–1. They are not even isomorphic, much less coincident. This surely involves substantial descriptive complexity.

In biology, at least, the picture is further complicated by another

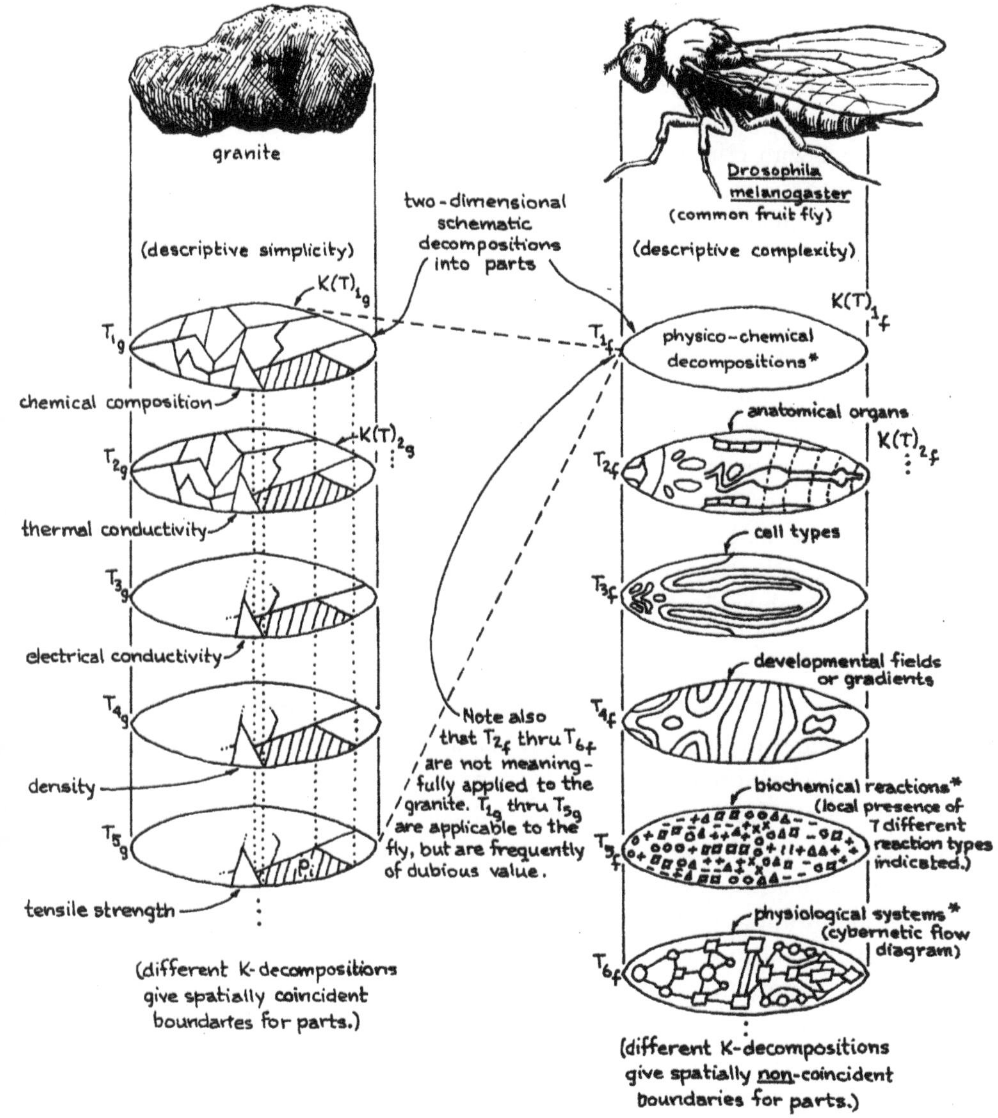

Figure 9.1 Descriptive simplicity and complexity. In the three asterisked cases, spatial localizability is not even clearly a manageable way of describing the relevant subsystems.

factor: different theoretical perspectives are not nearly as well individuated as in the physical sciences. Thus, anatomical, physiological, developmental, and biochemical criteria, not to mention paleontological information and inferences of phylogenetic relations and homologies, all interact with criteria of evolutionary significance in analyzing organisms into functional systems and subsystems. This borrowing of criteria for individuation of parts from different and diverse theoretical perspectives is one factor that can make functional organization in general and biology in particular a conceptual morass at times. This is further discussed in Wimsatt (1971a, chapters 6 and 7, and Wimsatt, 2002a).

Descriptive complexity has a point, largely because of what I call *interactional complexity*. This is a measure of the complexity of the causal interactions of a system, with special attention paid to those interactions that cross boundaries between one theoretical perspective and another.

Many systems can be decomposed into subsystems for which the *intra-systemic* causal interactions are all much stronger than the *extra-systemic* ones (see Figure 9.2). This is the concept of "near-complete decomposability" described by Simon and others (see, e.g., Ando, Fisher, and Simon, 1963; Levins, 1970a; Simon, 1962). Such systems can be characterized in terms of a parameter, ϵ_c, that depends upon the location of the system in phase space and is a measure of the relative magnitudes of inter- and intra-systemic interactions for these subsystems.[12] This notion will be called S-ϵ_c-decomposition, and the subsystems produced according to such a decomposition will be denoted by $\{s^i\epsilon_c\}$. A system is *interactionally simple* (relative to ϵ_c) if none of the subsystems in $\{s^i\epsilon_c\}$ cross boundaries between the different K-decompositions of a system, and *interactionally complex* in proportion to the extent to which they do (see Figure 9.2, b and c).

The importance of interactional complexity is as follows: The parameter ϵ_c can also be used as a measure of the accuracy of a prediction of the behavior of the system under a given decomposition if inter-systemic interactions are (perhaps counterfactually) assumed to be negligible. The larger ϵ_c is for that system under that decomposition, the less accurate the prediction. Alternatively, if a specific value of ϵ_c, say ϵ_c^*, is picked in order to achieve a certain desired accuracy of prediction of the behavior of a system, and the system turns out to be interactionally complex for that value of ϵ_c, then the investigator *must* consider the system from more than one theoretical perspective if he or she is to be able to make predictions with the desired level of accuracy.

Obviously, the value of ϵ_c^* is an important factor here. If a system is

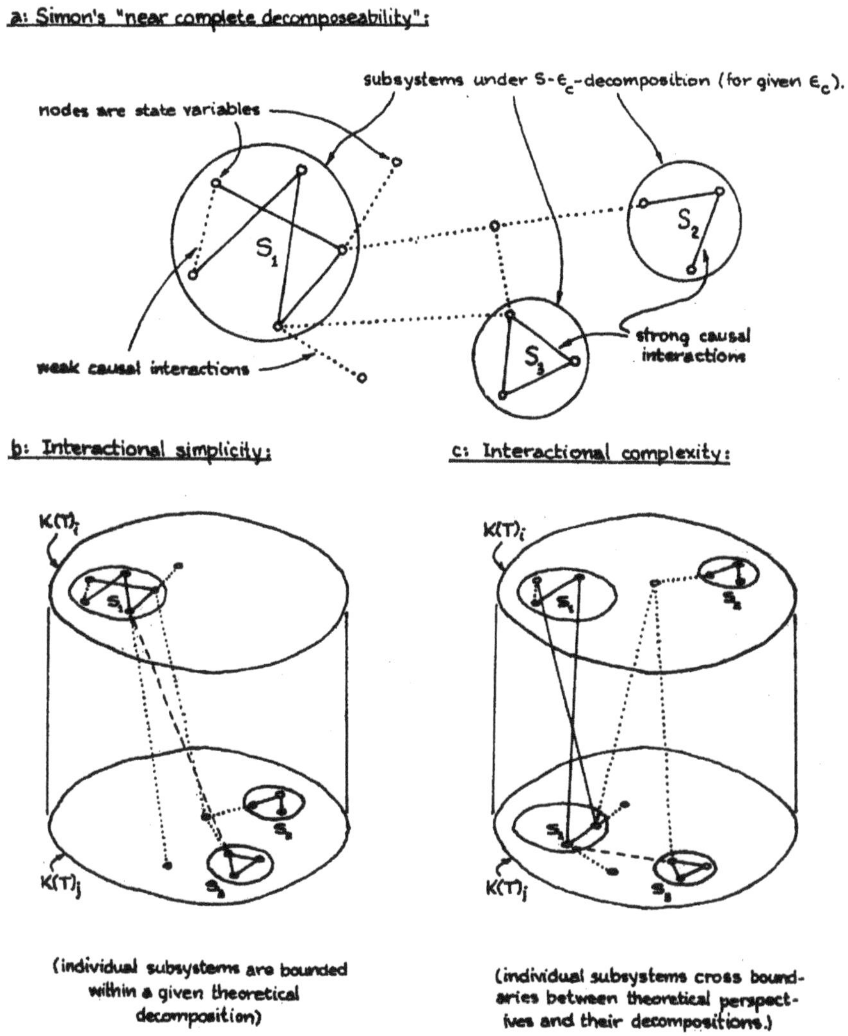

Figure 9.2. Near-decomposability and interactional complexity. Views in this figure represent decomposition into *sets of state variables* in different perspectives, whereas decompositions in Figure 9.1 were into *sets of parts*. These are related, but should not be confused. Thus, strong correlations among state variables would be a version of Campbell's (1958) "common fate" criterion for individuating objects or parts, though the other criteria he discusses (spatial proximity, similarity in other properties) would conflict, and the entities we would pick out represent a compromise.

interactionally simple for a given value of ϵ_c^* it will remain so for all larger values of that parameter, since larger values of ϵ_c^* denote lower standards of predictive precision. A system that is interactionally complex for one value of ϵ_c^* may be interactionally simple for larger values. Thus the interactional simplicity of a system also increases as the minimum value of ϵ_c^*, for which it is interactionally simple, decreases.

In any case of interactional complexity, the investigator is forced to attempt to relate the different K-decompositions in order to trace and analyze the causal networks in the system. This is a relatively straightforward task if the system is descriptively simple, since the spatial decomposition of the system into parts in one perspective automatically gives the spatial decompositions (but not all the properties!) for the other perspectives. But if the system is descriptively complex and is also interactionally complex for more than a very small number of interactions, the investigator is forced to analyze the relations of parts for virtually all parts in the different decompositions, and probably even to construct connections between the different perspectives at the theoretical level.

Many investigators of biological, social, and other complicated systems have claimed that no one perspective appears to do justice to their objects of study, or, somewhat more obscurely, that their systems are unanalyzable into component parts—or at least that there is no clear way to perform this analysis. Kauffman (1971) and Levins (1970a) both claim that in complex systems there are a number of different possible decompositions and often no way of choosing between them. Levins' remarks suggest something even stronger: "[For] a system in which the component subsystems have *evolved* together [the subsystems] are not even obviously separable. . . . It may be conceptually difficult to decide what are the really relevant component subsystems. . . . This decomposition of a *complex* system into subsystems can be done in many ways . . . it is no longer obvious what the proper subsystems are" (Levins, 1970a, p. 76). It seems reasonable to construe these claims in part as claims that the systems in question are interactionally and descriptively complex. Levins' claim about evolved systems raises further questions about the origins of complexity.

Evolution, Complexity, and Organization

Why should some systems be interactionally complex? If this question is examined in an a priori manner, it is perhaps more amazing that some systems should be interactionally simple. The systems, as analyzed to apply S-decompositions, are composed of state variables and causal relations between them. We would expect that different state variables, picked at random, would reflect properties of parts or systems in different theoretical perspectives. Unless causal relations or state variables were organized in a rather specific way relative to the different theoretical perspectives, one would expect that interactional complexity would be the norm.

But isn't this what successful theories do for us—that is, isn't it the mark of a successful theory of a range of phenomena that it unites and embraces the causally relevant parameters and state variables within a single theoretical perspective? This question suggests that if our theories are successful, then they should produce descriptions of systems according to which the systems are interactionally simple. I think that this would be to put the conceptual cart before the phenomenal horse. As the criterion (one of many) for the adequacy of a theory of a system, this statement seems correct but it is hardly sufficient. Also, one should not automatically assume that our *existing* theories are adequate theories of complex systems. The belief that they are is based largely on a still unfilled reductionist promise.[13]

It is true that our existing theories work well on simple systems—simple in part (but only *in part*) because these theories are constructed so as to render them interactionally simple. But one cannot assume that it is always possible to find a theory that will render a given system interactionally simple. As Ross Ashby (1956) said some years ago:

> Science stands today on something of a divide. For two centuries it has been exploring systems that are either intrinsically simple or that are capable of being analyzed into simple components. The fact that such a dogma as "vary the factors one at a time" could be accepted for a century shows that scientists were largely concerned in investigating such systems as *allowed* this method; for this method is often fundamentally impossible in complex systems. (p. 5)

Nonetheless, the a priori argument for the interactional complexity of systems given at the beginning of this section is intrinsically defective, for it ignores facts that every scientist takes for granted—namely, that systems *are* constrained and that state variables are not causally related at random. Thus, some of the arguments of "The Architecture of Complexity" (Simon, 1962) appear to suggest that there *are* physical constraints on evolving systems that would render them interactionally simple and that they are descriptively complex only after the manner of a multi-level theory, with many-one mappings of parts from lower to higher levels. It is an open question as to whether this is indeed the conclusion Simon intends, but in any case, it seems to me to be mistaken.

Simon makes an elegant case for the conclusion that all evolved systems containing many causally interrelated parts will be hierarchically organized. This is via an argument that for two systems of roughly equal complexity, each to be built out of simple components (and each subject to perturbations tending to cause decomposition), the one that arises out of the successive aggregation of individually stable subassemblies into larger subassemblies will have a much higher probability (or lower expected time) of formation than the one that does not. (See Simon's parable of the watchmakers, "Tempus" and "Hora" in 1962, pp. 90–95.) Thus, one would expect that at least the vast majority of complex evolved systems would be hierarchically organized.

But Simon's use of the concept of near-decomposability in the same article sometimes appears to suggest that he believes such hierarchical systems to be nearly decomposable in a nestable manner—with smaller subassemblies (at lower levels) having successively stronger interactions and no S-decompositions crossing boundaries between levels. Indeed, if the subassemblies that go to make up a hierarchically organized system are stable, isn't it the case that these subassemblies at all levels must be the subsystems that emerge for various characteristic values of ϵ_c in the different level S-decompositions of a system?[14]

This conclusion would fail to distinguish the decomposability or stability of the subassemblies before they aggregate from their decomposability or stability (in isolation) after they have aggregated—especially a long time afterwards, when they have had time to undergo a process of mutually coadaptive changes under the optimizing forces of natural selection. The optima and conditions of stability for a system of aggregated parts are in general different in a non-aggregative way from the optima and conditions of stability for its parts taken in isolation.[15]

Naive design procedures in engineering, in which the organization of the designed system was made to correspond to the conceptual breakdown of the design problem into different functional requirements, with a 1–1 correspondence between physical parts and functions, have given way to more sophisticated circuit minimization and optimal design techniques. These methods have led to increases in efficiency and reliability by letting several less complicated parts jointly perform a function that had required a single more complicated part, and, where possible, simultaneously letting these simpler parts perform more than one function (in what might before have been distinct functional subsystems). This has the effect of making different functional subsystems more interdependent than they had been before, and of encouraging still further specialization of function, and interdependence of parts. It is reasonable to believe that the optimizing effects of selection do just this for evolving systems, and if so, that hierarchically aggregating systems will tend to lose their neat S-decomposability by levels and become interactionally complex.[16]

This argument is buttressed empirically by considering what happens when natural organized systems are artificially decomposed into subassemblies that are the closest modern equivalents of the subassemblies from which the systems presumably came. Few modern men (or better, couples, for bisexual organisms) could survive for long outside of our specialized society. The same goes for mammalian cells—at least under naturally occurring conditions, even though multi-cellular organisms are presumably descended from unicellular types. Even many bacteria cannot survive and reproduce outside of a reasonably sized culture of similar bacteria. The current belief of some biologists is that mitochondria and chloroplasts originated as separate organisms, and acquired their present role in animal and plant cells via parasitic or symbiotic association. According to this view these once independent organisms (or subassemblies) are now so totally integrated with their host that only their independent genetic systems are a clue to their origin (Margulis, 1971).

With increasing differentiation of function in systems, different subsystems become dependent, not only upon the presence of other subsystems, but also upon their arrangement. Experiments with the transplantation of imaginal discs in the larvae of holometabolous insects demonstrates that the developmental fate of these discs (which develop into organs in the pupal stage) depends not only upon the disc and its "age," but in some cases also upon its location. Slime moulds are remarkable among organisms for their ability to function undisturbed if they are pushed through a sieve in their undifferentiated form. This is quite unusual for multi-cellular animals. An adult man or mouse would do considerably less well under similar conditions.

The end result, I think, is that one cannot expect hierarchical organization resulting from selection processes to be S-decomposable into different levels—or at least, not into the different levels of organization relevant at the time of aggregation. Nor could one expect that such organization would be interactionally simple when decomposed according to any other theoretical perspectives bearing no intrinsic relation to the selection forces acting upon it.

Nor, unfortunately, is there even any guarantee that functional organization in terms of the operation of selection mechanisms (Wimsatt, 1971a, 1972) is the road to a descriptively and interactionally simple analysis of such systems. The use of functional criteria might lead to more simplicity perhaps (I would argue strongly that they would!), but functional systems are still subject to physical, chemical, and biological constraints at a number of levels, and never completely lose the marks of the systems from which they have evolved—even down to the level of the basic chemical elements of which they are composed. These simultaneous constraints seem almost certain to result in interactional complexity.

Complexity and the Localization of Function

In the last section, I suggested how considerations of efficiency in evolution would lead to the co-adaptation and increased interdependence of parts of a functional system, and that this would lead to increases in the descriptive and interactional complexity of that system. One aspect of this increase in complexity is a trend away from 1–1 mappings between functions and recognizable physical objects. It seems plausible to suggest that one of the main temptations for vitalistic and (more recently) anti-reductionist thinking in biology and psychology is due to this well-documented failure of functional systems to correspond to well-delineated and spatially compact physical systems. (Richard Gregory's excellent and suggestive remarks [1961, 1962] on the problem of localization of function in the brain [and his humorous illustrations with engineering examples] offer too many riches to mine them superficially here.)

It is only too tempting to infer from the fact that functional organization does not correspond neatly to the most readily observable physical organization—the organization of physical objects—to the howling non sequitur that functional organization is not physical.[17] A tantalizing explanation for our tendencies in this direction is hinted at in Donald Campbell's (1958) suggestion that our willingness to "entify" things as real is directly proportional to the number of coincident boundaries we find.[18]

Thus, organisms count as systems because of the coincidence of a number of boundaries at, roughly, the surface of the skin or its functional equivalent. In addition to the relatively discontinuous change of a number of physical and physiological variables at what is taken to be the organism/environment interface, the systems thus picked out are usually relatively independent agents biologically, since what we call organisms usually live, die, metabolize, mate, and move relatively independently of one another. Indeed, we tend to marvel at the problem cases—eukaryotic cells, slime molds, and social insects—where one or more of these boundaries do not coincide, and have problems deciding whether a given unit is an organism, assemblage of organisms, or part of an organism.

But what holds true of an organism (that many boundaries coincide at its skin) need not hold true of its parts. Inside the complex system there is a hegemony of different constraints and perspectives and boundaries. If what Campbell says is correct, this hegemony leads us to be slow or dubious about objectifying the parts of such a system. What is unobjectifiable is to that extent unphysical, and so functional organization becomes a thicket for vital forces and mental entities. It is no accident that those systems for which vitalisms and mentalisms have received spirited defenses are those systems that are also paradigmatically complex.

The difficulties with the spatial localization of function in complexly organized systems suggest a more positive approach to at least one aspect of the psycho-physical identity thesis.[19] In 1961, Jerome Shaffer took account of the frequently discussed non-spatiality of mental events and proposed that the spatial location of corresponding brain events could, as a *convention,* be taken as the location of the corresponding mental events. Norman Malcolm (1964, p. 119n6) argued that as this would be just a convention, talk of the location of mental events would just be taken as a shorthand way of locating the corresponding physical events.[20] Malcolm and Shaffer's later discussions (Malcolm, 1971; Shaffer, 1965) raise other issues that lead off in other directions, but neither of them seem to take seriously the implications of the interactional and descriptive complexity of functional organization.

It is not merely that functionally characterized events and systems are spatially distributed or hard to locate exactly. That much can be said for bulk terms like *water,* or even more, like *fog* and *smoke.* But ordinary bulk matter, like ordinary fields, can be conceived of as homogeneous. The problem is that a number of different functionally characterized systems, each with substantial and different powers to affect (or effect) behavior appear to be as interdigitated and intermingled as the infinite regress of qualities-within-qualities of Anaxagoras' seeds. Furthermore, the high degree of redundancy and plasticity of the cortex—pointed to by the work of Lashley (1951) and his doctrine of "equipotentiality"—make it seem as if functional systems are not essentially located *anywhere*. The apparent contradiction of having a number of (functionally) distinct organized systems, each of which appears to occupy all or most of the same space, and at the same time none of it, leads to the tendency to deny spatiality at all,[21] or in less extreme forms, to invent special kinds of quasi-physical predicates. The non-physicalist and anti-reductionistic strains in the writings of Jerry Fodor (1965, 1968), Hilary Putnam (1967), and Margaret Boden (1970, 1972), who speak with a strange ambivalence about their "functional roles," "programs," and "internal models," reflect this no less than did nineteenth-century vitalism.

This tendency is, I think, at least partially explained by the "pathological" behavior of boundaries in complex systems and what this does to our normally workable criteria for spatial objectification. This way of putting it may be misleading for it could be argued that we conceived of the mental as non-spatial long before we had any idea (gleaned from neurophysiology) about the problems with localization of function in the cerebral cortex. But there are alternatives to suggesting that active awareness of these problems led us to conceive of the mental as non-spatial. One might suppose, for example, that spatial objectification is an *active* hypothesis that we apply to those groups of phenomena that tie up into sufficiently neat packages in the right ways. Based on this account, the mental realm is not denied spatiality, it just has not yet been added to the list.

The denial of spatiality as a category to mental entities has the ring of a philosopher's invention. Now that we see one reason why we might not have been able to attribute an exact location to mental events, we can wonder perhaps whether the common man need ever have been more than an agnostic about their spatiality. If he has asserted more, he has probably only succumbed to the seductions of philosophy—deifying as conceptually true a hypothesis that future writers may well decide is empirically false.[22]

CHAPTER TEN

The Ontology of Complex Systems

Levels of Organization, Perspectives, and Causal Thickets

Willard van Orman Quine once said that he had a preference for a desert ontology. This was in an earlier day when concerns with logical structure and ontological simplicity reigned supreme. Ontological genocide was practiced upon whole classes of upper-level or "derivative" entities in the name of elegance, and we were secure in the belief that one strayed irremediably into the realm of conceptual confusion and possible error the further one got from ontic fundamentalism. In those days, one paid more attention to generic worries about possible errors (motivated by our common training in philosophical skepticism) than to actual errors derived from distancing oneself too far from the nitty-gritty details of actual theory, actual inferences from actual data, the actual conditions under which we posited and detected entities, calibrated and "burned in" instruments, identified and rejected artifacts, debugged programs and procedures, explained the mechanisms behind regularities, judged correlations to be spurious, and in general, the real complexities and richness of actual scientific practice. The belief that logic and philosophy were prior to any possible science has had a number of distorting effects on philosophy of science. One effect was that for ontology, we seemed never to be able to reject the null hypothesis: "Don't multiply entities beyond necessity."

But Ockham's razor (or was it Ockham's eraser?) has a curiously ambiguous form—an escape clause that can turn it into a safety razor: How do we determine what is necessary? With the right standards, one could remain an Ockhamite while recognizing a world that has the rich multi-layered and interdependent ontology of the tropical rain forest—that is, our world. It is tempting to believe that recognizing such a worldview requires adopting lax or sloppy standards—for it has a lot more in it than Ockhamites traditionally would countenance. Quite to the contrary, I think that the standards for this transformation are not lax, but only different. Indeed, the standards that I urge are closer to our experience and arguably more fundamental than those used during the hegemony of foundationalist methods and values.

In the first section of this chapter, I discuss the criterion for what is real—what I call *robustness*—a criterion that applies most simply and directly, though not exclusively, to objects. In subsequent sections, I use robustness and other information about our world to delineate the major structural features—primarily levels, but with some comments on what I call *perspectives* and *causal thickets*—that dominate our world, our theories, and the language we use to talk about both. These are higher-level ontological features, *Organizational Baupläne,* related to the things that people usually talk about under the topic of ontology (things like objects, properties, events, capacities, and propensities) as paragraphs are to words and phonemes or morphemes. But they are there nonetheless, it is only our concern with the little things, motivated by foundationalist or reductionist concerns, which has deflected our attention from them. This ontology—of levels, perspectives, and causal thickets—is no less required for a full accounting of the phenomena of the physical sciences than it is for biology and the social sciences, but its obdurate necessity has seemed more obvious in these latter cases. This may now be changing. The increased interest in fractal phenomena and chaotic and, more generally, non-linear dynamics emerging from the so-called exact sciences has brought many noisy residua of the ontological scrap heaps of the physical sciences to the center of attention as theoretically revealing data, structures, and objects with new-found status. Most of these things have never before made it into theory—or if so, only into the "theory of observation" under the topic of "error analysis" where they lived in the ubiquitous error term. Messiness—or at least the right kinds of messiness—is now almost a virtue in many of the sciences, as the recent explosion of interest in complexity seems to attest.[1] Levels, perspectives, and causal thickets are major ontological players in these complex areas—domains with significant implications for how to approach many of philosophy's most refractory problems.

Because the aim of this chapter is ultimately taxonomic—to say what there is, or to describe some of the *bigger* things that are—the descriptive sections basically take the form of a list of properties, elaborated either to further explain ideas likely to be unfamiliar, or to explain relations among the properties that help to give the ideas of level and perspective their cohesiveness. Taxonomy may sound boring, but I hope to show you that the description of and relations between a family of newly discovered species can be an exciting task.

I. Robustness and Reality

Before I say what there is in this complex world, I should give my criteria for regarding something as real or trustworthy. Particularly among those of a foundationalist persuasion, it is common to start by providing some criterion, be it indubitability, incorrigibility, or other means of picking out things or assumptions whose veracity is not open to question. One then says that those things are real (true, indubitable, or whatever) if it is either one of these primitive things or if it is derivable from them via a valid series of inferences. Only things admitted in one of these two ways are allowed. I share the foundationalist's concern with securing reliability for our conceptual structures, but I don't think that there *are* any criteria that *both* give indubitability or render error impossible, *and* permit any interesting inferences from that starting point. Thus, I would rather give a criterion that offers *relative* reliability, one that you're better off using than not, indeed better off using it than any other, and that seems to have a number of the right properties to build upon. Rather than opting for a global or metaphysical realism (an aim that bedevils most of the analyses of "scientific realists"), I want criteria for what is real that are decidedly local—which are the kinds of criteria used by working scientists in deciding whether results are real or artifactual, trustworthy or untrustworthy, objective or subjective (in contexts where the latter is legitimately criticized—which is not everywhere). When this criterion is used, eliminative reductionism is seen as generally unsound, and entities at a variety of levels—as well as the levels themselves—can be recognized for the real objects they are, and traditional foundationalism and ontic fundamentalism are in trouble. They will survive, if at all, as a local kind of problem-solving technique of significant but limited usefulness. (But see Chapter 7, on dynamical foundationalism.)

Following Levins (1966), I call this criterion *robustness*. (Chapter 5 analyzes and reviews this concept and methodology; Wimsatt, 1980a, 1980b, has relevant case studies. Campbell's [1966] concept of "triangulation" captures many of the same ideas, and his classic work with Fiske [1959] on the "multi-trait-multi-method matrix" brought this methodology to the social sciences.) *Things are robust if they are accessible (detectable, measurable, derivable, definable, producible, or the like) in a variety of independent ways.* A related but narrower criterion (experimental manipulability via different means) has since been suggested by Hacking (1983), who draws a close link with experiment, and limits his discussions to the realism of entities. But robustness plays a similar role also in the judgment of properties, relations, and even propositions, as well as for the larger structures—levels and perspectives—described below (see also Wimsatt, 1981a, 1974, 1976a). Furthermore, independent means of access are not limited to experimental manipulations but can range all the way from non-intrusive observation or measurement to mathematical or logical derivation, with many stops in between. Experimental manipulation is just a special case. We feel more confident of objects, properties, relationships, and so forth that we can detect, derive, measure, or observe in a variety of *independent* ways because the chance that we could be simultaneously wrong in each of these ways declines with the number of independent checks we have.[2] We can only make the probability of failure decline—though it can get very small, it does not go to zero. This criterion does not give certainty. Nothing does. There are no magic bullets in science—or anywhere else, for that matter. But if that's so, then certainty is not so important as generations of philosophers have supposed.

The independence of these different means of access is crucial. Independence is often not easy to demonstrate, and failures of independence are often well hidden. Cases of pseudo-robustness, while not common, are not truly rare either, and invariably seem to involve unperceived failures of the independence assumption, or—relatedly—not sufficiently broad variation in the means of access.[3] (Wimsatt, 1980b, 1981a, discusses cases of spurious or pseudo-robustness in population biology and psychology, and Culp, 1995, gives a careful and enlightening dissection of degrees of independence and interdependence among experimental techniques in molecular genetics. See contrary arguments by Rasmussen, 1993, and Culp, 1994, about the use of robustness in the analysis of an artifactual "entity," the mesosome, in recent cell biology.) Indeed, if the checks or means of detection are probabilistically independent, the probability that they could *all* be wrong is the product of their individual probabilities of failure, and *this* probability declines very rapidly (i.e., the reliability of correct detection increases rapidly) as the number of means of access increases, *even if the means are individually not very reliable.* This gives us the requisite sense of independence for this criterion—namely, that the *probability of failure* of the different means of access should be independent. Of course, one cannot infer immediately from apparent physical independence of the means of access to their probabilistic independence. That is a further hypothesis that is sometimes false. Probabilistic independence represents a kind of mathematical idealization—a mathematical model of physical processes or, in more complex cases, of a system of interrelated physical, biological, psychological, and social processes.

Although nothing will guarantee freedom from error, robustness has the right kind of properties as a criterion for the real, and has features that naturally generate plausible results. Furthermore, it works reliably as a criterion in the face of real world complexities, where we are judging the operational goodness of the criterion—not its goodness under idealized circumstances. We are judging its performance as well as its competence, as it were. It even has the right metaphysical and epistemological properties. Thus, it is part of our concept of an object that objects have a multiplicity of properties, which generally require different kinds of tests or procedures for their determination or measurement. It follows that *our concept of an object is a concept of something that is knowable robustly.* Indeed, one of the ways in which we detect illusions is that appearances to one sensory modality are not borne out with the appropriate confirmation in the other sensory modalities—confirming, for a visual hallucination or mirage that what we see before us is not an object, not real (Campbell, 1966).

Robustness can wear two faces in a kind of epistemological figure-ground reversal that leads to a kind of almost magical appearance of bringing yourself up by your own bootstraps. Particularly in the early stages of an investigation, we may use *agreement* of different means of detection, measurement, or derivation to posit an object or an objective property or relation that is the common cause of these various manifestations. At a certain stage, we will accept the existence of the entity or property as established—however corrigibly—and begin to use the *differences* observed through the diverse means of access to it as telling us still more about the object. (It is after all *that kind of thing or property that is detectable via these diverse means,* and *shows itself differently through them.*) We will at the same time use these differences to tell about the means of access to the object. (This *one* thing or property appears in these diverse ways *through these different means of access.*)

In this latter stage, we may compare the performance of the different means on a variety of target objects. In so doing, we are both calibrating each means against the others, and learning about their respective limitations.[4] This kind of switching back and forth can lead to considerable successive refinement both in our knowledge of the object(s) in question, and of the characteristics and limitations of the tools we have for accessing them.[5] The fine tuning and power of the refinements are increased if the objects in question turn out to form a class of diverse entities that can all be studied via the same means—as genes did for the Morgan school (Wimsatt, 1992).

Robustness has had a surprising history—it seems to be always there, but seldom noticed. Thus, seventeenth-century philosophers made a distinction between primary qualities (shape, extension, impenetrability, etc.) that they held were really in objects, and secondary qualities (color, taste, sound, etc.) that they held were induced in us by our interactions with the primary qualities of objects. Descartes took the primary qualities of objects as the fundamental properties of matter from which he tried to explain all else through derivation, and it was a general feature of such theories to try to explain secondary qualities in terms of primary qualities. This kind of relationship between primary and derived things became central to and emblematic of deductive and foundational approaches. The ironic fact, not noted at the time, is that the properties that Descartes and others following him chose as primary qualities were all knowable in more than one sensory modality, whereas the secondary qualities were known in just one sensory modality.[6]

Thus, in modern jargon, *the primary qualities are robust and the secondary qualities are not.* The explanatory principle of that period translates as: *Explain that which is not robust in terms of that which is—or, by extension, that which is less robust in terms of that which is more so.*[7] This is still a good principle, and one that is generally followed—it serves equally well in foundationalist and in non-foundationalist camps. It is different from, independent of, and if anything, more basic than anything else in the foundationalist methodology. Ironically then, we see that the paradigm of foundationalist approaches is simultaneously a paradigm use of robustness as a criterion for the real, and that the best applications of the deductivist paradigm occur when the foundational assumptions, objects, or properties are robust.

This indicates a coincident starting point for deductivist and robustness paradigms. There are other ways—elaborated in Chapter 5—in which they diverge. Thus, on the deductivist paradigm, the length of derivations doesn't matter (as long as they are finite), and additional derivations of the same conclusion through different means are redundant and unnecessary. But if overall reliability is the primary concern, and one has at each stage a small but finite chance of misapplying valid inference rules, then the length of serial deductive arguments does matter. Furthermore, in a world where failure is possible, multiple derivations of a result by different paths are no longer otiose as a way of checking or providing further support. One can stray still further from foundationalist values: *with parallel independent means of support available and the net reliability of the conclusion as the only concern, there is no longer any reason to limit inferences to truth-preserving ones, and the use of good inductive, abductive, or more generally, heuristic principles may have a place in the construction of exemplary arguments—in philosophy as well as elsewhere.*[8] Indeed, robustness as a criterion of superiority among arguments can and should cast a very broad and long epistemological shadow, once we get away from the unrealistic assumptions about human reasoning that have anchored 350 years of foundationalist thought.

I intend to apply these methodological lessons right here. Throughout this chapter, I not only use the concept of robustness as a tool in the analysis, but I also employ it in the structure of the argument by using multiple concepts and arguments that individually have a heuristic character—having less than deductive analytical force. There are lots of characterizations that represent strong tendency statements, which can be cashed out in terms of statistical rather than universal claims. This is data that can't by the nature of the objects be formulated or used in arguments that require necessary and sufficient conditions. Attempts to tighten them up would only render formulations that are too narrow in scope or fail to capture most of the interesting phenomena. It is suggestive of the situation for which "fuzzy set theory" was invented, though the present character of that theory makes no allowance for the systematic character of biases and exceptions (Wimsatt, 1985, 1992). This is a common pattern for entities, regularities, mechanisms, and explanations involving complex systems, yet we shouldn't refuse to discuss them for that reason. They are too important for their reality to be denied, or rendered suspect by false simplifications or idealizing assumptions. We should value *for that reason* an analysis that recognizes the centrality they have in everyday life.

In a way, then, this analysis has something in common with folk psychology and some of the basic assumptions of ordinary language philosophy—like them it takes for granted that the world we see, live in, respond to, and act upon is too important, too central to our way of being, to be dismissed. But this much is not just anti-scientific sloppiness (ordinary language philosophers went much further). For all of the ontological radicalism of quantum mechanics, Niels Bohr felt the need to postulate his "correspondence principle"—that an adequacy condition for quantum theory was that it had to produce (in the right limits) the macroscopic phenomena we observe everyday. The approach advocated here proceeds more like Bohr (in spirit, if not in content), and less like ordinary language philosophy in trying to suggest the outlines of a more realistic scientifically motivated epistemology and metaphysics for approaching these problems. But before attending to the ordinary phenomenology of this new taxonomy, a bit of abstraction is necessary to see where we are going in this new philosophical landscape.

Ontologically, one could take the primary working matter of the world to be causal relationships, which are connected to one another in a variety of ways—and together make up patterns of causal networks. (I won't address problems with causality in this chapter. Those who favor "Humean skepticism" will also find lots else to object to here, and can stop reading now unless they want to see how far you can get without it!) These networks should be viewed as a sort of bulk causal matter—an undifferentiated tissue of causal structures—in effect the biochemical pathways of the world, whose topology, under some global constraints, yields interesting forms. Under some conditions, these networks are organized into larger patterns that comprise *levels of organization,* and under somewhat different conditions they yield the kinds of systematic slices across which I have called *perspectives.* Under some conditions, they are so richly connected that neither perspectives nor levels seem to capture their organization, and for this condition, I have coined the term *causal thickets.* Much of psychology and the social sciences, for all the appearances of local order and local approximations to levels and perspectives, when looked at more globally and once the various idealizations of our theories are recognized, seem to be in this third state, or in a hybrid mixture that contains elements of all three. These three kinds of structures are rich in methodological and philosophical consequences for understanding the strengths and limitations of different approaches to studying problems and phenomena in systems characterized by one of them. We now turn to the first of these *Organizational Baupläne*—levels of organization.

II. Levels of Organization

The analysis presented here elaborates on parts of two earlier papers on reductionism and levels of organization (Wimsatt, 1976a, and Chapter 9). There has been a fair amount of work on levels since, in which they are taken to mean an astounding variety of things. Much of it, though relevant to the analysis of some complex systems, leads in the wrong direction for present purposes. Thus, I agree with McClamrock's argument (1991) that Marr's (1982) three levels (algorithmic, computational, and hardware) are better viewed as levels of analysis or of abstraction, or as kinds of functional perspectives on a system, than as compositional levels of organization. This conflation is apparently a common kind of mistake among philosophers of psychology.

More generally, people sometimes talk as if the material, psychological, and sociocultural realms constitute monadic levels (e.g., as in Popper's first, second, and third worlds). These rough distinctions are of major importance because they delimit regions where different major concepts, theories, methodologies, and explanatory strategies dominate, but they are larger heterogeneous aggregates spanning multiple levels and including also other less well-ordered structures rather than single individual levels of organization. Thus, by *any* criteria, there are obviously multiple compositional levels of organization within the material realm: elementary particle, atom, molecule, macro-molecule, and so forth, or, within the biological realm, as units of selection, for example, selfish genes (transposons), some kinds of supergenes (chromosome inversions), selfish gametes (the t-allele case in mice), selfish cells (cancer), selfish organisms, and selfish groups—all of which would fit into the material realm, traditionally conceived.[9] Similarly, most current cognitive theories recognize multiple levels of a compositional character[10] within the mental realm: structural representations of belief or planning, linguistic structure, or hierarchical representations of features in a classification system. Atomic families, small groups, mobs, speakers of a local dialect, social classes, sectors of the economy, and citizens of a nation-state are all obviously social, or sometimes sociocultural units at diverse levels of organization—whose interactions follow diverse dynamics.

By level of organization, I mean here compositional levels—hierarchical divisions of stuff (paradigmatically but not necessarily material stuff) organized by part-whole relations, in which wholes at one level function as parts at the next (and at all higher) levels. Though composition relations are transitive (so one could collapse the highest level systems to the smallest parts), levels are usually decomposed only one level at a time, and only as needed.[11] (Thus, neurons are presumably composed of parts like membranes, dendrites, and synapses, which are in turn made of molecules, which are in turn made of atoms, and so forth down to quarks, though to the connectionist modeler, neurons are adaptive modules with properties like incoming and outgoing connections and thresholds, and which might as well be indivisible atoms for all of the use that is made of their still lower level properties.) Most of what I say below relates to material compositional hierarchies and levels, because I utilize constraints characteristic of the physical world—which also includes the physics of biological, psychological, and social objects.

Nonetheless, this is not a reductionist analysis in the sense in which a philosopher might use that term. (I would urge, however that it *is* reductionist, or at least broadly mechanistic as those terms would be understood by most scientists. See Wimsatt, 1976b, 1979.) Nor should it be taken as implying, either in evolutionary history, or in current state-of-the-art genetic engineering, that usually or always, the preferred, most effective, or (stepping back to punt) even a *practically possible* way of making a given upper-level object is by assembling a bunch of lower-level parts. This over-extension of what I have called (1976a) the "engineering paradigm" is one of the things that have given reductionism and materialism bad names. (I remind the reader that the paradigms of genetically engineered molecules are not examples of *ab initio* constructions, but rather examples of the conversion of naturally occurring organic factories to the production of other products.) There is some assembly to be sure, but it is assembly of the jigs on the production line, and sometimes rearrangement and redirection of the line—not construction of the factory. To believe otherwise is to mistake arguments *in principle* for arguments *in practice*. (For the limitations and interpretation of such in principle claims, see Chapter 11.) Ultimately, we sometimes just have to stop promising and deliver the goods.

One of the reasons that it is important to look at material compositional levels more closely is that a number of properties of higher-level systems, which are treated as if they were emergent in some nonreductionist sense, follow directly from rather general properties of purely material compositional levels.[12] Thus, there is nothing intrinsically mentalistic (or social or cultural) about multiple-realizability, or the dynamical autonomy of upper level phenomena, or the anomalousness of higher-level regularities relative to the lower-level ones. Though each of these traits has been taken by some philosophers to be characteristic of the mental, they are actually characteristic of any move from a lower compositional level to a higher one. That goes for the theory of chemical bonding relative to fundamental quantum-mechanical theories of the atom no less than for "the" relation between the neurophysiological (*which* neurophysiological level?) and "the" cognitive (*which* cognitive level?). *These traits are features that always accompany the emergence of a new stable level of organization.*

As a kind of reductionist, I want to get as much as I can about higher levels from the properties of lower ones. As a kind of holist, it is tempting to try to do the reverse. For evolving systems, it is not controversial to argue that the arrangement of lower-level parts (and consequently the appearance of certain higher-level phenomena) is a product of higher-level selection forces (Campbell, 1974b). And you can do both at the same time (and we do) as long as you don't commit yourself to saying that the system you study is to be exhaustively characterized by one approach or the other, but regard them as complementary. So it is possible to be a reductionist and a holist too—but not *any* kind of reductionist, or holist. Unlike an eliminative reductionist, I think that we add knowledge of both the upper level and the lower level by constructing a reduction. We add to the richness of reality by recognizing these linkages—not subtract from it. Eliminativists generally worry too much about the possibility of error at the upper level, and not enough about how stable and resilient—how robust—most upper-level phenomena are, a fact that can make the upper-level details more revealing under some conditions than the lower-level ones.

The notion of a compositional level of organization is presupposed but left unanalyzed by virtually all extant analyses of inter-level reduction and emergence. A pioneering and important attempt to deal with levels of organization (and even more with the naturally resulting concepts of hierarchy) is Herbert Simon's (1962) classic "The Architecture of Complexity," which contains both useful conceptual distinctions and arguments of absolutely central importance. The views expressed here show Simon's influence strongly, but go further in other directions. I urge a view that Simon would share: *that levels of organization are a deep, non-arbitrary, and extremely important feature of the ontological architecture of our natural world, and almost certainly of any world that could produce, and be inhabited or understood by, intelligent beings.* (This gives levels an almost Kantian flavor.) *Levels and other modes of organization cannot be taken for granted, but demand char-*

acterization and analysis. If I am right (Wimsatt, 1976a), compositional levels of organization are the simplest general and large-scale structures for the organization of matter. *They are constituted by families of entities usually of comparable size and dynamical properties, which characteristically interact primarily with one another, and which, taken together, give an apparent rough closure over a range of phenomena and regularities.* (For anyone who still believes in "necessary and sufficient conditions" style analyses, I note at least five qualifiers in this sentence—all apparently necessary—that would be difficult at best to deal with, and the referents of these qualifiers are also often disturbingly general, and correspondingly unclear. Note also, that I said that levels are "constituted by," not "defined in terms of." Definitional language is notoriously unhelpful in contexts like these. Broad-stroke characterizations, focused with qualifications and illuminated with examples, are more useful.)

Levels are in many ways the ontological analogues of conceptual schemes—though without the difficulties said (e.g., by Davidson, 1973) to attend the supposition that there is more than one of them. We live in or at one, and most of our important everyday interactions are with other entities at our level of organization—i.e., with people, tables, chairs, cars, dishes, or computers. We don't normally interact with a person's cells, or with a computer's memory chips. Persons and computers are designed to be opaque with respect to the operation of their lower-level hardware—we don't usually "see" such hardware details unless they cause a macroscopically observable malfunction, or unless we take the deliberate and special additional steps to allow us to observe things at different levels. Most of the explanations of the behavior of an entity, and most of the means for manipulating, causing, or modulating its behavior, will be found and most naturally expressed in terms of entities, properties, activities, and regularities at the same level. Our level is our common world of folk psychology, or more broadly, of the objects that populate Sellars' "manifest image" or its scientific same-level descendants.[13]

A number of other levels are also accessible to us—in part because their effects occasionally leak up or down to our level (through those few interactions that fail to be characteristically level-bound),[14] and in part because we have actively searched for and exploited these few direct connections with other levels to enrich and expand our awareness of and control over these other domains of phenomena within and around us.[15] [Author's note, 2003: In doing so, we are "extending our senses," a particularly apt description since our senses at the one end—

and developmental adaptations of cognition and physiology at the other—already are designed to stretch the range of size and time scales over which we can perceive changes in and act upon nature. See the discussion of environmental grain in Chapter 12 for further explication of these ideas.] Because any complex material objects can be described at a number of different levels of organization, identity, composition, or instantiation, relations must hold between descriptions of the same object at different levels. These provide additional important means of accessing the different levels and calibrating relations between them, and the inspiration for explanatory reductionist mechanistic theories of the behavior of the systems in question.[16]

At lower levels of organization (those of the atom and molecule) we tend to have well-defined types of definitely specified composition and, at least in principle, an exhaustively specifiable range of possible states. At higher levels of organization (from our anthropocentric perspective, but definitely middle-range on a cosmological scale) levels become less well-defined in terms of size scale and other properties (see the top row of Figure 10.1). Higher-level types of entities may no longer have crisp compositional formulae,[17] but cover a range and, in some cases, composition may no longer be a primary individuating characteristic.[18] They must do so for two connected reasons: (1) the disparately composed entities at a given level may nonetheless show multiple similarities in their behavior under similar conditions—all to be covered by multiple regularities (thus engendering at least rough multiple-realizability as the rule rather than the exception), and (2) these similar entities found at higher levels, despite their similarities, become occasions for an increasing number of exceptions to whatever regularities we can construct (see Wimsatt, 1972) because of the increased richness of ways entities have of interacting with one another (due in part to the increasing number of degrees of freedom and of emergent properties).

As the richness of causal connections within and between levels increases, levels of organization shade successively into two other qualitatively different kinds of ontological structures that I have called, respectively, "perspectives" (Wimsatt, 1974) and "causal thickets" (Wimsatt, 1976a). Objects whose mode of organization is characterized by the three distinct types of structures (levels of organization, perspective, and causal thickets) have interestingly different consequences for the methodology of sciences that study them. Below I describe some properties of levels of organization, and then say rather less about perspectives and causal thickets. These remarks are intended less as an analysis (in terms of necessary and sufficient conditions) than as a characterization of some of their most important properties (many of which are discussed further in Wimsatt, 1976a). The complex interplay of these various criteria and forces that mould levels of organization is one of the main things that give the complex sciences their richness and texture.

Levels of organization have a variety of properties that make them rich in ontological and epistemological consequences. Taken individually, these properties seem to be almost accidentally associated—important but merely empirical or contingent properties. Looked at more closely, their *merely empirical* status is probably more a product of the fact that they haven't yet been taken seriously by any of the dominant philosophical views. In fact, these properties of levels are closely connected in ways that make the features of levels and their analysis not just a contingent empirical matter. (For further discussion of some topics not found below—including the role of first- and third-person perspectives in an account of levels of organization and further remarks on the degree to which levels of organization are inevitable features of nature and of our conceptual scheme—see Wimsatt, 1976a.) In the following section I discuss these contingent properties, tying them together with a network of further empirical and conceptual facts as I go.

1. Compositional Levels of Organization: The Role of Size

a. *Successive levels of organization represent a compositional hierarchy.* If one entity is a part of another it is characteristically at a lower level of organization than the other, though in some cases and for some purposes, parts of roughly commensurate sizes as the whole system are treated as being at its level. Entities at the same level of organization are usually of roughly the same size, though there tends to be greater size variance (even proportionally) at higher levels of organization, largely due to the increasing number of degrees of freedom and ways of interacting characteristic of larger systems. With the "engineering paradigm" (Wimsatt, 1976a)—that we normally assemble complex systems out of simpler parts, a process that can be iterated—entities at successively higher levels of organization tend to show roughly geometric increases in size (see also Simon, 1962).

b. *Size and surface/volume ratio, which is a function of size, are major factors in determining which physical forces are most central to the explanation of behavior* (see Haldane, 1927), *so the size of characteristic objects at a level is not an accidental feature of this analysis.* Changing size is a necessary consequence of compositional hierarchies (given the old saw about how two [simple] objects can't occupy the same place at the same time), but changing size is also central to how different level entities get their different properties. The size-scaling factor between adjacent levels is not arbitrary—if so it would have a simpler solution. To see this, let's suppose it were arbitrary. Why not arbitrarily pick, for example, a binary aggregation scale in which every time two similar (same-level) objects are aggregated, it involves going up a level of organization? This would surely be both possible and preferable if levels were determined by convention, or by a search for the most algorithmically economical generating relations.

Nor is it entirely without a physical basis. Binary aggregation seems natural for the architecture of computer memory, and binary doubling is naturally inherent in cell replication. In fact, starting with the same elementary particles, this scheme would produce an organizational hierarchy of all nature as regular as a giant fractal lattice. (This would be both simpler and far more elegant than what actually happens.) But, pursuing the cell-division example for a minute, this line does not produce natural vertebrae in the search for nature's joints for more than the first few cell-divisions past the zygote. Then differentiation begins, and other properties become more important, such as which cells are inside and which are on the outside of the developing cell-mass. Cell divisions in different lineages lose their synchrony fairly quickly in most metazoans. Some cell-types die and are continuously replaced by others of the same type, while others go on dividing with no significant mortality in their lineages. Consequently, organisms with a large number of cells show no tendency greater than random to have their cell-numbers be at or close to integral powers of two, and the relevant functional units don't show bottom up binary regularities either. The basic problem with binary aggregation is that this aggregation mode does not track the regularities found in nature—the entities thus produced would seldom be those with any broad natural significance.

This idea of binary aggregation was introduced as an aggregative mode which—despite occasional significant pairing—is so obviously not an architectural principle for the natural world to demonstrate that the problem has a natural rather than a conventional or purely formal solution. (One might ask social constructivists why this is so!) Although size scale is an important causal determinant of levels of organization, it is not the only one. The relevant (and highly variable) geometric scaling factor between successive levels is itself a complex function of the interplay of different physical forces on relatively stable structures at the different levels, and the kind of system in question.

c. Size is a relevant, and in many cases a good criterion because a number of causal interactions characteristically become significant or insignificant together for things in a certain size range. *Size is thus a robust indicator for many other kinds of causal interactions.*[19] *This should be one of the reasons why physics has so many straightforward and simple applications to aspects of our macroscopic world.* Dust particles and bacteria are not *prima facie* good choices to be functional analogues for anything, but their common size and mass range nonetheless create strong similarities across whole arrays of their behavior. They both make excellent Brownian motion particles—and indeed the discoverer of Brownian motion made the plausible assumption that all such particles were alive. (After all, how else could small entities move around apparently actively in an obviously inert fluid!?) Size has further consequences for the design of means of locomotion in bacteria that have to deal with the fact that at their size scale, it is not a trivial matter to move in ways that are not both reversible and reversed—and thus for their movements to actually take them anywhere (Purcell, 1977)!

d. *Size is not a sufficient indicator of level*—consider bacterium-sized black holes. These definitely would not exhibit Brownian motion, at least not for conditions found in our part of the universe because they would be incomparably too massive. This is not (just) a philosopher's silly hypothetical example, though it may have been a physicist's game. An extended series of letters in the journal *Nature* in 1974–1975 discussed the existence and properties of black holes in the size range of 10^{-2} to 10^{-4} mm. in diameter. Cosmological debates had suggested that the creation of such microscopic black holes in the early history of the universe was a possibility. The discussion in *Nature* considered whether one of them could have caused the gigantic explosion over Tunguska in Siberia in 1908 (the standard candidate is a meteor some 40–50 meters in diameter). Debate ceased when it was pointed out that on the black hole hypothesis there should have been a comparable exit hole and explosion in the Baltic Sea shortly thereafter. Such a black hole (1) would not show Brownian motion, or behave in any other way like a Brownian motion particle; and (2) things around it would respond to it in a bulk, aggregate, or an "average" way—for example, the rate at which it will accumulate mass and emit radiation is a function of the net disposition of mass around it, not of the detailed organization of that mass or how it is grouped into particles or chunks. (It is so much more massive than they that its trajectory and relative rate of mass accretion—over short periods of time—is also virtually independent of them and their velocities, but only depends on where its trajectory passes relative to them. However, the objects close to the black hole are dominated in their behavior by its presence—they behave to it as an individual: individual details of its motion, size, and so on do matter for them.)

e. The example of the dust particle–size black holes suggests a natural criterion in addition to composition for ordering entities by level of organization—probably a sufficient criterion, but alas not a necessary one:[20] *Of two entities, if one relates to the other's properties as part of an average, but the second relates to the first as an individual, then the first is at a higher level of organization than the second.* This is of somewhat broader applicability in characterizing levels than compositional relations because it enables one to order entities that are not above and below one another in the same compositional hierarchy. It indicates a kind of individuation asymmetry relating to scale that is generally true of things found at different levels in compositional hierarchies, but is not limited (as the part-whole relation is) to things in the same hierarchy. In addition, it seems plausible to say that two things that relate to one another as individuals are at the same level, and two things that relate to each other as parts of averages are both embedded in larger systems, but may vary relative to each other with respect to level.[21]

2. Levels and the Simplicity of Stratification: A Layered Tropical Ontology and the Consequent Development of Language Strata

f. *Levels of organization can be thought of as local maxima of regularity and predictability in the phase space of alternative modes of organization of matter.* This is the closest I will come to a definition, because this characterization has rich connections with a number of other important properties of levels. The levels must be viewed as occupying a remapped space of reduced dimensionality relative to this enormous phase space of all physically possible states of matter, since in the levels-oriented ontology, there are strong interactions and similarities among quite diverse kinds of systems.[22] Because they are compositionally very diverse, these systems will tend to be far apart in the embedding phase space, but because they are similar in terms of the variables appropriate to the levels description, they must be close together in the reduced-dimensional projection of that space in terms appropriate to that level.[23] Almost all entities are at levels. *Since most direct interactions of something at a level of organization are with other things at that same level of organization, regularities of behavior of that entity will be most economically expressed in terms of variables and properties appropriate to that level.*[24]

In talking about these as local maxima, I mean to imply that entities with modestly larger or smaller values of key properties (think of size) would show messier regularities than and key into fewer regular relationships with the other entities and each other than is true for the entities we have. The larger number of regularities or stable patterns involving the larger number of relatively stable entities, both concentrated at or near levels of organization, makes the characterization of levels as local maxima of regularity and predictability correct. This is analogous to a kind of "fitness maximization" claim for ontology, springing from a deep embeddedness of our world in a spectrum of different equilibrating and selection processes acting on different size and time scales (see also Dennett, 1995, for convergent "deep" claims about an evolutionary ontology and dynamics).

g. The fact that most direct interactions of something at a level of organization will be with other things at that level means that detectors of entities at a level will be or will have parts that are at the same level as the target entity, and that will interact with it via properties characteristic of that level. This has several direct implications:

1. The theory of instruments for *us* to detect properties or entities at level *x* will involve causal interactions, mechanisms, objects, properties, generalizations, and regularities of level *x*.
2. If we are at a different level, this theory of instruments will also involve causal interactions, mechanisms, objects, properties, generalizations, and regularities at *our* level, since we need to be able to detect and record their output. For these reasons, and for others, eliminative reduction is often not possible, necessary, or desirable—our very instruments anchor us at our level, as well as at the level we are observing. Such instruments are *inter-level transducers.*
3. The entities of a level will be multiply anchored through causal interactions to other entities at that same level, and will therefore show substantial robustness at that level.
4. Many of the properties attributed to entities at a given level (or sometimes attributed to the instrument used to detect them) will in fact be disguised relational properties—properties of the interaction between target entity and instrument. (This, or something like it, should be the correct move for the classical secondary qualities, but it also occurs for many other theoretical properties—perhaps most notoriously fitness, which is a relational property of phenotype and environment, but is misleadingly attributed without qualifications to organisms, traits, and genes.)

5. Many of the apparent ontological paradoxes characteristic of different level accounts of a system—paradoxes which may appear to require the elimination of upper-level properties and entities to a zealous reductionist—arise from forgetting this relational character. In Eddington's "two tables" paradox, there is nothing contradictory in saying that this table is both continuous, colored, and solid (when using my fingers and eyes as probes) and at the same time mostly colorless empty space (when using a beam of electrons as a probe.

h. Theories come in levels (to analogize an observation of John Dillinger) because that's where the entities are. Simpler theories can be built with those entities (and their major interactions) than with slightly larger or smaller or otherwise different ones. On this account of the theorist as bank robber (or forager, or economist), theories of entities at levels provide the biggest bang for a buck. These entities will be theoretically fruitful because of their many causal interactions, and the appropriate choice of entities at levels will more often produce naturally segmented systems that are nearly decomposable—which "cut Nature at its joints" (Wimsatt 1976a). Thus language (in which concrete nouns—entity words—are learned first) and theories constructed using and refining this language are in this way *responses to* rather than *determiners of* the structure of the world.[25] A causal asymmetry is asserted here that runs counter to most recent linguistic or social-relativist views of the world. During the heyday of linguistic philosophy one might almost have had the impression that nature came in levels because language came in strata—a kind of theory dependence or conceptual scheme dependence of our ontology.[26] For most of the natural world, this has it exactly backwards: language is a tool for dealing with problems in the environment (including the human environment, and including the environment of different levels of organization accessed by our ever-further-reaching and multi-faceted instrumentation). For the most part, language has the macroscopic structure that it does because of the structure of the environment, and only relatively rarely is it the other way around. If most of the robust entities are at levels (as they are),[27] then the levels will themselves be robust—they will be relatively stable and multiply detectable. Theories are tools for representing, explaining, and dealing effectively with Nature. If they deal whenever possible with objects and properties that are at levels, they will be simpler, and will deal with things that are stabler, and (for that reason), also more common and persistent.

3. The Coevolution of Levels and Their Entities

i. Richard Levins (1968) argues that *organisms evolve in such a way as to minimize the uncertainty in their environments.* This is an important truth—but only half of the story: organisms will try (1) to be as *unpredictable* as possible to their predators, while (2) trying to render the behavior of resources they need, including prey, as *predictable* as possible! This selection for unpredictability (together with selection to respond adaptively to energetically tiny informational cues in the environment) introduces a level of predictive complexity in aspects of the detailed behavior of biological systems that seems to have no parallel in the inorganic world.[28] These kinds of interactions should lead naturally to positive feedbacks, non-linear dynamics, and chaotic behavior. *This interdigitating web of designed predictabilities and unpredictabilities, together with the consequent selection for heightened sensory acuities, probably serve more than anything else to make the regularities of the biological natural order so conditional, so context-sensitive, and so complex.* It leads to the exploitation of sources of information, good predictors of fitness-relevant parameters, wherever they can be found—including at other levels of organization. *Thus organisms, just like human scientists, sometimes have reasons for developing interactions that are not level-bound, and these opportunistic inter-level connections make higher-level phenomena less well-defined with respect to level, and levels themselves more diffuse.* The fact that these trans-level interactions for such things as functional organization (Wimsatt, 2002) can themselves sometimes be described in a systematic way that is not level-bound is ultimately what makes what I describe as *perspectives* below so important for the analysis of biological systems.

j. More generally, considering Levins' original insight, *as stable foci of regularity and predictability, levels should act as attractors for other systems changing under selection pressures.* These evolving systems will do so by plugging into regularities in as many levels as are accessible to them—in effect by matching levels, where possible, with their environments.[29] When they do so, then their own regularities of behavior become part of the context to which other organisms adapt. This insight is a major feature in most or all concepts of the ecological niche (see Schoener, 1989, for a review), and is further generalizable.

k. *Levels themselves evolve over time,* with higher levels becoming occupied and lower levels becoming more densely occupied, while the biological objects comprising them and their interactions change on still faster dynamics. The temporal course of levels thus mimics the ecological phenomena of succession, and the stratified and rich ontology of the tropical rainforest rather than that of a Quinean desert. This is a perspective seemingly more appropriate to modern cosmology (which is a story of the successive occupation of higher and higher of the lower "physical" levels up through the atomic and molecular scale—and paradoxically, the differentiation of lower and lower of the higher physical levels on the astronomical scale) than it is to modern ontology, but it is also profoundly evolutionary. The level of organization is more like an ecosystem than a species—it evolves as a product of the evolutionary trajectories of the entities that compose it, and provides selection forces that guide their evolution (by affecting what is stable). *From the evolutionary perspective, levels define niches for their composing entities, but these are coevolving niches that are products of the entities that make up the levels.* (Compare the "constructional" view of the relationship between organism and environment of Levins and Lewontin, 1985; the concept(s) of the ecological niche by Schoener, 1989; and, for an important and instructive extension of the concepts of niche and species to the evolution of theories and research traditions, see Allchin, 1991.)

Note—as philosopher Chuck Dyke has urged upon me—that this last observation places an important constraint on the ways in which levels or their entities can be regarded as compositionally defined. In Section II I noted that while levels were compositional, this should not lead one to the mistaken view that the best way to make a higher-level entity (according to the engineering paradigm) was to assemble it out of lower-level parts. On the view advocated here, within the organic and social realms (I won't speak for large "merely physical" aggregates), levels are for many purposes co-evolved, generated, or developed, rather than aggregated. It is still true that in a relevant sense, any higher-level entity will be composed (without remainder—I still believe in the conservation of mass) of its lower-level parts, but it will be a (mechanically explicable) non-random generated complex of those or other lower-level parts, which may have required a diversity of "chaperones" (as molecular biologists call other molecules designed to facilitate a given reaction) and other same and higher-level co-generating complexes for its construction or development. But if this is true for many of the entities at a level, and if the entities at a level act as coevolutionary forces on one another, it is also true for the level itself, and the description of the level as a compositional entity will—to that extent—be misleading.

4. Levels, Robustness, and Explanation

1. There is a general *level-centered orientation of explanations* that can be explained in terms of the greater stability and robustness of entities at levels of organization, and probably more globally, in terms of the consequent robustness of levels themselves. This is a general and important meta-principle for the organization of explanations that is usually taken for granted and seldom commented on. It facilitates explanatory clarity, but occasionally misfires (see the discussion of perceptual focus in the last two sections of Wimsatt, 1980b, where I discuss the biasing effect of the tendency to refer group phenomena down to the individual level of description in the units of selection controversy). The robustness of levels tends to make them stable reference points that are relatively invariant across different perspectives and therefore natural points at which to anchor explanations of other things. *Explanations of the behavior of between-level entities tend to be referred upwards or downwards in level, or both—rather than being pursued in terms of other between-level things. Even the fine tuning of the exact "altitude" of the between level entity—its size and thus the distance it is above the lower and the distance it is below the upper levels—is motivated by concerns originating at one or the other of the levels. The robustness of levels makes the level-relativity of explanations a special case of the phenomenon referred to in the preceding section—the explanation of that which is not robust in terms of that which is robust.* I will consider the case of Brownian motion as a between-level phenomenon, which, by its very nature requires very special relations to the level below and the level above. (For a more technical exposition of some of the details, see Jeans, 1940.)

A good Brownian motion particle must be small enough that sampling error effects in molecular collisions produce temporally local imbalances in change of momentum between colliding molecules and the particle—giving net random fluctuations in the motion of the particle. In effect, it is enough larger than the colliding molecules that it jiggles relatively slowly (the law of large numbers works pretty well), but not so much larger that it works perfectly (that the jiggles are too small to detect). In a gas, the colliding molecules are moving at a mean speed equal to the speed of sound (of the order of 1100 ft./sec. in air at room temperature at sea level—so-called standard temperature and pressure). The Brownian motion particle must be enough larger than the gas molecules that individual collisions do not move it too fast or far before the next collision (or actually, the next significant failure in local averaging of collisions), so that we can continue to track it visually. Increased size of a particle (relative to its molecular drivers) acts in four ways to facilitate tracking: (1) it slows down motion in response to a collision with a particle of a given momentum; (2) the larger cross-section gives more collisions per unit time, giving temporal averaging in a shorter distance and decreases the expected absolute path length (or time) until the next perceived change in direction; (3) the increased size also decreases its *relative* path length (the ratio of path length to diameter), increasing the perceived *relative* stability of its position and motion—an important variable in our perceptual ability to track it; and (4) the Brownian motion particle also has to be large enough to reflect light in the visible spectrum, or else we couldn't see it (but if the particle is too large, it will not move enough for us to be able to detect the motion).

Individual "jaggies" in the Brownian motion particle's trajectory do not generally correspond to individual molecular collisions, but rather to local imbalances in collisions that force a distinguishable change in its velocity in times short enough to be perceived as instantaneous. Our visual system reifies paths between these super-threshold changes as straight-line trajectories, with piecewise constant velocities, but the value of that threshold is a complex function of illumination level, our static and dynamic angular resolving power, flicker-fusion frequency, and the wavelength of the reflected light—not to mention the magnification and optics of any instrumentation we use to watch it. (It is this fact that is responsible for the frequent claim that Brownian motion is a fractal phenomenon: changes in the magnification of the scene, or of the motion sensitivity characteristics of the detector will change the length scale over which velocity changes are detected.) If there are entities causing the changes in direction that *we* notice, *as we reify these changes,* they are *clusters* of collisions, rather than individual collisions, and the character and size of the clusters that we will reify as a group is a function of our perceptual parameters. (Other organisms would see it differently—possibly resolving a fractal pattern on a different scale determined by the relevant parameters of *their* visual systems.)

The colliding molecules are below the Brownian motion particle in level, and we are above it, but *there are no levels in between for the Brownian motion particles to occupy.* If anything is at its level, it is these clusters of molecules, whose grouped collisions cause noticeable changes in velocity or direction of the particle. We do not recognize these clusters as entities for at least two reasons: (1) the perceiver-dependent and thus subjective time and size scale fractal characteristics of the Brownian motion—changes in which would change the temporal boundaries of the relevant clusters, and (2) the lack of unity of the cause of these motions—because the clusters are mere temporary assemblages that have no stability—they don't make "good" objects.[30] Explanations are, as here, referred downwards and upwards in level.

Another revealing indicator that Brownian motion particles are between levels is that they are given no intrinsic characterizations—as is indicated by the fact that things as diverse as dust motes and bacteria can all be Brownian motion particles. *Between-level entities tend to be defined functionally rather than in terms of their intrinsic properties—it is almost as if they have no intrinsic properties to use in such a definition.*[31] If so, this suggests the paradoxical conclusion that we may recognize the intrinsic properties of things, at least in part, due to characteristic interactions they have with other same-level things, since only levels have the intensity of different kinds of interactions among entities to fix unique sets of intrinsic properties as being causally relevant.[32] Multiple realizability in between-level contexts washes out the causal salience of most specific intrinsic properties.

m. It is also true that in our world, the dominant methodology is reductionist—we tend to explain features of the behavior of an entity in terms of its internal features, rather than how it relates to its environment. This implies a kind of explanatory priority, that *things not explicable at a given level are to be referred to the next lowest level, rather than to the next highest level.* This is a contingent, but very deep feature of our methodological world—sufficiently so that we tend to be suspicious when we are called on to explain phenomena by going up a level (as with functional explanations), or even by staying at the same level (as with phenomenological causal theories). These suspicions are frequently unjustified, and there are situations where explanations in terms of other same-level or higher-level entities are exactly what is required. Different aspects of the reasons for and character of this bias are discussed at length in Wimsatt, 1976a, part III; Chapter 11 in this volume; and Wimsatt, 1980b (the section on reductionist problem-solving heuristics and their biases), and I will not discuss them further here.

5. Time Scales, Multiple Realizability, Stability, and Dynamical Autonomy

n. As noted by Simon (1962), *processes at higher levels* (with a few important exceptions) *tend to take place at slower rates than processes at lower levels as measured by their "relaxation times"—the time it takes a reaction to go a certain fraction (usually one-half) of the distance to equilibrium.*[33] This phenomenon would certainly follow from the fact that it takes longer for causal effects to propagate larger distances. The coupling of size and time scale might look suspiciously like an application of relativity theory to physical processes, but it is not that simple. Most causal effects propagate at speeds that are a negligible fraction of the speed of light—governed by different processes that have more to do ultimately with quantum mechanics than relativity (the rate of propagation of disturbances of various energies in various solid, liquid, and gaseous media). Even if these processes are rooted in quantum mechanics, they would be so via pathways that—at least in the organic realm—are sometimes torturously indirect. (Consider the rate of propagation of membrane depolarization pulses in nerve fibers, and locomotion speed in all types of animals—both of which *increase* for larger structures, but in ways that lead to decreases in the frequency of repetitive actions for larger animals. Thus, an elephant runs much faster than a mouse, while its legs move at a much lower frequency. Bearing this in mind, I was astounded to discover that my expensive SLR camera did not have a lens speed fast enough to stop an ant in motion!) The net effect is to make one chary of any simplistic explanation for this probably very heterodox phenomenon.

o. *The multiple-realizability of higher-level properties or types is a general fact of nature, and applies to any descriptions of entities at two different levels of organization.* (it is thus entertaining to see philosophers of psychology act as if this characteristic is a special property of the mental realm). Multiple-realizability is entailed jointly by (1) the astronomically larger number of possible distinguishable micro-states than possible distinguishable macro-states—a ratio which (assuming that micro- and macro-variables have equal numbers of allowable states) grows roughly as an exponential function of the ratio of sizes of characteristic entities at the two levels, and (2) the numerical identity of the upper-level system thus described with the lower-level system thus described. Given that relatively many states at the micro-level must (because of the numerical identity) map into relatively few at the macro-level, the multiple-realizability of the few by the many follows (Wimsatt, 1981a).

p. More importantly, the *dynamical autonomy* of upper-level causal variables and causal relations—their apparent independence of exactly what happens at the micro-level—is entailed by this multiple-realizability and two further facts: (3) the relative stability of macro-level features (which persist for a characteristically longer time than micro-level features as a joint result of longer relaxation times and multiple realizability—items *n* and *o* above) in the face of (4) a constant flux of micro-level changes on a smaller size and shorter time-scale. (These items can be collapsed into a single assumption by taking the relative character of the stability claim seriously.) *The stability of macro-states in these conditions further entails that the vast majority of neighboring (dynamically accessible) micro-states map into the same or (more rarely) into neighboring macro-states.* To suppose otherwise would require at least a tremendously convoluted and radically improbable mapping from micro-states to macro-states—if it were even consistently possible. It is dynamical autonomy, more than anything else, which makes room for higher-level causal phenomena and theories, and the causal effectiveness of macro-level manipulations.

q. *Dynamical autonomy in turn entails that most (and in simple multi-level systems, an astronomical majority of) micro-level changes don't make a causal difference at the macro-level,* and that, except for cases of causal divergence (such as are found widely in chaotic dynamical systems, but are still presumably relatively rare since they would be selected against in most circumstances), most macroscopically causally efficacious factors will correspond to major global and often structural differences at the micro-level. The possibility of micro-level chaos shows that most macro-systems that show stability (or the respects in which they show stability) are tuned in such a way that the micro-level changes do not cause deviation amplifying (and therefore unpredictable) changes at the macro-level in those respects. In many simpler systems (for example, the mappings between micro-states and macro-states for a gas under conditions in which it does not show turbulence) we get this easily, but it applies to more complex systems as well if the systems are to show distinguishable macroscopic order.

An example may help, and we have a particularly important one at hand, for *the genetic system is a paradigmatic example of a striking kind of paradox frequently found in evolving systems. It is systematically tuned (as a matter of design) so that small differences can have effects on a variety of size scales including the very large, in which context dependence of effects is a common phenomenon, but where it is crucial that most differences do not have significant effects most of the time.* (I suspect that most people used to inter-level relations of the sort characteristic of classical statistical mechanics, where "law of large number" averaging is a reasonable mode of moving from one level to the next, will find the complex interplay of sensitivities and regularizing equilibrations of the relations between genotype and phenotype to be quite remarkable.)

Consider the following: We are given the genetic variability at many loci characteristic of virtually all species of organisms, and the scrambling effects of genetic recombination, so that each offspring is essentially without precedent in the specification of its genotype. Offspring of the same parents (save for identical twins) should characteristically differ at thousands of loci. Furthermore, we know that small genetic changes can and often do have large effects, and that interaction between genes in producing their effects is the rule rather than the exception.

Given these facts, if we didn't know any better, it would be plausible to expect almost no correlation in phenotypic properties between different members of a species (within the range of properties defined by that species), and between parents and their offspring. Yet offspring commonly inherit their parents' traits, as well as their fitnesses—not perfectly, but much better than random. The stability of the phenotype at many levels is essential for the heritability of fitness required for the evolutionary process to work. Not only must elephants breed elephants, humans humans, and *Drosophila Drosophila,* but the variability and systematic and independent inheritance of individual survival-relevant characters from parents to offspring within each species must be preserved—not glued together with a thicket of epistatic and linkage interactions—if temporally and spatially local adaptation to changing environments is going to be possible. We are constantly told by geneticists of cases where a single base change in a gene or a single amino acid change in a protein has enormous consequences for adaptation and function at a variety of higher levels of organization. But this has to be the exception rather than the rule for evolution as we know it to be possible. (Sickle-cell anemia remains the classic case here, and there still aren't many cases known as yet, though these should increase with our knowledge of developmental genetics.) Nonetheless, the plain fact remains that most genetic changes that happen under biologically normal conditions have no readily discernible effects (see Lewontin, 1978, on "quasi-independence," and Wimsatt, 1981b, for further discussion). Wagner (2005) provides a superb review and analysis of this "designed neutrality" at multiple levels of organization.

Therefore, most small micro-state changes do not make a difference at the macro level—even in systems that are characteristically sensitive to small changes. The converse does not follow: as pointed out above, closely related or identical macro-states may be realized by widely disparate kinds of micro-states, as illustrated by the Brownian motion of dust motes and bacteria!

r. *For instantiations of stable macro-level properties, in a sense there is no micro-level explanation for why they have happened, since changes in these properties, even if characterized at the micro-level, are macroscopic in scope.*[34] In giving extensive micro-level detail in an explanation, there is an implication that the detail matters—that the event or phenomenon in question would not have happened but for the cited details, that if just one detail were different, the outcome would have been significantly different. But if a process shows multiple realizability and dynamical autonomy this is just what is denied for the relation of most microscopic events to their macroscopic descriptions. There is, however, a crucial related question—namely, *why are these macroscopic states, properties, and relations stable?* This question will require an answer that is at least partially anchored in lower-level mechanisms—though not in large numbers of context-sensitive micro-level details. (If selection processes are involved in the explanation, it may also require reference to events at higher levels as well.)

s. The operation of *evolutionary and differential selection processes should tend to expand the scope of dynamical autonomy—increasing the range of multiple realizability—still further in cases where a macro-level property contributes positively to fitness.* Mutations will accumulate, which make its realization more likely and easier (this is a kind of generalized "Baldwin effect" response to selection). (Now, a decade after this was written, Wagner [2005] has provided robust empirical and theoretical support for this conclusion.)

Dynamical autonomy begins with the stability of properties of physical systems, but as the systems get larger and more complex, and their behavior more potentially variable, selection can breed stability of these usually more complex and contextual properties. Even in cases where the environment is unstable, making different properties desirable for fitness in different environmental contexts, evolution should select for context-sensitivity and conditional developmental programs—which tend to make the right things in the right contexts—all thereby increasing the heritability or stability of fitness across different environments (Wimsatt, 1986a). The only fly in this ointment is the increasing capabilities of the predators, parasites, and competitors of each species—referred to in item i above, and enshrined in Leigh van Valen's

(1973) Red Queen hypothesis—that even though each species is evolving, because of the co-evolution of others, you have to run as fast as you can just to stay in the same place. This should simply serve to generate increasing complexity and context-sensitivity of at least some organic interactions, and ultimately lead to the breakdown—through interpenetration and demodularization—of well-defined levels, and the emergence of other modes of organization in the ontology of complex systems.

One might think that one could go up indefinitely, successively aggregating and composing larger and larger systems into entities that occupy still higher levels of organization, but—whether as empirical fact, robust statistical regularity, or nomic necessity—other things emerge as salient cuts on natural processes and systems as these systems become more complex. My best guess is to think that the systems for which these other relevant modes of organization emerge are all products of biological or cultural evolution, since these are processes that tend to produce complex, contextually conditional, systematic, and characteristically adaptive behavior (see item i above), which has to simultaneously meet a variety of constraints at different levels of organization. But in lieu of more robust arguments for this conclusion, we must beware of overgeneralizing from the cases that our theories (and our interests) have given us the greatest reasons to consider. In the next section I try to characterize the conditions leading to the breakdown of well-defined levels and the emergence of perspectives. For an important review and analysis of levels and explanation in neuroscience that leads in complementary directions, see Craver (2007).

6. *From Levels to Perspectives: The Breakdown of Levels*

As long as there are well-defined levels of organization, there are relatively unambiguous inclusion or compositional relations relating all of the things described at different levels of organization. In that case, inter-level identificatory hypotheses are an important tool of explanatory progress in localizing and elaborating lower-level mechanisms that explain upper-level phenomena (Wimsatt, 2006a). There are relatively unproblematic assignments of all entities and properties with respect to level, and often systematic theories of phenomena at the respective levels. At this stage, theories are either directed to phenomena at specific levels or (for inter-level theories) acting to tie levels together by elaborating inter-level mechanisms or connections (see Maull, 1977; Wimsatt, 1976b; Darden and Maull, 1977). But, conversely, when neat compositional relations break down, levels become less useful as ways of characterizing the organization of systems—or at least less useful if they are asked to handle the task alone. At this point, other ontological structures enter, either as additional tools or as replacements. These are what I have called *perspectives*—intriguingly quasi-subjective (or at least observer, technique, or technology-relative) cuts on the phenomena characteristic of a system, which needn't be bound to given levels. Since the discussions of perspectives in Wimsatt (1974, included here as Chapter 9), and of the relation of levels, perspective, and causal thickets in Wimsatt (1976a), an even broader diversity of different perspective-like things have appeared in the literature of the last 30 years, and have been invoked to solve a similarly broad range of problems. This characterization of perspectives is tentative, incomplete, and still unsettled even on such major questions as to whether they are a unitary kind of thing. Nonetheless, there is a class of such things that do have a lot in common. Below I provide a tentative list of properties of these strange objects, and a set of examples suggesting some of their differences as well as their similarities. Further refinements will have to await another occasion.

The transitions suggested here—from levels to perspectives to causal thickets—characterize systems in terms of increasing complexity and context-dependence, and lower modularity and degree of regularity. This is an ordering in terms of kinds of complexity. It is not a natural evolutionary trajectory for systems, or any other kind of natural dynamical transition. Although, if I am right, systems later in this sequence first appear after systems earlier in the sequence (as a result of the continuing action of biological and sociocultural evolutionary and developmental processes), there are specifiable circumstances in which selection processes favor simplicity, modularity, near-decomposability, increased regularities of behavior, and well-defined compositional relations. Thus, with few exceptions, the order given here should be regarded as taxonomic, rather than temporal. Given the taxonomy, we may later wish to argue about temporal trends.

t. As higher levels get more complex (they have more degrees of freedom), they get more diffuse, and they overlap more in size scale and other related properties with neighboring levels, and engender perspectives and thickets. With more molar properties at the higher levels, and each one a potential pathway for causal interaction with entities that have or respond to that property, there are more ways to "plug into" a level. With more degrees of freedom, higher-level objects get potentially richer in their budget of properties, more multi-dimensional. At their best, they should thus be capable of higher degrees of robustness than lower-level entities. (There should be more ways of interacting with a spouse than with a quark!) There should *also* be more ways of being *not* very robust, of being only marginally connected to the causal processes of a level, and also more ways in which objects could interact simultaneously with or *bridge* two or more neighboring levels. These last two kinds of cases would increase the diffuseness of the levels associated with the entities. Thus, as levels get higher and more complex (up, roughly through the level of the ecological community or ecosystem, and perhaps on up to the biosphere), we should expect them to get more diffuse, for levels to overlap more, and for it to get more difficult to localize an entity or phenomenon by level unambiguously and for all contexts (see Figure 10.1).[35]

u. As objects find new ways to bridge levels, fluctuations at the lower level, which without the bridge average out at the upper level, are now transmitted directly (as we can observe Brownian motion with the aid of a microscope, and through that, the effects of micro-level events), generating the possibility of macro-level amplification of these micro-level events, creating a kind of sensitive dependence on initial conditions that will tend to increase the number of circumstances under which macro-level regularities will break down. Thus, we should expect that the maximum degree of regularity of upper-level phenomena for complex organized systems would be less than that for simpler systems composed of more homogeneous parts. This is the complement to the "diffusion" of levels: as they come to span a broader range of sizes, the maximum predictability decreases, almost as if the area under the level waveform for each level is a constant[36] (see level c in Figure 10.1). One must remember that small differences—fluctuations or signals—make a difference when they are detected by a system designed to respond to them, and for which the pattern is significant. The human eye can detect a single photon—a micro-level event to be sure, but not yet a pattern. The number of photons, if appropriately distributed in space and time, necessary to convey information is larger than this (probably of the order of 10), but still astoundingly small.[37] Detectable information can lead to macroscopically major (and, with modern technology, even further divergent) behavior.

v. At the same time we should also expect (ultimately, for reasons of increased dimensionality) to find more frequent, obvious, and severe context-dependence of the behavior of our entities at higher levels of organization. This would most often be expressed via systematic and not so systematic exceptions to simple generalizations involving these entities. This is one of the reasons why it is better to think of regulari-

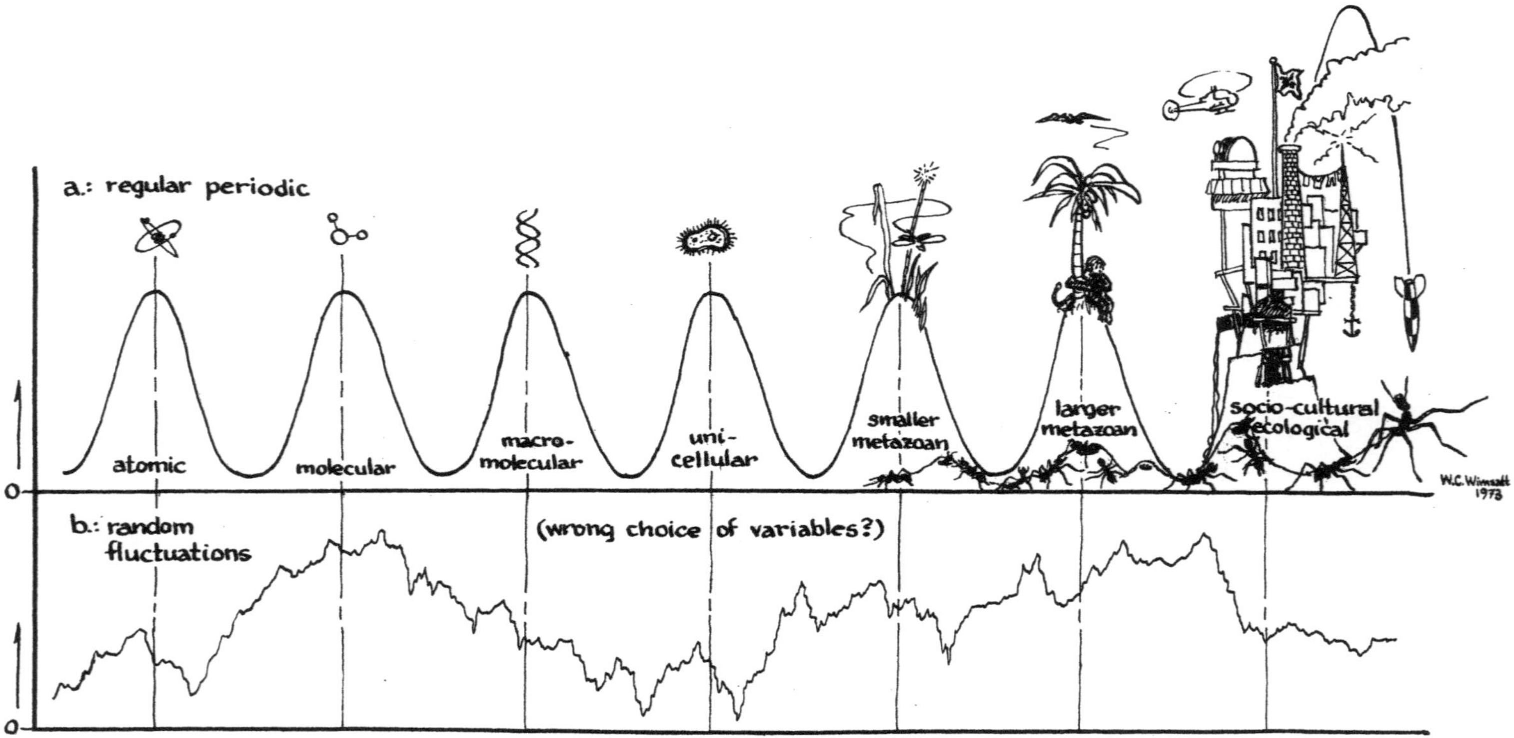
a.: regular periodic
atomic
molecular
macro-
molecular
uni-
cellular
smaller
metazoan
larger
metazoan
socio-cultural
ecological
0
b.: random
fluctuations
(wrong choice of variables?)
0
W.C. Wimsatt
1973

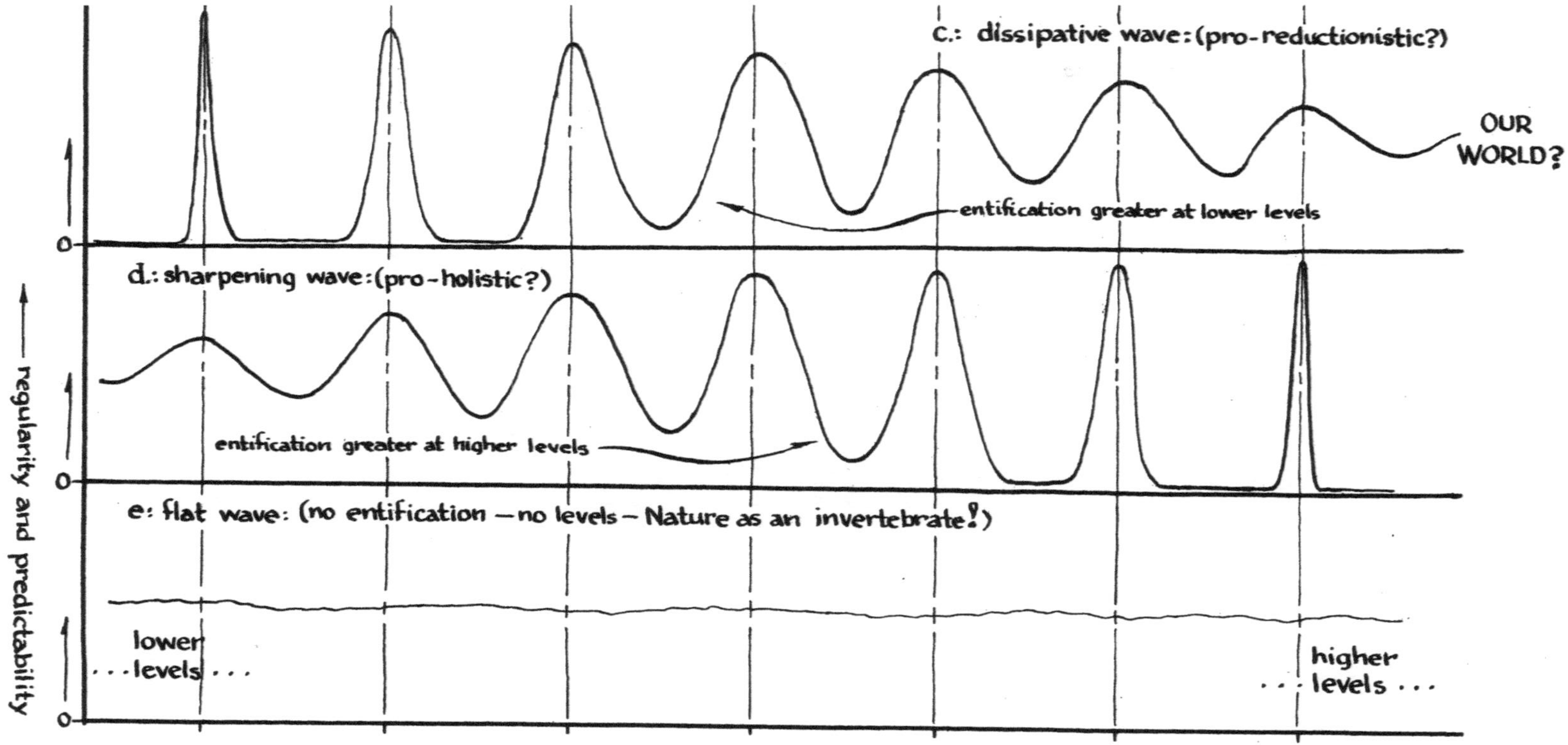
c.: dissipative wave: (pro-reductionistic?)
OUR WORLD?
entification greater at lower levels
d.: sharpening wave: (pro-holistic?)
entification greater at higher levels
e: flat wave: (no entification — no levels — Nature as an invertebrate!)
lower
. . . levels . . .
higher
. . . levels . . .
0
0
0
regularity and predictability

ties in complex systems in terms of mechanisms rather than laws. The latter, but not the former, suggests a search for exceptionless generalities and explanatory completeness, whereas the former fit naturally into a scheme that is satisfied by providing a characteristic *ceteris paribus* qualified articulation of causal factors (see Chapter 11).

w. Finally, as it becomes more common for entities to interact directly with other entities only through a subset of the properties that are causally relevant at that level, with different entities responding to different subsets, the notion of a *niche* (derived originally from ecology, cf. Schoener, 1989, and also as applied to research programs and theories by Allchin, 1991) becomes more relevant to the characterization of their behavior. This notion of a niche makes it clear and naturally explicable how different systems could act upon and react to the "same" environment in fundamentally different ways (Wimsatt, 1976a). The fact that the niche must be characterized relative to the organism—it is mutually defined by the organism and its "objective" environment (the

Figure 10.1 Waveform representation of compositional levels of organization as they might occur in different conceivable worlds—not all of which are physically *possible* worlds. In each row, the vertical axis is the degree of regularity and predictability—or, in more modern terms, the degree of pattern—for objects of different sizes. Size is represented logarithmically along the *x*-axis, so that regular periodic maxima would represent patterns found at geometrically increasing size scales. (Such scales would be expected if objects at each level were aggregates of roughly commensurate numbers of objects from the level immediately below.) It is argued in the text that the diagrammatic top row (a) and the second row below it (c) are the best representations of levels of organization in our world—(a) for its periodic character spilling over in an unruly fashion increasingly at higher levels, suggesting (c) for the greater diffuseness of the higher levels of organization (in the middle range of size scales that we occupy). The levels diagrammed here are really only the middle ones. Presumably, quantum mechanics renders the very small again diffuse, and astronomical scales again produce well-defined objects interacting in a relatively limited number of well-defined ways. I believe that waveforms (d) and (e) are *not* found in our world. As discussed in Wimsatt (1976a), a waveform like (d) would favor holistic over reductionist methodologies, and non-periodic forms like (b) or (e)—where there are no levels of organization—are ruled out by Simon's arguments concerning the role of evolution via stable subassemblies. Given the obvious existence of levels of organization over the range sampled and the random excursions in the sampled variable, the reasonable assumption for (b) is an incorrect choice of variables, or perhaps diagnosis of a causal thicket.

configuration of physical and biotic factors affecting its evolution)—introduces a feature of subjectivity that I explore further in the next section.

III. Perspectives: A Preliminary Characterization

What I call *perspectives* is a diverse range of things that nonetheless appear to have at least some of the properties of being "from a point of view" or to have a subjective or quasi-subjective character. In spite of that, perspectives differ substantially in terms of their other properties, and in terms of their relative objectivity. Their subjective character is because of the properties that they do share, which are discussed below. (The parenthetical remarks in these paragraphs are usually further elaborations of how this is so.)

1. *Perspectives involve a set of variables that are used to characterize systems or to partition objects into parts, which together give a systematic account of a domain of phenomena, and are peculiarly salient to an observer or class of observers because of the characteristic ways in which those observers interact causally with the system or systems in question.* (So far, this does not distinguish a perspective either from a methodological approach, or from the ecological niche of a species—two things that both have a kind of observer-relativity, and also have the curious objective-subjective duality I think characterizes a broad range of perspectives.)

2. *The set of variables in question is recognized not to give a complete description of all aspects of the systems that they are used to investigate.* Thus there is an explicit denial of a closure clause. (If this captures an important aspect of subjectivity, which I think it does, it is the recognition that it makes no sense to speak of something as subjective [or as objective] without the other category—which at this stage [from the subjective side] involves at least the recognition that there is something outside of the boundary of the subjective.)

3. In spite of this, there may be a restricted closure of the following sort: *there is a reasonably well-defined class of problems that can be solved without bringing in information from outside the perspective.* These are treated as paradigmatic problems for that perspective. These may also be problems that cannot (or cannot plausibly) be solved in any other way. So there are paradigmatic anatomical, physiological, and genetic problems, though (cf. (2) above), no one believes that these approaches individually exhaust what may be said about the organism. (This suggests a kind of unity and systematic problem-solving utility to

the subjective. There are things one can accomplish wholly within the subjective perspective, and things that only can be plausibly solved from within the subjective—or a particular subjective—perspective.) In effect, this says that perspectives partition problem-space in a nearly decomposable fashion.

4. Indeed, it is commonly taken for granted that *multiple perspectives can be applied to different aspects of the behavior of a system.* (Without this, there is not yet a recognition of the objective—a recognition of the robustness of the system accessed by the different perspectives.) I refrain at this stage from saying that the objective requires the existence of other subjectivities (thus perhaps characterizable as the intersubjectively accessible, or interpersonally robust), or merely the applicability of other perspectives (which could still be true in a Robinson Crusoe universe with plenty of external robust objects but no other persons).

I won't say any more here about the personal, interpersonal, and material realms, but to note that robustness, levels, and the idea of a perspective, together with an account of what it is to have a shared perspective are useful tools in characterizing our objective, mental, and social worlds. (In the last section of Wimsatt [1976a] I note and exploit parallels between the kinds of access we have to things at our own level and the less direct access we have to things at other levels, and the dichotomy between first-person and third-person perspectives.) But that is for another time and place. I want to consider particularly the kinds of complexities that make levels break down. The next two properties of perspectives were described more fully in Chapter 9 and in Wimsatt (1976a).

5. *Simple systems as well as complex ones can be described from a variety of perspectives, but will differ in the degree to which they have problems that are trans-perspectival—which require the use of information from more than one perspective for their solution.* Simpler problems are bounded and solvable with the resources of a single perspective. Simpler systems have more of their problems (or more of their problems for the purposes at hand) bounded within individual perspectives. Note that since problems usually arise out of purposes, a system can be simple for some purposes, and complex for others.

6. The complexity of trans-perspectival problems also varies from simpler to more complex with whether they decompose systems in ways that (a) are spatially coincident (in which case the different perspectives must also be either at the same level, or span the same range

of levels); (b) are hierarchically rationalizable relative to one another, so that the parts of one perspective are all whole systems in another (in which case the perspectives are related to one another as different level descriptions of the same system); or (c) overlap in arbitrary ways. The last case produces an enormous increase in complexity, but is common in the biological, psychological, and social worlds. (This is called *descriptive complexity* in Chapter 9, and the preceding kind of complexity is called *interactional complexity*.)

7. *Note that levels come out as a kind of special case of perspectives on this analysis—a class of perspectives that map compositionally to one another so that their entities are related without cross-cutting overlaps in a hierarchical manner.* It is tempting to say that we need to require also that the entities/parts at levels are especially robust, though that may come out for free given that hierarchical (and modular) compositionality will tend to require or entail substantial robustness of the systems and parts at all levels. Note that thus far, I have introduced nothing that a hard core materialist could not accept. (Indeed, I believe that all that I have introduced so far a hard-core materialist *must* accept.) Given this, hierarchical compositionality suggests a number of further interesting (but at this time still speculative) connections: (a) The "nearly sealed" aspect of living at a level of organization (the fact that level-leakage is relatively rare), and the comparatively torturous and indirect paths to systematic access to another level can at least help to explain qualitatively the first-person/third-person dichotomy between subjective and objective modes of access indicated in Wimsatt (1976a); and through that (b) it may suggest naturally how subjectivities can be seen to be anchored in a natural world. (c) Also, if "level leakage" is just a variety of "perspectival leakage," it suggests that and how modest amounts of comparability or leakage between subjectivities may be essential both to the recognition of other subjectivities and the reality anchoring of our own (necessitated by the private language argument). It also predicts (d) that, how, and why the breakdown of levels with increasing complexity can come to create problems for the localization and bounding of subjectivities as well as for the bounding of well-defined perspectives. This latter problem I take to be connected to new wave contextual embedded and distributed theories of consciousness.

I now wish to consider perspectives that are not levels. They may fail to be levels either because they are too small, they are located mostly at levels, or because they aren't of sufficiently broad span to count as levels. Or they may fail to be levels because, in a way, they are too big—they cross-cut levels: they are transverse sections that do not include more than a small fraction of the phenomena at any given level, but span phenomena at more than one (usually at several) levels. It is these two basic kinds of entities that allow us to go beyond levels to importantly different kinds of entities.

8. *The smaller kind of perspectives are those things that look most subjective, since they are most explicitly keyed to the point of view of a particular kind of organism or observer.* When objectively characterized without regard to other than physical or biological properties, I call these *niches,* because I think that the ecological niche of a biological species is the prime exemplar of this[38] (on niches, see Schoener, 1989). When characterized explicitly cognitively and subjectively, with respect to the cognitive and sensory capacities for and from the point of view of an animal, I call this the *subjective niche,* or *Ümwelt,* to use von Üexkull's (1934) term (von Üexkull and Nagel, 1974, are the best, and remarkably close, exemplars of this position). This notion of perspective naturally suggests further subdivisions that are psychological or cultural rather than biological in character, and how to make these further subdivisions (and how many) is an important question, although I do not address it further. Is there a paranoid schizophrenic's perspective? One or many? Is there a female perspective (is it cultural or biological)? Is there a feminist one? An upstate New Yorker's or a Manhattanite's perspective? An only child's (first child's, second child's, etc.) perspective? Does each new interest group or reference group individuate a perspective or a component of a perspective? Does every person? Does every life stage? How has my perspective changed since I was an assistant professor? Got married? Became a father? Learned how to program in *Pascal?* Even we must follow Quine's desert intuitions for ontology here in recognizing that there may be too many potential perspectives standing in the doorway! So how should we decide? This domain is the topography of many of the most important battles in the social sciences.

9. What I called perspectives in my 1974 paper is not usefully captured by any of the notions of perspective discussed so far. It is the *larger* variety of perspectives promised above. It is a more robust ontological category than they are, since it is not essentially defined by the relationship of a single kind of entity with its environment. Perspectives in the 1974 paper (a) spanned more than one level, and thus could not

be ordered as higher and lower or more primary and secondary than one another; (b) gave criteria for decomposing systems into parts using the properties and tools appropriate to that perspective; (c) were manifestly incomplete descriptions of their objects; (d) were such that different perspectives (for complex systems) could cut up systems in quite different ways that were not easily comparable to one another; (e) had a class of problems that they could solve in isolation; and (f) (for complex systems) had other problems that could not be solved without bringing in the resources of another perspective or perspectives.

Anatomy, physiology, and genetics are different perspectives on an organism in this sense. Perspectives may sometimes correspond loosely to disciplines, but need not. They may be either larger or smaller. Thus, the adaptationist perspective, in which the parts of an organism are all analyzed in terms of their evolutionary function—those aspects of behavior responsible for their selection, elaboration, and maintenance—which is larger than a discipline, unless disciplinary lines are drawn extremely broadly to include it; the discipline of evolutionary biology, for example. Fate maps also seem plausible as perspectives in this sense, in which the cells of a developing embryo (or layers, or regions—so this is not confined to a level) are marked to indicate what they will become are a specialized representational tool within classical developmental biology, and thus much smaller than a discipline. There are specialized tools for revealing these (such as radioisotope labeling, which can give an iconic representation of the fate of a cell, layer, or region through development).

If I were to rename this kind of perspective now (as I probably should), I would call them *sections*—short for cross sections (or perhaps sometimes transverse sections in messier cases!)—views chosen by architects, engineers, and anatomists to give particularly revealing aspects of their complex structures; *views that can cross-cut one another in various ways, and at various angles; views that are individually recognized as incomplete; views that may be specialized for or better for representing or for solving different problems; and views that (like perspectives) contain information not only individually, but also in how they articulate.*

10. Important ontological features of perspectives are captured in Figure 10.2,[39] which indicates that *perspectives cannot be ordered compositionally relative to one another—you cannot say that the objects or parts of one perspective are "really" composed of the objects or parts of another—or if you could do so, that a corresponding claim could be*

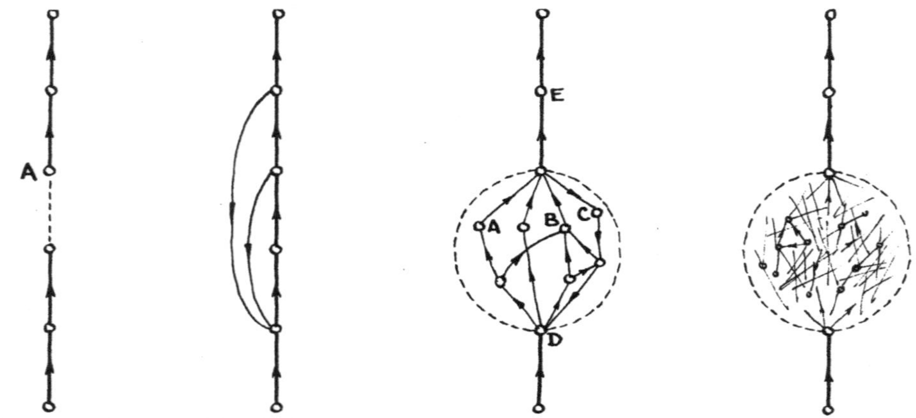

(a) simple reduction
(b) emergent at A.
(c) selection and explanatory "feedbacks"
(d) descriptive and interactional complexity
(e) defineable perspectives break down.

f) divergent branching yielding diverse products.

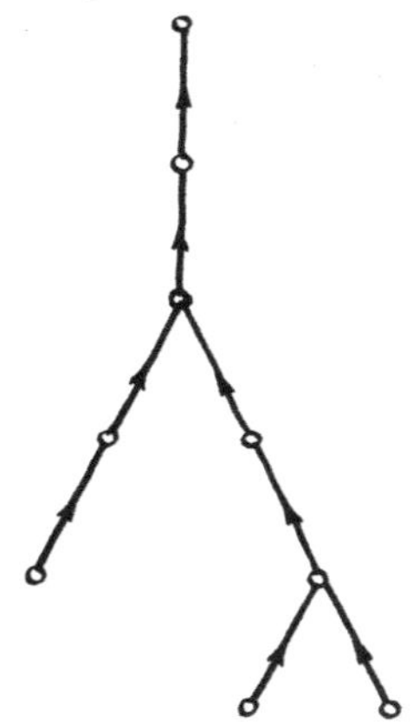

g) convergent branching yielding heterogeneous products.

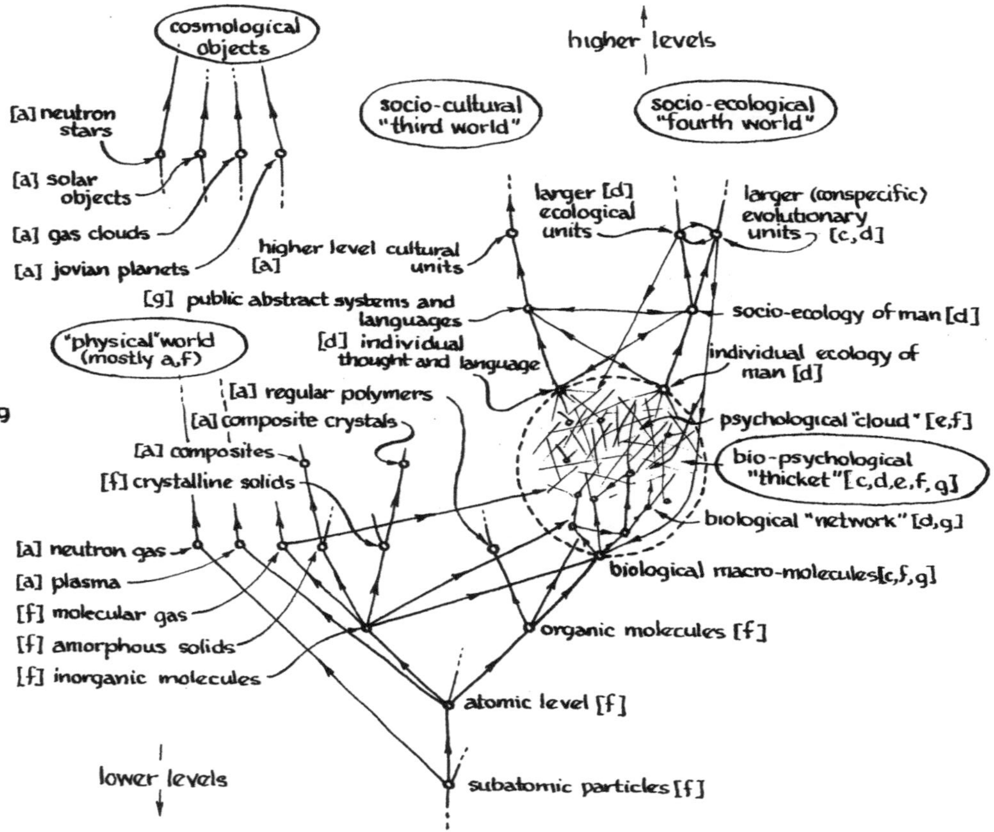

(h) A reductionistic (?) illustrative phylogenetic ontology of our world as we see it.

made in the other direction with equal justice. (Are anatomical features composed of physiological processes or conversely? The question doesn't make sense, but information from each perspective is relevant to the solution of at least some problems in the other.) But compositional talk is not forbidden within perspectives, even putting levels

Figure 10.2. Complex orderings of levels and perspectives. This figure depicts modes of composition of aggregate and complex systems, ordered in terms of the direction of explanatory relations. (Nodes are levels or perspectives; arrows give direction of explanation.) In simple systems this follows compositional relations, with behavior of the wholes explained in terms of the properties and relations of the parts. Thus, the simple reduction picture of the "unity of science" movement is given by (a), in which each level explains the one above it—a picture which, as Roger Sperry complained, "seeks to explain eventually everything in terms of essentially nothing" (quoted in Wimsatt, 1976a). The classical picture of emergence (as a failure of reduction) introduces a gap, as in (b). (This account is rejected in Chapter 12.) Explanatory feedbacks from higher to lower levels are introduced by selection processes (Campbell, 1974b), diagrammed in (c). Complex organization of the phenotype (as a product of selection processes) builds on explanatory feedbacks from higher to lower levels, creating further ordering problems with the emergence of perspectives (d), and increased interactional complexity producing cross-perspectival problems, and ultimately breakdowns of and ambiguities in the boundaries between perspectives, resulting (e) in "causal thickets" (Wimsatt, 1976a).

Section (h) of the figure is a compound diagram illustrating the composition modes of various kinds of physical, biological, psychological, and social systems (letters in square brackets refer to the local character of the network around that node). It is illustrative: I will not argue for its detailed architecture, and it may be wrong in representing complex physical systems in the biological, psychological, social, and cultural realms. I see no obvious errors, however. The biological organism (a developed language using socialized human) has perspectival structure (actually at its lower levels of biological organization, merging continuously with causal thicket structure as we get into the internalized psychological and social realm). Two ontological lineages emerge from this: those of cultural objects (abstract objects, presumably also viewable as abstract relational properties of objects in the second lineage), and socio-ecological objects (kinds of complex material systems having the whole range of social, ecological, biological, cultural, and psychological properties). I believe that the connectivity patterns relating these various realms inside and outside the individual are much more complex than represented here. Thus, social institutions obviously are complex hybrids of objects at a variety of levels from both of these lineages. There are causal thickets above and outside the individual interacting rather directly in various ways with causal thickets inside the individual, and an embodied socialized theory of consciousness is required.

aside as a special case. For a perspective, you may (and usually will, if you are a materialist!) be able to find lower-level objects (indeed, a greatest lower bound, or *GLB*, of largest common parts) such that all the entities in the perspectives are composed of them—de facto atoms, as it were. There may also similarly be higher level objects (correspondingly, a lowest upper bound, or *LUB*, of smallest common systems), such that the objects in the perspective are all parts of those objects (e.g., organisms), but for the regions in between the GLBs and LUBs, there is at most local orderability of compositionally ordered parts *within* each of the perspectives. If they exist, GLBs and LUBs of a set of perspectives are rich in implications. The GLB and LUB decompositions of embedded and embedding systems will both be robust, because they will be level-descriptions and will be orderable relative to one another. Given their unambiguous robustness and status as entities at levels, there will be a tendency to regard them as more important or as ontologically more central than descriptions derived from the perspectives in between. Reductionists will tend to favor the GLB descriptions on down through lower levels of organization, and functionalists or holists will tend to favor the LUB descriptions, and possibly on up through higher levels. If we accept the objectivity of the GLB and LUB descriptions, this will *tend*[40] to fix all the perspectives between them within the objective realm (or, more generally, to give them any ontological properties common to the two bounding levels). If so, these properties will be aggregative rather than emergent properties for that class and within that range of descriptions of systems (see Wimsatt, 1986a, and Chapter 12 in this volume).

The most interesting thing about perspectives follows from this ontological feature:

11. If compositional ordering relations break down, as they may between descriptions of the same object in different perspectives, then above a GLB and below an LUB traditional formulations of materialism are inadequate for ontological reasons *because you can't say what is composed of what,* although your complex system contains nothing immaterial. If this is right, then in that interesting size range in between atoms and organisms (or perhaps in many regions in between atoms and societies) you will often find a situation for parts or properties where neither type-identities nor token-identities appear to be of much use (Wimsatt, 1976a). Token identities aren't of much use anyways, beyond expressing advocacy of a token materialism. Nancy Cartwright said in a recent lecture[41] that token identities are too weak—they ignore the systematic regularities that are there, even in messy

cases. The problem (as she also noted) for type identities (and also for laws as they are normally conceived of by philosophers) is that the systematic regularities aren't exceptionless either. And you can't make them exceptionless without introducing so many qualifications as to make them essentially useless (I urged similar views in 1976—see Chapter 11). But we can't even get to this juncture if we can't specify composition relations, and in this interregnum of multiple partial incomplete perspectives, we can't. This might seem to be the death knell for any possible reductionisms—as it clearly is for any formalist or deductive accounts of reduction. It is also clearly at least highly problematic for any identity-based accounts like that urged in Wimsatt 1976a and 1976b (see also related discussions of what happens when localizations break down in Wimsatt, 1974, and in greater depth—using connectionist models as an example—in Bechtel and Richardson, 1993) But note also that this breakdown occurs without really doing anything to compromise the spirit of materialism because we can understand *in materialistic terms* why compositional relations are problematic, and a variety of general structural and methodological features about the situation, and can do so without admitting that there are any phenomena (or regularities) that we cannot explain. This is a remarkable situation, but one that characterizes, for at least some problems and properties, all naturally evolved systems.

12. But even with the varieties of incomparability suggested above, *organisms can share dimensions of niches in that some causal factors can be causally important to all, or to an important subset of them. This makes these dimensions or causal factors particularly important in explaining their behavior, and also particularly real, objective, and robust.* One way for perspectives to emerge (in the sense of sections, above) would be around causal clusters of variables that are robust niche dimensions—as sets of descriptive variables whose analysis generates adequate solutions to classes of correlative problems. The primary qualities would be good examples here, and statics (for physical structures) and anatomy (for biological ones) stand as good correlative perspectival theoretical structures. Within biology, similar or shared niche dimensions may be important causes of convergent evolution.

Finally, insofar as a theory deals with only a subset of the causally relevant properties of an object, it has a perspectival character, but if the properties it deals with are sufficiently robust and fruitful, it may be easy to forget this fact. It is worth considering (on another occasion) whether and when theories (in general or in particular; folk psychology or Kuhnian paradigms) should be viewed as being or as providing a perspective.

13. How do we judge whether perspectives are real? I think that there are two ways. First, when there is agreement across perspectives in identifying or saying things about objects they access in common, this judgment not only recognizes the robustness of the object, but—indirectly—confirms the means of access. Second, we can treat the perspective as object, rather than as means of access to other objects. But then the same criteria of robustness should apply—the extent to which the perspective is multiply detectable, in this case by being articulatable with other perspectives, affects the degree to which it is real to us. We can do this in various ways, with different ways appropriate to the kind of perspective it is. I will mention only one here, because it is already commonly recognized in methodological discussions in the social sciences. It is this activity we are practicing when we practice *Verstehen* (seeing the act in the way the agent did, and judging it to be rational or otherwise explicable from their perspective) to understand action. The target here is not action, or its justification, but the explanation of the action. And we can provide an explanation by putting ourselves in the other agent's shoes, and see that the action *is* rationalizable from their perspective. (Of course, it doesn't justify the action, the perspective could be that of a heinous fellow.) If we understand the action, in the sense of explaining it, then by taking on the perspective, and successfully practicing *Verstehen,* we have not only explained the action, but also confirmed the existence of that perspective, and its salience to the action.

IV. Causal Thickets

I noted above that each perspective will tend also to contribute to the solution of some problems that it cannot solve by itself—and that for more complex systems, this would tend to happen more frequently. With increases in the complexity of objects, and in their number and variety of degrees of freedom, they can interact with one another in more varied and complex ways, and more problems involving their behavior require the use of two or more perspectives for their solution. Sometimes, when there is a range of problems that can characteristically be solved using two or three particular perspectives or disciplines together, a new subdiscipline gets formed (e.g., psycholinguistics, or even developmental psycholinguistics). Sometimes problems are fought over by practitioners from two or more different perspectives. And sometimes problems appear to be big enough, or generally enough stated (e.g., the mind-body problem), that they seem to be intrinsically multi-perspectival. Since a perspective maintains its identity in part by having problems that its corresponding discipline can characteristically solve by itself, the characteristic identification of important problems with certain perspectives and the identity of perspectives tend to break down simultaneously. When the relative frequency of such problems gets too high (either as a function of the way the world is or as a function of the inefficiency of our conceptualizations in organizing our problem-structures), the boundaries of perspectives begin to break down and it becomes more difficult to decide which perspective (or perspectives) a problem belongs to. (Correspondingly, as the preceding parenthetical remark might suggest, it becomes harder to tell when we are talking about our world and when we are reflecting only or primarily on our own conceptualizations. Thus the "perspective," and many of the claims of the new deconstructionists and sociological relativists are in a way predictable and explainable in this situation—which I remind you, is still characterizable within a broadly materialistic perspective!)

This breakdown of boundaries induces competition among the different methodologies associated with the different perspectives, and so we should expect that methodological disagreements would proliferate, along with disputes about how to fragment systems into parts and how to best define key terms. As the boundaries break down this far, not only is it true that others' perspectives intrude on the one you wish to argue for, but also that your perspective can seem to reach legitimately to the horizon. Paradoxically, as the perspectives weaken in their own domain, they don't retreat, like good scientific theories, but their generality appears to increase without bound. (Deconstructionism is not the only banner to have claimed the whole field—witness methodological individualism under the banner of rational decision theory [fighting mostly prisoners' dilemmas], or the self-reinforcing behaviorisms of a generation ago.) At that point, philosophers may rush in where scientists fear to tread—or perhaps have done so and stubbed their toes! Here, if anywhere, philosophers may be useful if they know the lay of the land.

Perspectives have now degenerated into a *causal thicket*. This term is intended to indicate a situation of disorder and boundary ambiguities. Perspectives may still seem to have an organizing power (just as viewing a thicket or shrub from different sides will reveal a shape to its bushy confusion), but there will be too many boundary disputes. Claims may be made that phenomena are at a given level, or are to be viewed from a given perspective, and any level of analysis or perspective that has successful associated theories will attempt to claim disputed territory. But that is just the point—there will be a lot of disputed territory—and the disputes will often turn on how the system is to be cut up for analysis—or even (to those of a holistic persuasion) whether it can be cut up for analysis at all. (Some connectionists seem to expect that local analysis will fail for all interesting mental properties, which will therefore be holistically distributed, while others are busy denying that we will have to recognize any mental properties because they don't find them at any locations!) Most complex biological problems involve levels, perspectives, or a combination of both—except in neurophysiology and some areas of developmental biology. The neurophysiological, psychological, and social realms are mostly thickets, which are only occasionally well-ordered enough for local problems to be treated as perspectival or level-relative problems. All of this enormously complicates talk of reduction, because with such multiply connected entities, and the failure of the ability to say what is composed of what, it may now be almost impossible to determine what is being reduced, what is doing the reducing, and what even is the proper scope of the system under analysis and the problem we are being asked to solve.

The proliferation of disputes of this form involves an unusually large proportion of conceptual issues, methodological arguments, and boundary disputes. This phenomenon is predictable simply from looking at the form of complexity such systems take, and the form disputes should take when boundaries break down. Some of these disputes are likely to indicate sources of genuine disagreement, but this can't be determined when so many things are up for grabs. Moreover, the natural tendencies of most theorists toward expansionist territorial claims, and of all of us to understand the merits of our own positions better than those of our opponents, makes frequent disagreements seem inevitable where there are boundary ambiguities. Localization of problems with the existing conceptual structures, and of disputes to the right trouble spots will have to await the development of conceptual structures, methodologies, and new explanations of mechanisms in terms of them. If this explanation for their occurrence is correct or nearly so, an unusually large fraction of the disputes should be resolvable as people from the different groups learn and work out how to talk with one another, if (and it is a big sociological *if*) they maintain a commitment to try to understand one another rather than bloating their reputations by taking cheap shots at the opposition. This is perhaps the deepest pragmatic commitment of science—that it is in one's interest to come to understand differences, and then to resolve them. This yields an ultimately realist picture only because the world has an indefinitely large number of constraints for acceptable theories, if you know where to look. But you'd better get an overall sense of the geography before you decide on your colonizing strategy. This has a lesson as well, of which eliminativists should beware: you don't make friends with the natives (folk) by denying their legitimacy (psychology), and you can't tell what's in the territory without a native guide. You can play imperialist without heeding these warnings, but it usually requires more resources, costs a lot more, and takes a lot longer. And you may end up having to grant them autonomy anyway!

So far, we seem to have defined causal thickets as a kind of wastebasket category. They needn't be. On a priori grounds, considering the possible connectivities of causal networks, shouldn't causal thickets be the norm, and relatively insulated levels or perspectives the rare cases? Wouldn't causal thickets be, as it were, the high entropy or generic states of the causal structure of the universe—sort of an ontological primal slime? This is to exchange assumptions of simplicity and order in the universe for assumptions of randomness in causal connection—a kind of structural disorder. An absurd view, one might say, but not a priori absurd. To be sure, we wouldn't exist, and couldn't survive in such a universe, but considering it provides a useful kind of change in perspective. One of the remarkable things about our universe is the degree of order we find in it. To be sure, it is not an exceptionless static order—crystalline without flaw. There are regularities at all levels, and mechanisms tying them together, and perspectives that give cross-sectional cuts on the phenomena for a range of problems. And then there are some things that are just too multiply-connected to fit exhaustively into any of these ontological categories. And we can say something about the conditions in which we expect each of these to arise, and their methodological consequences. This looks a lot more complex than the old story, but it provides tools and ways of thinking and talking that seem a lot closer to the truth. And, as I've been trying to tell you, that's the way the world is.

Reductive Explanation

A Functional Account

Philosophical discussions of reduction seem at odds or unsettled on a number of questions and concerns:

1. Is reduction a relation between real or between reconstructed theories, and if the latter, how much reconstruction is appropriate (Ruse, 1971; Hull, 1974)? Or is reduction best construed as a relation between theories at all (Maull, 1974; Wimsatt, 1976a)?
2. Is reduction primarily connected with theory succession, with theoretical explanation, or with both (Nickles, 1973; Wimsatt, 1976a)?
3. Is translatability *in principle* sufficient, or must we have the translations in hand, and if the former, how do we judge the possibility of translation when we don't have one?[1]
4. What is the point of defending the formal model of reduction if it doesn't actually happen (Hull, 1974; Ruse, 1971), or if the defense has the consequence that if reductions occur, they are trivial and uninformative (Hull, 1974), or merely incidental consequences of the purposeful activity of the scientist *qua* scientist in devising explanations (Schaffner, 1974b)?
5. At least in biology, most scientists see their work as explaining types of phenomena by discovering mechanisms, rather than explaining theories by deriving them from or reducing them to

other theories, and *this* is seen by them as reduction, or as integrally tied to it.[2]

6. None of the philosophers currently writing on this topic are suggesting inadequacies in the kinds[3] of mechanisms postulated by molecular geneticists for the explanation of more macroscopic genetic phenomena.
7. Nonetheless, Ruse (in 1971, though no longer) and Hull (1974) seem to suggest that there is no reduction (only a replacement), and Schaffner (1974b) suggests that a reduction is occurring, but is a merely incidental consequence of the activity of these scientists.

What possibly can explain this wide disagreement between scientists who appear to take reductive explanation seriously and to regard it as perhaps *the* important consequence of their work, and philosophers who are attempting to faithfully characterize their activity and its rationale? Can reduction be as unimportant (or nonexistent) in science as these philosophers seem to suggest? I think the answer must be "no," and that there are four main factors that are responsible for the present philosophical confusion on this point:

1. Philosophers have taken the "linguistic turn" and talk about relations between linguistic entities, whereas biologists are more frequently unabashed (or sometimes abashed) realists, and talk about mechanisms, causal relations, and phenomena. Though not necessarily vicious, I think that the linguistic move has lead philosophers astray. Here I defend a realistic account of reduction.[4]
2. While virtually everyone agrees that a philosopher, by the nature of his task, must be interested in doing some rational reconstruction, doing so serves different ends in different contexts. A failure to distinguish these ends and how they may be served contributes to the apparent defensibility of the formal model of reduction.
3. No real competitor to the formalistic (or more generally structuralist) account of reduction has been forthcoming. Therefore, there has been a tendency to regard *informal reductions* [5] as either nonreductions or as *deficient* reductions, which can be remedied by becoming formalized. I will outline some aspects of a functional account of reduction that suggests that informal re-

ductions are the proper end of scientific analyses aiming at reductive explanations.

4. An emphasis on structural (deductive, formal, logical) similarities has led to a lumping of cases of theory succession with cases of theoretical explanation, with the result that discussions of reduction, replacement, identification, and explanation (which have radically different significances in the two contexts) have become thoroughly muddled.[6] A functional account of these activities yields important clarifications of their nature.

I wish to say something more about item 2, before turning to my analysis of reduction, which concerns primarily item 3 and 4. The first point enters mainly by implication.

Two Kinds of Rational Reconstruction

There are at least two (and probably more) contexts where talk of rational reconstruction seems appropriate in connection with plausible and useful activities of philosophers of science.

Rational$_1$: An Optimal Strategy

One might want to abstract from the often-irrelevant details and sometimes mistaken moves of the actual practice of science to reconstruct the significant patterns of scientific activity.[7] Insofar as these patterns can be claimed to be a relatively efficient, or even an *optimal,* way of achieving or trying to achieve the ends of such activity, the reconstruction could claim to be a rational reconstruction in the sense of rational decision theory—that it represented the way one ought to do that activity. As such the philosopher of science is a *therapist with respect to scientific strategy.*

Rational$_2$: A Canon of Logical Rigor

A physicist (and nowadays with increasing frequency, a biologist) might ask a mathematician for formal help. He might wish to prove a mathematical conjecture whose truth or falsity he is uncertain of and that has important implications for his work. Or he may have an argument that he can formulate more informally, but desires more rigor either to buttress the argument or to determine more precisely the conditions under which it holds. As such, a mathematician is a *therapist with respect to formal argument,* logic, and critical thinking. These are also roles that could legitimately and usefully be played by a philosopher of science.

In either case the philosopher of science would be analyzing or criticizing an activity in terms of how well it served the ends of the scientist, and in each case, the activity itself and the analysis of it further these ends.

Note that the functions of the philosopher of science in these two cases are, at least *prima facie*, not equivalent. It is not at all clear that improvements in rigor, per se, are a rational and efficient way to do science—say, for finding explanations—nor even that the ultimate end state of science will be to improve the rigor of theories *that are otherwise adequate* (i.e., after their other problems have been solved). Improvements in rigor are sometimes useful, but not always. Philosophers of science have sometimes talked as if improvement in rigor is a scientific-end-in-itself, but no one here is doing so. I believe that the sort of confirmation and troubleshooting suggested above is the main function of rigorous argument in science, and that rigor is not a scientific-end-in-itself.

One effect of logical empiricism (with its emphasis on the logical structure of laws, theories, explanations, predictions, and experiments) has been to blur—even to obliterate—the distinction between these two senses of *rational reconstruction.* This conflation has had a disastrous effect upon the analysis of reduction, proceeding as it has in terms of the formal model. Schaffner's (1974b) thesis of the "peripherality" of reduction suggests that any successful defense of the formal model would win a pyrrhic victory. In terms of the above distinctions, I would describe this peripherality of the formal model as follows: It is not $rational_1$ to view formal (i.e., $rational_2$) reduction as a scientific-end-in-itself because science then becomes an inefficient and ineffective way of pursuing known scientific ends (such as explanation). And although the formal model of reduction is by definition a $rational_2$ model, it is not even an effective *means* to some end because it is not the answer to a request for formal (i.e., $rational_2$) assistance that anyone has made or would be likely to make! Thus, although early discussions of formal reduction seemed to hold out the hope that it would perform the functions of both kinds of rational criticism, it is my impression that more recent sophisticated discussions (such as Schaffner's) have given up on both claims. But these claims are not peripheral and readily dispensable. They represent one of the major motivations for pursuing either a formalistic or a reductionist strategy in science. If they must be given up, one's claim to be analyzing reduction as that concept is used in science must be suspect.

Paradoxically, if a non-formal (or perhaps, partially formal) account of reduction is allowed, it can be seen to be a rational activity in both senses: It is an efficient (rational_1) way in which to proceed, and it proceeds by using logical instruments for the critical (rational_2) evaluation of theoretical and observational claims. Because it is a partially formal model, the use of formal methods (as discussed by Schaffner and Ruse) is to be expected on this model also, and it derives confirmation from the cases they adduce to support the formal model. It does not require total systematization, however, which has *not* been exemplified in any of the cases they discuss and which formal reduction requires (see, e.g., Schaffner, 1976, p. 614).

How do we get such an alternative to the formal model of reduction? Just as a characterization of logical *structure* (a rational_2 reconstruction) suggests and is suggested by a formal model of reduction, the view of scientific activity as purposive suggests *a functional* analysis and characterization—a rational_1 reconstruction—of reduction. Such an analysis may distinguish between activities having similar structure in some respects,[8] while pointing to and explaining further structural differences that are ignored on the formal approach. Most importantly, a functionalist approach shows why the research aims of the scientist *contribute to* (in the sense of moving in the direction of) fulfilling the aims of the formal model, but are in fact *different from* and even *inconsistent with* actually getting there. Then a stronger version of Schaffner's (1974b) peripherality thesis is justified:

(P1) Not only is progress toward formal reduction incidental, but

(P2) It also seems to be epiphenomenal, since this progress toward formal reduction appears to have no *further* consequences.

(P3) Finally, if (as I believe) getting there is inconsistent with the real aims of science, this "progress" is bound to remain incomplete.

Successional versus Inter-Level Reductions

The functional viewpoint is perhaps best developed by expanding upon and modifying Schaffner's (1967) model, which has many useful features, though the end result will be quite different (see figures 11.1 and

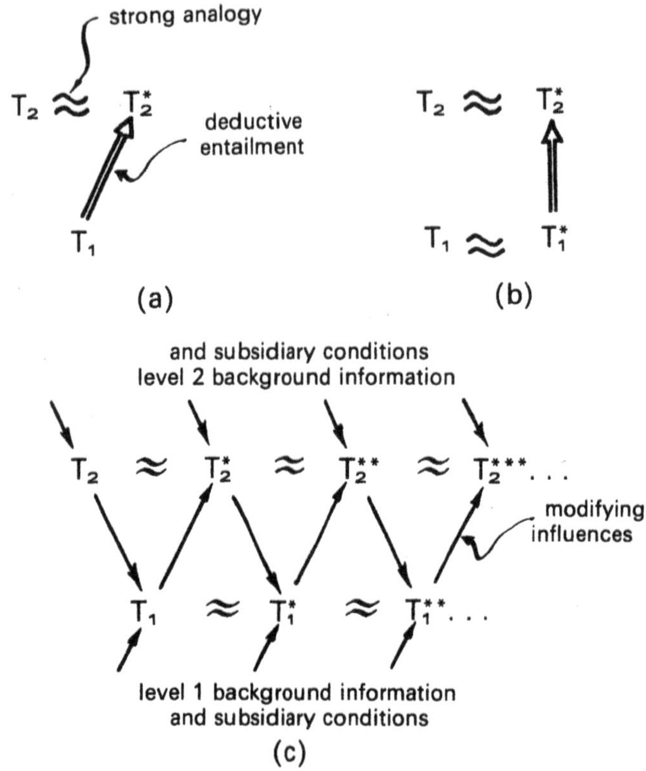

Figure 11.1. (a) Theory reduction, from Schaffner (1967). T_2: reduced theory; T_2^*: corrected reduced theory; T_1: reducing theory. (b) Theory reduction, from Schaffner (1969). T_1^*: modified (corrected?) reducing theory. (c) Co-evolution of theories at different levels, from an early draft of 1976a).

11.3). Most importantly, Schaffner distinguished between and included both a derivability condition between the reducing theory (T_1), and a corrected version of the reduced theory (T_2*), and a condition of strong analogy between T_2* and its uncorrected predecessor, T_2. These two relations are prototypic of two distinct relationships, each of which has been called *reduction.*

Schaffner's condition of strong analogy is closely related to Nickles' "reduction$_2$" (Nickles, 1973, pp. 194ff.) and to what I elsewhere (Wimsatt, 1976a) and below call *successional* or *intra-level* reduction. Nickles' account, emphasizing transformational and possibly non-deductive relations between successive competing theories affords an important partial explication of "strong analogy." A functional account of this activity explains many of the structural features Nickles proposes, and others that he does not mention.

What is not clear on Schaffner's model, but implicit in Nickles' is that reduction$_2$ (which is a kind of "pattern matching" problem and could also be regarded as *demonstrating and analyzing* the "strong analogy" between T_2 and T_2*)[9] is neither automatic nor self-evident. It has a point, involves work, and is performed for reasons separate from the functions of the "other" reductive relation. Nickles suggests that reduction$_2$ performs heuristic and justificatory functions vis-à-vis the uncorrected older T_2.[10]

I believe that reduction$_2$ is fundamentally connected with theory succession (of T_2 by T_2*) and performs rather more functions than Nickles makes out. *It is most immediately a transformational operation whose function is to localize and analyze the similarities and differences between T_2 and T_2** that in turn serve a variety of further functions. Most interestingly, because none of these functions are served by making comparisons other than between T_2* and its immediate predecessor, T_2, and in any case, similarities and differences become *less* localizable as changes accumulate, successional reduction would be expected to be *intransitive,* and to behave as a similarity relation.[11] *Thus the intransitivity of successional reduction is an explicable feature, not a given, on the functional account of this activity.*

For further analysis of the specific uses made of these localized similarities and differences between T_2 and T_2* and diagrammed in Figure 11.2 (see part II of Wimsatt, 1976a). However, the following contrasts between "successional" and "explanatory" reductions are noted here.

1. *Successional reduction is* and must be *a relation between theories* (since it is these that exhibit the similarities and differences), unlike *explanatory reduction* which *is not,* in any but degenerately simple cases.

2. *Replacement* occurs only with *the failure* of successional reduction—failure to localize similarities and differences among successive competing theories. Replacement and successional reduction are opposites. But for explanatory reductions, replaceability is closer to and is by many treated as a *synonym* for reducibility. A failure of T_1 to reduce T_2 (perhaps derivatively, by reducing T_2*) would make T_2 and its successors *emergent* and *irreplaceable* relative to T_1. *Replacement obviously has two different meanings here.*

3. *Successional reductions are intransitive.* A number of them "add up" to a replacement. *Explanatory reductions are transitive.* (It is this last fact that raised the hopes among advocates of "unity of science" for

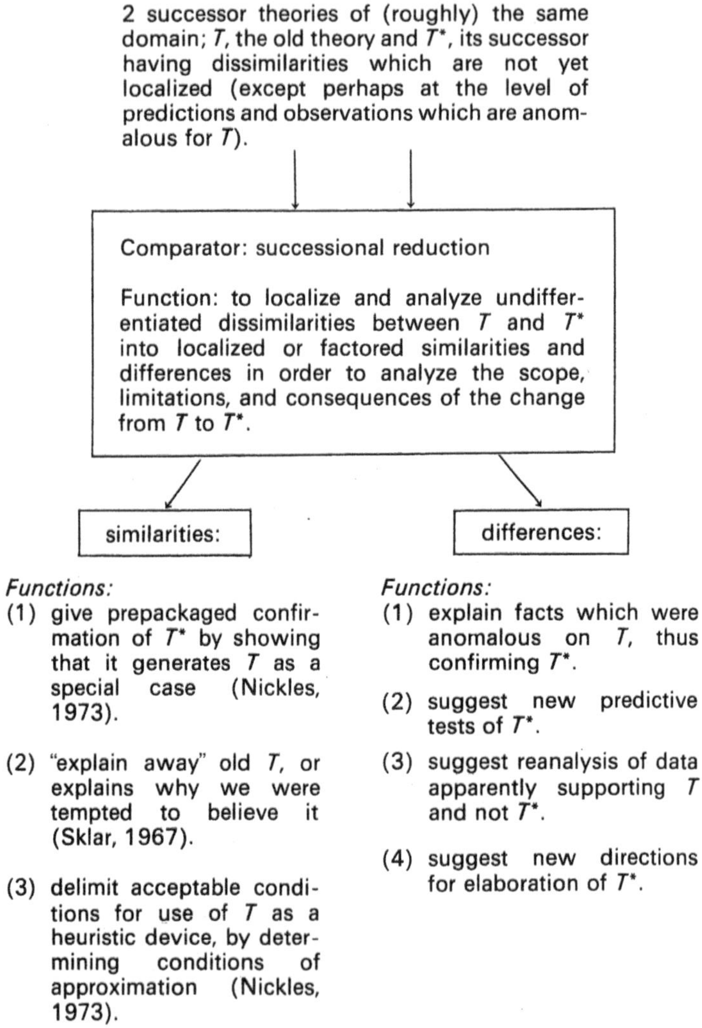

Figure 11.2. Functions of similarities and differences in successional reduction.

Nickles (1973) also suggested that "reductions$_2$" may be done in various ways. This makes sense if the point of the transformation is how best to factor out similarities and differences.

Successional reductions may be possible "locally" (for parts of theories) even when not possible globally (for the whole theory).

Differences in meaning among key terms may be regarded as irrelevant, so long as they are localizable to allow fixing praise and blame on specific components of T and T^* in comparatively evaluating them. See also Glymour (1975). Thus, the "meaning change" objection is avoidable.

great ontological economies through reduction, about which I have more to say below. See also Wimsatt, 1976a.)

4. Talk about elimination might be appropriate for the posited entities of corrected and replaced theories if the new theory is sufficiently different that there is no significant continuity between old and new entities. But such talk is frequently illegitimately extended to contexts of *explanatory reduction.* This is often motivated by talk of ontological or postulational simplicity in the light of supposed translatability and deducibility (discussed further below), but in at least some cases looks suspiciously like treating reduction and replacement as opposites. Thus, in arguing that the formal model of reduction doesn't fit the relation of Mendelian to molecular genetics, Hull and Ruse[12] each suggest that it looks more like a case of replacement. As I suggested in item 2 above, the opposition between reduction and replacement is appropriate for successional reduction, but *not* for interlevel or explanatory reduction. Their claim is thus misplaced if it concerns the relation between T_1 and T_2. Though intelligible if construed as concerning the relation between T_2 and T_2*, I would disagree on the facts of the case, and agree with Schaffner (1976) and Ruse's (1976) more recent view that there is no replacement, but a reduction. To explain why, I must say a great deal more about explanatory reductions to which I now turn.

Levels of Organization and the Co-Evolution and Development of Inter-Level Theories

Rather than talking directly about reductive relations between theories, the approach I have taken (Wimsatt, 1976a) is the realistic one of regarding levels of organization—features of the world—as primary, and defined in such a way that it is natural that theories should be about entities at these levels of organization. The notion of a level implies a partial ordering, such that higher level entities are composed of lower level entities. In a universe where reductionism is a good research strategy, the properties of higher-level entities are predominantly best explained in terms of the properties and interrelations of lower-level entities.

But I argue further that levels of organization are primarily characterized as local maxima of regularity and predictability in the phase space of different modes of organization of matter. Given this, selection forces (and at lower levels, the stability considerations into which these shade) suggest that the majority of readily definable entities will be found in the (phase space) neighborhood of levels of organization, and that the simplest and most powerful theories will be about entities at these levels.[13]

Nothing in this approach entails that levels defined as local maxima of regularity and predictability must always be well-defined and delineated, or strictly linearly orderable (although they usually are for simpler systems), and certain conditions can be suggested (in *this* world) where these assumptions are false (see Chapter 9 and Wimsatt, 1976a, part III). These are conditions where neat composition relations cannot be specified for all (or perhaps even for any) of the entities in these different *perspectives*. (Level talk *requires* the possibility of specifying composition relations, so I talk about perspectives when this condition is not met.) This failure of orderability leads to the "intertwining" of theories mentioned by Schaffner (1974b) in discussing the operon model (see also Schaffner, 1974a) in support of his thesis of the "peripherality of reduction," and to the much more extreme situation suggested by Maull (1974) in her penetrating analysis of the same case—which she sees as the development of an inter–(multi-)level theory rather than the tying or merging together of preexisting theories.

These sorts of complexities are ignored in discussions of the standard model of reduction, and Hull's (1974) discussions of the difficulties of translation just begin to characterize one of their major effects. Nor is this problem limited to genetics. Fodor's (1974) discussion supports the view that the standard model is of substantially greater scope and provides a careful analysis of problems that arise for the standard ("type reduction") account of reduction in these areas. But the standard model just looks so right that it is hard to see how it *could* be wrong. In this light, claims like those of Hull and Fodor seem almost counterintuitive, and it becomes easy to give them short shrift. There are several sources of bias in favor of the standard model that contribute to this appearance:

1. There is a general tendency to characterize the lower-level theory (T_1) as "more general" and "more explanatory" than the upper-level theories (T_2 and T_2^*), trading on our general reductionist prejudices in favor of using compositional information (rather than, e.g., contextual information) in an explanation. This has complex sources that I have discussed elsewhere (Wimsatt, 1976a), and has as one of its effects the tendency to assume that lower-level theories correct upper-level theories, but not conversely.[14]

2. Another important source of bias leading to this error is the distinction between contexts of justification and contexts of discovery, and the attention paid to the former at the expense of the latter. We pri-

marily worry about justifying edifices—theoretical structures that have already undergone substantial revision and selection, and that we have begun to presuppose in a variety of other areas and are thus loath to revise in any substantial way. We discover and propose models tentatively and usually without much commitment. We give them up or modify them easily because little else depends upon it. For reductions (or at least for those that look much like they will come close to satisfying the formal model), the lower-level theory is already well into the edifice stage, and it is thus not surprising that lower-level corrections are less visible, having for the most part already occurred (this is entrenchment, in the sense of Chapter 7).

3. Another bias toward the standard model is introduced via the view that explanations involve giving laws, rather than citing causal factors or giving causal mechanisms. How this is introduced (laws suggest greater systematization than do causal factors) and avoided (by accepted Salmon's account [1971] of statistical explanation) is discussed below.

4. Discussions of translatability tend to revolve around those cases where it looks easiest to give a translation, and it is often easier for properties than for objects (which are characterized by a variety of theoretically relevant properties if they are important objects). It is easier for objects if they are not functionally defined (or are fallaciously *treated* as if they were not) since function makes features of the *context* highly relevant. (As linguists know, a context-dependent translation is an incomplete translation.) Functionally defined processes can be the most difficult, since they will often be associated with a number of objects that will also be involved in *other* functional processes (see Chapter 9), and can be realized in different ways. This is the domain where the functional localization errors induced by the aggregativity biases discussed in Chapter 12 have the largest effects.

Discussions of reduction in genetics have not even approached the translation of some of these terms. Terms from population genetics like "heterosis," "additive (multiplicative, non-additive, non-multiplicative) interactions in fitness," and Lewontin's "coupling coefficient" (1974, p. 294), represent things we look for and find mechanisms for, but general or context-independent translations at a molecular level seem absurd—both impossible and pointless. Context-dependent translations are easy to come by, of course. Discovering the mechanisms in specific cases *gives us* that. But that won't do for the formal model: for those purposes a *context-dependent translation is not a translation.*

What would a new view of inter-level reduction look like?

Schaffner's (1969) modifications to T_1 in order to affect the reduction (Figure 11.1b) is a step toward the picture I would draw: *Theoretical conceptions of entities at different levels coevolve and are mutually elaborated* (particularly at places where they "touch"—where we come closest to having inter-level translations)[15] *under the pressure of one another and "outside" influences* (see Figure 11.1c). In this picture, both successional reductions (or replacements) and explanatory reductions are occurring in an intricately interwoven fashion. Very roughly, all corrections in theory get packed into a "successional" component (because Leibniz's Law applied to inter-level identities ferrets them out of the other component), and all unfalsified explanatory and compositional statements get packed into the "explanatory reduction" component. Theory at different levels progresses by piecemeal modification, in a manner paradigmatically exemplified by Maull's discussion of the operon theory (1974, chapter 2, and 1976) (see Figure 11.3 for the following discussion).

Three things should be noticed about these modifications.

1. Their form may well be deductive or quasi-deductive in character, but if so, the arguments are usually both enthymematic and riddled with *ceteris paribus* assumptions. Typically, it is decided that a T_1-level mechanism cannot accommodate a T_2-level phenomenon without modification to T_1^*, in which case inferential failure of T_1 is the source of the change; or from T_1 and appropriate boundary conditions, we infer, predict, or deduce that a phenomenon that is incompatible with T_2, but not with a T_2^* and observed results should occur, in which case an inferential success of T_1 and its associated mechanisms is the source of the change.

2. The modification occurs without a total deductive systematization, or often even an informal recodification of the theories. The new theories are characterized in terms of the changes from the preceding theories, but because they were similarly characterized there is hardly ever a thorough systematization.

3. The important difference of this picture from Schaffner's is that it is primarily the *changes* in theories that result from deductive arguments. Seldom if ever is any even sizeable fragment of a theory deduced wholesale from another, and seldom if ever is even a single theory sufficiently systematized to meet the conditions for applying the formal model. Furthermore, it is so clearly unnecessary and irrelevant to the search for explanations.

Schaffner's own accounts (1974a, 1974b) and that of Maull (1974) are beautiful confirmation of this highly efficient but formally highly

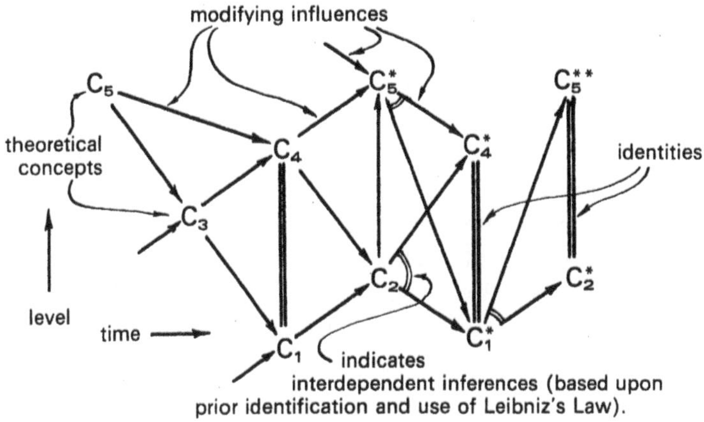

(a)

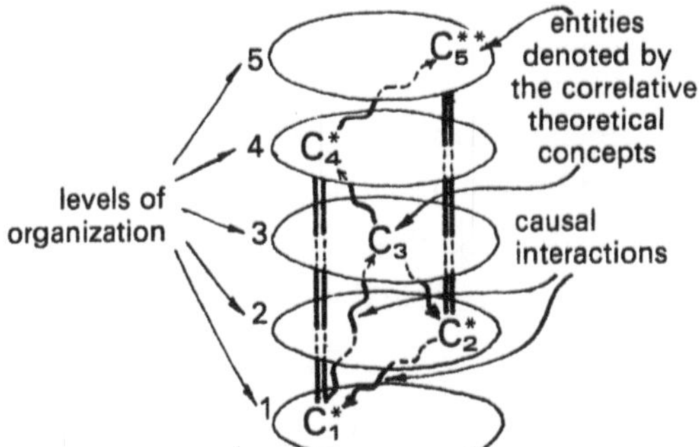

(b)

Figure 11.3. An extension of the model of Maull (1974) involving the use of identities as proposed in Wimsatt (1976a) in the co-evolution of concepts in the development of an inter-level theory of the operation of a causal mechanism. Strong analogy between concepts and their descendants ($C^{n\,*}_{m}$, C^{n+1*}_{m}) is assumed generally (but not necessarily universally) to hold, but is not represented here, in order to simplify the diagram.

(a) Inference structure of the development of the theory. (b) Resultant causal structure of the mechanism according to the theory; development of an inter-level theory.

confusing strategy of theory evolution. These suggest that the vertical arrows *not* be interpreted as total entailments between theories (or reductions, where upwards arrows are concerned), but as single rough deductions or inferences from attempts to match the structure of causal mechanisms as described at different levels resulting in changes in *parts* of theories. There is, to be sure, use of deductive argument, and lower-level explanation of upper-level phenomena. The examples of Ruse (1976; hemoglobin and sickle cell anemia), Maull (1974), and Schaffner (1974a, 1974b) are marvelous. But as Hull (1974) points out, they do *not* touch the issue of whether a total deductive systematization is occurring since such cases would also be expected on the view of reduction advanced here. So, then, why should one bother to attempt to characterize reduction along the lines of the formal model? There just seems to be too big a gap between principle and practice for the principle to be very interesting.

Aside from philosophical predilections of an eliminative sort, there seem to be two reasons for holding onto the formal model of reduction:

1. The belief that as the fit gets better between upper-and lower-level theories, their relationship asymptotically approaches the conditions of the formal model of reduction.
2. The belief that even if the fit never asymptotes, or if it does, doesn't converge on the formal model, the latter represents an aim of scientists.

While Schaffner (1974b) has questioned whether trying to accomplish the reductionist program per se is a good scientific strategy, I suspect that he (and perhaps many scientists) believe that it is at least a secret hope or end. I want to examine the grounds for this latter belief, and suggest an alternative interpretation that is more consistent with scientists' actual behavior. This interpretation also raises serious questions about the first assumption.

Finally, the formal model would not be nearly as tempting if there was not, for each philosopher talking about "translating away" upper-level vocabulary, a scientist talking about "analyzing away" upper-level entities. It thus looks as if a claim about words can be "cashed in" for a claim about entities; a claim that many scientists appear to accept. This claim will be analyzed from another perspective in the next chapter—what does it mean to say that a property of the whole is "nothing more than" properties of the parts?

So the formal model appears to have direct support in the talk of many scientists of the "nothing more than" persuasion. But of what are they persuaded? Are the translations or analyses like those promised by Schaffner *immediately* forthcoming? Usually not. No one actually ground them all out, but that's said to be just a practical difficulty. It is *in principle* possible. But in principle claims have been failing, only to be replaced by new ones, since the time of Democritus. Given their history, such in principle claims could not plausibly be treated as self-warranting. But then what else warrants them? How can we evaluate these in principle claims, to distinguish good ones from bad ones? Or perhaps these in principle claims are not the claims they seem to be, to knowledge the claimant cannot have. I suggest rather that they are important tools in the task of looking for explanations. Before discussing this, I must consider the views about *explanation.*

Two Views of Explanation: Major Factors and Mechanisms versus Laws and Deductive Completeness

I accept Salmon's (1971) account of explanation as a successful search for "statistically relevant" partitions of the reference class of the event being explained, with two provisos. First, I will make some modifications (explained below and in the appendix) to bring it into line with a view of science as an activity conducted according to cost-benefit considerations. Second, I assume that in finding statistically relevant partitions, we are doing so with the aim of partitioning the reference class into *kinds of mechanisms,* or kinds of cases involving a given mechanism (I am thus giving a realist interpretation to his model). In a *reductive* explanation, these mechanisms or factors are at a lower level of organization than that of the phenomenon being explained.

One of the intriguing features of Salmon's account is his move from constructing (statistical) *laws* to a search for statistically relevant *factors.* Laws suggest the need for a complete account of the conditions under which they apply and are correct, and the connection of explanation with laws thus naturally suggests the sort of exhaustive search for factors and conditions that would go along with a complete translation of terms or a complete deductive reduction. By contrast, a search for factors (especially a search for the *major* factors—enter cost-benefit considerations!) ties in naturally with a view of explanation as a search for the mechanisms that produce a given phenomenon, and as an account of how they do it. This search stops short of an exhaustive deductive account by sticking much of the initial and boundary conditions and many background assumptions into a *ceteris paribus* qualifier on the explanation *because they are too unimportant or insufficiently general to be accounted as part of the "mechanism."*

The deductivist or formal account *can* give superficial recognition to such differences of importance by different labeling (laws, boundary conditions, initial conditions, etc.) of different parts of the deductive basis. However, in looking first for a valid deduction, the formal account treats all such information as if it were fundamentally alike because it is all equally necessary for the deduction to go through. It thus rides roughshod over realistic intuitions as to differences in the roles and importance of these different kinds of information. Hull (1974) is sensitive to this in arguing that a single molecular mechanism can lead to different Mendelian traits, for which he has been criticized by Ruse (1976) and Schaffner (1976). Neither Hull nor I nor the scientists who would agree with us are anti-reductionists or anti-determinists. We are simply responding to widespread and reproducible intuitions as to when a change in the total state-description is counted as a change in the mechanism, and when it is not.

This judgment and its reproducibility are explicable on a combination of realistic, evolutionary, and cost-benefit considerations about the nature of scientific theorizing: A mechanism is a "kind," and cost-benefit considerations on the complexity of the theory introduce a "crossover point" beyond which a phenomenon or state is too infrequent or unimportant in a theory to be reified as a kind. There will thus be cases involving the same mechanism with different outcomes that will be attributed to differences in the (more variable and less central) initial or boundary conditions, or to violation of the nebulous *ceteris paribus* clause.

The deductivist also makes and must make such judgments of relative importance, but the baggage of having to construct a valid deduction and of having to treat the correspondences between lower and upper levels as "translations" leads to dangerous misdescriptions of what is going on in several respects.

1. It is only too easy to assume that variations in the boundary conditions are predictively negligible because they are treated as of negligible or lesser general explanatory importance. A failure to include them as part of the mechanism as Hull (1974) has done indicates the latter, but in no way implies either that the same mechanism always produces the same output, or that this failure indicates that the same

total state of the system is on different occasions yielding different outcomes. These are mistaken interpretations that become tempting when Hull's discussion of mechanisms is read as if it were about state-descriptions, and when the only differences of importance are assumed to be differences of deducibility or predictability.

2. Schaffner's claim (1976, pp. 624–625) that Hull's discussion of mechanisms misconstrues the logic of the formal model is double-edged. He would in effect substitute talk about state descriptions. But if the scientists are interested in *mechanisms* and Hull's point is defensible in terms of the way we investigate and reason about mechanisms (as I think are so), of what relevance is Schaffner's probably correct claim that the formal model is defensible if we translate from talk about mechanisms to talk about state descriptions? If scientists aren't interested in state descriptions, Schaffner has apparently defended the formal correctness of his model at the cost of showing its irrelevance to how scientists talk and reason about reduction. Schaffner's claim about the peripherality of reduction begins to look more and more as if it applies more modestly and correctly to *the formal model of reduction.*

3. An equally dangerous move accompanies Schaffner's account of the relation between micro- and macro-descriptions as "translation." Schaffner (1976, p. 630n 25) *assumes* the constancy of the environment and unstated initial and boundary conditions over a range of different cases in constructing his "translation" for the dominance relation. This is done "for reasons of simplicity and logical clarity" (ibid.). But while this is an appropriate defense of simplifying assumptions in a model or idealization, it is not an appropriate move in defense of a "translation" that is to be used in the way that his are. Thus *one thing his assumption does is to mask the real context-dependence of his "translation" by artificially assuming that the context in constant!* But if one is trying to establish that context-independent translations can be given (a necessary move if one is to use these translations as general premises in a deduction over a range of cases in which the context changes), this move is to beg the question; it is to hide deductive incompleteness by trading it for translational incorrectness or equivocation. Schaffner *cannot* do so (see Schaffner, 1976, pp. 622–623).

Schaffner would not assume this constancy if it were admitted or discovered that there were an important variable (or "part of the mechanism") contained in that set of things assumed constant. He would then attempt to delineate that variable, and include it in the translation. Thus the boundary between what is in the translation and what is assumed constant is fixed by the same judgments of importance used in delineating "mechanism" from "background" on the model that I (and I believe Hull) would defend. But what is not in the translation (or mechanism) is not thereby constant. It is quite variable, in fact, and *its very variability is one of the reasons for not including a detailed specification for it in the general theoretical account.* Its variability makes it unimportant for theory construction, and often for selection as well,[16] though it can often produce divergent predictive results and frustrate attempts at translation.

Although Salmon is probably not considered to be a scientific realist, his account of scientific explanation is a natural ally of realistic accounts of science because of its natural structural affinities for such explanations in terms of major factors and mechanisms, in general, and lower-level mechanisms in the case of reductive explanations (see Shimony, 1971; Boyd, 1973, 1980; Campbell, 1974a, 1974b; Wimsatt, 1976a).[17]

Levels of Organization and Explanatory Costs and Benefits

Suppose that the primary aim of science and of inter-level reduction is explanation. We wish to be able to explain every phenomenon under every informative description by showing, first if possible, how it is a product of causal interactions at its own level, but barring that, how it is a product of causal interactions at lower levels (a micro-level or reductive explanation), or least probably and desirably in our reductionist conceptual scheme (but absolutely unavoidably in a world of evolution driven by selection processes), how it is a product of causal interactions at higher levels (most commonly, a functional explanation).

This order of priorities in the search for an explanation follows naturally from the account of levels as local maxima of regularity and predictability, together with acceptance of a weakly but generically reductionist worldview, and the assumption that the *search* for explanatory factors is also conducted according to some sort of efficiency optimizing or cost-benefit considerations. The rationale for this is discussed more fully in Wimsatt (1976a) and is roughly as follows:

1. The characterization of levels of organization as local maxima of regularity and predictability implies that most entities will most probably interact most strongly with (and most phenomena will be most probably explained in terms of) other entities and phenomena at the same level.

2. A reductionist conceptual scheme (or world) is at least one in which when explanations are not forthcoming in terms of other same level entities and phenomena, one is more likely to look for (or find) an explanation in terms of lower-level phenomena and entities than in terms of higher-level phenomena and entities.
3. If a search for explanatory factors is conducted along some such principle as "Look in the most likely place first, and then in other places in the order of their likelihoods of yielding an explanation," then the above order of priorities is established.[18]

Salmon's account (1971) of explanation will be generally presupposed here, but with a cost-benefit clause added to it: not only are "statistically irrelevant" partitions products of a choice of explanatorily irrelevant variables (as he points out), but "statistically negligible" partitions are similarly products of explanatory negligible variables. This change is consonant with the remarks of the preceding section on recognizing the different roles and importance of mechanisms, boundary conditions, and the like in an explanation, but also has important further ramifications. Most crucially the intuitive sense of what it is for one variable to "screen off" another changes (as described in the appendix).

The idea that there can be explanatorily negligible partitions of the reference class of the event or phenomenon being explained suggests an asymmetry of explanatory strategy for cases that do and cases that do not meet macroscopic regularities or laws. When a macro-regularity has relatively few exceptions, redescribing a phenomenon that *meets* the macro-regularity in terms of an *exact* micro-regularity provides no (or negligibly) further explanation. All (or most) of the explanatory power of the lower-level description is "screened off" (Salmon, 1971, p. 55, but see the appendix below) by the success of the macro-regularity. The situation is different, however, for cases that are anomalies for or exceptions to the upper-level regularities. Since an anomaly does not meet the macro-regularity, the macro-regularity *cannot* screen off the micro-level variables. If the class of macro-level cases within which exceptions occur is significantly non-homogeneous when described in micro-level terms, *then* going to a lower-level description can be significantly explanatory, in that it may be possible to find a micro-level description partitioning the cases into exceptional and non-exceptional ones at the macro-level. We would then have a micro-explanation for the deviant phenomenon.

Thus, for example, the ideal gas law (or its corrected phenomenological successor), as a relationship between macroscopic causal factors, is explanation enough for occasions when gases obey it. Going to the micro-level in such a case is not (or negligibly) more explanatory. Of course, if all of the molecules go to one corner of the container, the micro-level must be invoked since the macro-level law does *not* apply, and in that case partitions in terms of micro-variables will be statistically relevant.

So one reason to look for information at lower levels is to explain exceptional cases at the upper level. The other main reason is to explain upper-level regularities. But part of explaining exceptional cases involves explaining why they are exceptional in a way that is consistent with the patterns found in the motley of cases explained by the upper-level law (*qua* set of interrelated causal factors). This usually involves explaining exceptional and motley cases in terms of a single class of mechanisms or micro-variables. This requires that the relevant kinds of micro-descriptions necessary to explain the exceptional cases *also* be usable in generating the upper law as a "special case" or "limiting" or "approximate" description. It thus leads to an explanation of a revised version of the upper-level law.[19]

But what is a law, and why bother to explain one if, as I have argued, mechanisms and major factors bear the primary role in explanations of events that laws have been thought to do? The answer that suggests itself in the cases I have looked at where laws are being explained in terms of lower-level factors and mechanisms is that *laws are regularities involving distributions of cases characterized at the macro level.* They are explained as the product of the interaction of the mechanisms and major factors invoked at the micro-level with the micro-level distributions of initial and boundary conditions. They are not *mere* regularities (or "accidental generalizations" as Nagel [1961] characterizes the infirm statement of law-like form) because they are exhibited as the product of *causal* interactions of micro-level mechanisms, factors, and initial and boundary conditions. Such law-statements thus support the appropriate counterfactual and subjunctive conditionals. Indeed, when a macro-regularity is explained in this manner, an understanding of the micro-level mechanisms and conditions that generate the macro-level distribution and how they do so give a much richer structure of counterfactuals expressible in terms of micro-descriptions than before.

I am not sure whether this characterization of a law is generalizable. It might seem limited to cases where the phenomena of a law admit of meaningful redescription at a lower level. But, at least in those cases where this characterization applies, and this would appear to cover all cases of (inter-level) reductive explanation, a law should be explicable in the same general way as an event. The only difference would be that instead of talking about individual constellations of mechanisms, factors, and conditions, we are talking about assumed *distributions* of the above.

The reduction of thermodynamics to statistical mechanics would provide useful examples of explanations of this sort (see, e.g., the much discussed explanations of the second law of thermodynamics). But so also would the history of the assumption of the "purity of the gametes in the heterozygote" that Hull (1974, 1976) makes much of in arguing that molecular genetics replaces, rather than reduces, Mendelian genetics. I believe that Hull is incorrect in his conclusion, and that an illustration of how this "law" is explained reductively helps us to see how much real continuity there is between Mendelian and molecular genetics.

An Example: The Assumption of "the Purity of the Gametes" in the Heterozygote

This assumption began life as Mendel's (1866/1956) "law of segregation"—to explain the fact that some apparently lost characters ("recessives") reappeared apparently unchanged in successive generations. Mendel's explanation was that in the company of certain alleles ("dominants") the factors did not express themselves as characters, *but that they were transmitted to offspring unchanged (by their allelic factors or anything else)* to express themselves in future genotypes in which they were homozygous or dominant.[20]

In the Mendelism that Castle attacked (Castle and Phillips, 1914), with his belief that the allelic genes "contaminated" one another in the heterozygous state, it was accepted that genes affecting a given character came in pairs (were alleles), but Mendel's other law—of "independent assortment" (that non-allelic genes assorted independently of one another in the offspring)—was being challenged in the early years of the twentieth century, both experimentally and theoretically, by Bateson and others, including Morgan and his students.

The "linear linkage" model of the Morgan school explained some of Castle's results (gradual changes in coat color conformation in rats) by the gradual accumulation through selection of so-called modifier genes at *other* loci (presumably linked on the same chromosome) that modified the *effect* of the genes identified as producing coat color, *without modifying the allelic genes themselves.* There was thus no need to suppose (in this case) that allelic genes "contaminated" one another in the heterozygous state. Castle's supporting claim that these modifications were irreversible was successfully contested experimentally (see Carlson, 1966; Wimsatt, 1992).

The Morgan model supposed that the genes were linearly arranged on chromosomes, with allelic genes on corresponding places on the homologous paired chromosomes. According to this model, homologous chromosomes would, at a certain part of the cell cycle, wind around one another forming "chiasmata," break, and exchange segments. This was called *crossing-over* and *recombination.* A central feature of the model was that genes on the same chromosome would tend to assort together, constituting linkage groups. This was in contradiction to Mendel's law of independent assortment. A prediction of the linear model and the mechanisms of recombination was that the probability of recombination between two points along the chromosome was a monotonic increasing function of the distance between points (being approximately linear for small distances and approaching 50 percent [or random assortment] for large distances).[21] These also were experimentally confirmed. Furthermore, *in the absence of any "atomistic" assumptions* (placing a lower bound on minimum distance between recombinations), *this model would predict a finite frequency for crossing-over within genes of any finite size.*

A gene has a size, and members of the Morgan school recognized this, though different ways of estimating it produced different results (Carlson, 1966, pp. 83, 85, 158ff. reviews this and the other issues of this paragraph; see also Wimsatt, 1992). Although it was usually assumed that the genes behaved like "beads-on-a-string" (or independent atoms) as far as recombination was concerned, Muller, a Morgan student, questioned whether these "atoms" were the same for recombinational and for mutational events. Other observed phenomena (like "position effect") also raised questions about the beads-on-a-string model. It also was generally supposed that genes had an underlying molecular nature, though it was unknown what this was, and how it produced the properties manifested by genes, so the idea that genes had a molecular infrastructure was not new. Indeed, the atomicity of the genes was clearly believed, to the extent that it was only with respect to the genetic or biological properties of the genes.

The details of how the molecular account of the gene explain "position effect" and the possibility of differences between recombinational, functional, and mutational criteria for individuating genes are well known (see, e.g., Hull, 1974, or any modern genetics text) and uncontroversial here. All of these have the effect of compromising the view of genes as monolithic, monadic atoms with respect to some of their biological properties. If there are any "atomic" units of DNA, it is the individual base pair—again not because smaller changes are impossible, but because if they occur, they are not counted as *genetic* changes. While this would show that there were no "atomic genes" of the size Morgan and his school had assumed, and that their different criteria of individuation picked out *different* larger compound assemblages of bases as genes, it is not necessarily a disproof of their genetic "atomism." It could just as well be taken as a demonstration that their atoms were smaller than they had thought (see note 24) and (being of at that time unknown constitution) had some unexpected properties that explained others that the genes had been thought to have.

How does the assumption of the "purity of the genes in the heterozygote" fare? This becomes a question of the possibility of intra-genic recombination—but not a simple question: we must ask not only what happens, but also, what an experiment detects. We can now explain in terms of the design of the recombination experiment why, even if they should occur readily, it was very difficult to find intra-genic cross-overs and recombinations. We can do this in terms of the molecularly characterized gene, *but there is no need to do so.* Morgan could have done so himself, as it is an obvious consequence of the classical model of the genome.

1. On this model, there were a large number of genes on each chromosome. Muller estimated in 1919 that there were at least 500 genes on the X chromosome in *Drosophila,* and we now know that to have been at least a four-fold underestimate.[22]
2. It was taken as a given then as now that any individual gene has a very high stability, which would have applied either to intragenic recombination or to any other mutational event.
3. The design of a recombination experiment involved looking at a small number of marker genes spaced along the chromosome in order to see how frequently they (or more accurately, the traits that signal their presence) stay together in offspring. The usual number of marker genes was two, though Sturtevant occasionally used three and four to detect multiple crossing-over.[23] Supposing even that one could detect any intra-genic recombination occurring in any of the marker genes (see item 4), the very small fraction of the genome being used as marker genes renders it very

probable that recombinational events will not occur in any of the markers, but will occur elsewhere along the chromosome, separating whole the marker genes on either side of the break.

4. We now know that intra-genic recombination would produce a non-functioning gene. This would have been scored by the Morgan school as a "loss" or "mutation" of a gene, rather than as an intra-genic recombination, so they probably did *not* detect any such events that did occur. (Only with later work on intracistronic complementation were the classical techniques sufficiently refined to detect such intra-genic events. But it is worth emphasizing that the problem was a technical one, and not a conceptual one for the classical approach.)

The net effect of this is twofold:

1. The classical model itself predicts that if genes are as small and as numerous as they had to be (and they were smaller and more numerous), intra-genic recombination would be hard or impossible to detect; even if virtually all recombinational events were intra-genic.
2. What was *seen* in recombination experiments was whole (marker) genes separating from one another untouched.

The first fact might have produced caution. It did not. The second observation led to an extrapolated assumption that recombination occurred *between genes, generally,* rather than just *between the observed genes.* But the first fact means that the new molecular picture is *not* that different from the old model. By analogy with the old model:

1. Crossing-over should be a monotonic increasing function of the length of the DNA involved.
2. The probability of crossing-over should be very near 0 for lengths of DNA of the order of functional genes—e.g., cistrons.
3. Individual base pairs, *at least,* still have the "atomistic" status of the bead-like genes of the old model, since crossing-over cannot meaningfully be said to occur within a base.
4. The linear arrangement of the genes on chromosomes (preserved in the linearity of the primary structure of the DNA molecule) is unchanged in the modern account, and plays a central role in accounting for the high stability of the genes, the high reliability of

> the segregation mechanisms (without which genetics would be impossible), and the low frequency of "contamination" in the heterozygote.

But intra-genic recombination is assumed to be possible on the molecular account, and not on the beads-on-a-string model. Does this make the molecular theory a "neo-contaminationist" theory rather than a neo-classical one?[24]

Castle (1919a–1919c) had no well-worked-out mechanism, only a set of experiments that purported to show that classical (pre-Morganian) Mendelism did not work. There was little in Castle's work from which "neo-contaminationists" could claim descent. The purported phenomena of Castle's experiments for "contamination" turned out to be non-existent or to admit of Morganian explanations. His explanations had no important connections with the explanations a molecular neo-contaminationist would give for his neo-contamination phenomena, but Morgan's did. Thus, without a theory, a mechanism, or a set of phenomena persisting through time to call their own, there is no "Castlian genetics," and there are no molecular neo-contaminationists.

The kinds of connections between the two accounts clearly support the claim that the mechanism of the Morganian and molecular theories (especially when looked at with the time and size scale appropriate to the Morganian account—a move appropriate to showing that T_2 and T_2^* are strongly analogous) are indeed strongly analogous. I thus agree with Schaffner and Ruse on this issue.

Indeed, there has been so little change, and what has changed has done so with such continuity that it is tempting not to describe this as a case of successional reduction at all. It is very tempting to say that Morgan's gene *is* the molecular gene, at a different level of description, and conversely. But to make this identification in the same breath with a claim of strong analogy is to invite confusion of identity by descent of concepts in successive theories (which is a similarity relation) with referential identity of different level descriptions of the same object (which is an identity relation). The former notion requires no further attention now, but the latter concept and its role in reductive explanations and analyses is radically different on this account from that suggested by the formal model. Furthermore, the much better fit of this account, of the role and uses of identity hypotheses with actual scientific practice, is one of the strongest arguments for this account and against that of the formal model. (Discussion of changing concepts of the gene has continued through multiple discoveries since. For a more recent review see Beurton, Falk, and Rheinberger, 2000.)

Identificatory Hypotheses as Tools in the Search for Explanations

In its earlier formulations, the classical model of reduction had nothing to say about the role of identifications in reduction. Thus, Nagel (1961) suggested that bridge laws or correspondence rules might be grounded in definitions, conventions, or empirically discovered correlations or hypothesized identifications, as if one was as good as another. The widespread instrumentalism and mistrust of identifications as metaphysical, and as going beyond the evidence, has perhaps led many writers away from asking why scientists might prefer to make one claim rather than another. In the one area where this has been hotly debated (and where postulating identities or postulating correspondences is seen as making a metaphysical difference that bears immediately on matters of importance), philosophers of mind appear to almost universally believe that identity claims are a solely metaphysical and evidentially unsupportable extension beyond the evidence of observable correspondences (see Kim, 1966, for a representative and influential view). Not until the 1970s (see, e.g., Causey, 1972) did philosophers of science find a necessary role for identities in reduction. I wish to suggest an unexplored and absolutely central role for hypothesized identifications as tools in the search for explanations which, among other things, explains a number of features concerning their use that have been considered to be unjustified, unjustifiable, or otherwise anomalous (some aspects of this analysis are discussed more fully in Wimsatt, 1976a, 2006a).

I will assume that we are faced with some upper-level explanatory problem: some phenomenon for which we have no micro-level explanation, or perhaps something that lower-level accounts would lead us to expect at the upper level, but which has not been observed. Such an explanatory failure suggests inaccurate compositional information, or none. How do we discover the source of these inaccuracies, of the locus of our incomplete information? An identity claim, with its subsequent application of Leibniz's Law, provides the most rigorous detector of possible error or of a failure of fit of applicable descriptions at different levels: *Two things are identical if and only if any property of either is a property of the other.* If there are properties apparently had by one but not by the other, then either the identity claim is false (as many are) or else *there are as yet undiscovered translations between descriptions at the different levels* that show that the relevant properties are indeed shared.

Thus, in principle translatability (or analyzability) is a corollary to and the cutting edge of an identity claim. The identity claim is in turn a tool to ferret out the source of explanatory failures which, by its transitivity, allows one to delve an arbitrary number of levels lower if need be to pinpoint the mismatch, or by its scope, to any properties—however diffuse and relational—to detect a relevant but ignored interaction. (For this reason, I do not share the view of some writers that Leibniz's Law should be weakened in all sorts of ways for intensional contexts, and the like.)

Several interesting features follow from this account:

1. It would be expected that identity claims and claims of translatability should be honored more in the breach than in the observance. They function primarily as templates, which help us to locate and to focus upon *relevant* differences—differences that can help us to solve explanatory problems—in order to remove these differences and thereby to make more accurate identity claims. Thus the warrant for claims of in principle translatability, which was questioned earlier, is the same as that for making the identity claim from which it flows.

2. The warrant for this claim is in part the warrant for using a good tool appropriately: that its employment at this time and in this place may help us to discover a description or suggest a redescription that will allow us to explain some heretofore unexplained phenomenon. There is *no* warrant for using the claim if it is *known* to be false. The strength of the claim, which makes it such a sensitive template, renders it easily falsified, and like any strong claim, its negation carries no or little significant information. Thus, if one of the standard defeating conditions for identification, such as causal relation or failure of spatiotemporal coincidence is known to obtain, the claim is dropped, though perhaps in favor of a correspondence claim (Wimsatt, 1976a, part II).

3. This kind of warrant can, however, apply early in the stages of an investigation, and explains behavior that seems irrational and unjustifiable on a more inductivist account of the making of identity claims. Identity claims are often made on the basis of correspondences between or explanations of only two or three properties, often together with some subsidiary background information of a non-correlational nature. This was in fact true for the early identifications, by Boveri (1902) and by Sutton (1903), of Mendel's "factors" with the chromosomes. To the inductivist, this would look like a wildly irresponsible claim: a projec-

tion from two or three properties of a pair of entities to *all* properties of those entities. Moreover, to add insult to injury, the burden of proof after the making of such a claim is not upon its maker (as one would expect on an inductivist account), but upon those who *doubt* the claim to come up with a counter-instance. Only then is the maker obligated to respond to the putative counter-instance, either by elaborating and defending the claim, or by giving it up, as the case seems to demand. Sutton and Boveri proposed a number of new correspondences on the basis of their identifications, and these were later observed, though subsequent conceptual modifications and clarifications led to an elaboration of the identification claims by Morgan and his students, and the generation of many new predicted correspondences (Darden, 1991; Wimsatt, 1976a, 2006a). The early stages at which identities are proposed; the fact that they seem to provide the basis for, rather than be made on the basis of claims of correspondence; and the location of the burden of proof after the making of an identity claim all support this account of the role of identity claims against the inductivist, who should expect the opposite in each case.

4. The fragility and falsifiability of identity claims are hidden by the "open texture" of our concepts (Waismann, 1951), and in more severe cases, by the same tendency to claim identity by descent of our concepts that makes successional reduction possible. With successional reduction, the similarities *and* differences in the successive theories are analyzed critically and used. Only afterwards is the similarity implied by the possibility of performing a successional reduction invoked to maximize the apparent continuity in this identity-by-descent of theoretical concepts. Similarly, with inter-level identifications, the similarities are used critically to ferret out the differences, and only afterwards are the newly assimilated differences reified after the fact into the original identification. The fact that it has become more specific, more detailed, and sometimes has undergone outright changes is hidden from us, so that we see only the continuity of "identity by descent" in our concept of the specific identifications we have made.

5. This analysis suggests that scientists should prefer identity claims to claims of correspondence when there is no specific reason (such as the violation of one of the identity conditions mentioned in item 2 above) to prefer correspondence. They should do so because they prefer the stronger tool, and not for reasons of "ontological simplicity" (or whatever) as suggested by Kim (1966). From a specific identification, after all, one can generate all necessary correspondences, including new

ones that might arise as new properties and relationships are discovered at one level or another. But from the set of correspondences one might derive from an identification given what is known at a given time, one could *not* (without covert reintroduction of the identification) know how to generate new correspondences to fit the new information as it comes in. Identifications are an effective guide to theory elaboration. Correspondences are not. Thus one can understand not only why identity claims might be made early in the course of an investigation, but also why the metaphysically more conservative strategy of making correspondence claims instead will not work. In a static view of science, identity claims and corresponding claims of correspondence only may be empirically indistinguishable. But in a dynamic view of science, only identity claims can effectively move science forward (this is substantially elaborated in Wimsatt, 2006a).

The analysis of reduction and of correlative activities proposed here has differed from most extant analyses in two important respects. First, it has been primarily functional, with the aim of deriving and explaining salient structural features (including some not explained by the standard model) in terms of their functioning in efficiently promoting the aims of science; most notably, explanation. Second, it has aimed at a dynamical account of science, in which optimally efficient change and elaboration are the primary process, and in which stasis is either an artificial construct, a temporary blockage that must be explained, or an end state that we are not likely to reach in the foreseeable future. I believe further that it supports realistic conceptions of the nature of theoretical entities, and of the functions and roles of scientific theory, and does so while being truer to the ways in which scientists *actually* behave than the extant analyses of these activities deriving from the structuralist, static, and often instrumentalist logical empiricist tradition. Finally, it fits into a broader generically evolutionary account of man and his activities, and encourages me to believe that biology may soon be a source for paradigms and analyses that will inform philosophy and philosophy of science generally, rather than being little more than the backwards field for the brushfire skirmish in which philosophical imperialists moving out from the "hard" sciences stop to try their weapons. The latter time is now fast receding into the past, but it is not yet so far that we cannot remember it.

Appendix: Modifications Appropriate to a Cost-Benefit Version of Salmon's Account of Explanation

Salmon (1971, p. 55) defines what it is for one variable to "screen off" another as follows:

> D screens off C from B in reference class A if and only if:
> (i) P(B/A.C.D) = P(B/A.D) [C adds nothing to D.]
> (ii) P(B/A.C.D) ≠ P(B/A.C) [D adds something to C.]

Thus, on this interpretation, microstate description D in statistical thermodynamics *screens off* the macro-state description C from B (a macrostate in accordance with a phenomenological macro-law) in A (a macroscopically characterized assumed-ideal gas). This is so because of those fluctuations from the equilibrium state predictable from D, but not predictable from C, which generates the inequality in (ii).

Note how this definition handles an upper-level anomaly (say, a macroscopically unpredictable fluctuation). Since it would be true that:

(1) $$P(B^*/A.C.D) = P(B^*/A.D)$$

(2) $$P(B^*/A.C.D) \neq P(B^*/A.C)$$

where all is as before except that B* is a macro-state violating phenomenological macro-laws, it is clear that according to the above definition, D screens off B* from C in A.

It is the consequence and intent of Salmon's definition that any strict improvement in information requires saying that the variables generating the improvement screen off any other set of variables that they represent this sort of improvement upon. *This is so no matter how small the improvement and how great the cost resulting from adopting the new set of variables.* It is another consequence of accepting a view of scientific method appropriate to LaPlacean demons.

Scientific practice and good sense suggest the value of a different notion of screening off, which, because of its obvious connections with cost-benefit analysis, might be called the "effectively screens off" relation:

C *effectively screens off* D from B in reference class A if (and perhaps not only if):

(a) $$P(B/A.C.D) = P(B/A.D)$$

(b) $$P(B/A.C.D) \approx P(B/A.C)$$

[D improves the characterization only a little.]

(c) $C(D) >> C(C)$

[D is enormously more expensive information to get than C.]

(c') D is a *compositional redescription* of C.

Some comments are in order about conditions (c) and (c'), which are probably alternatives, or nearly so. The second condition comes closer to capturing the intended application of the effective screening off relationship in the present context, since I am here considering inter-level explanatory reductions, where the lower level is a compositional redescription of the upper level. Furthermore, at least empirically, the truth of (c') appears to guarantee the truth of (c), at least for those kinds of cases we are likely to regard as interesting compositional redescriptions, and thus for all of those cases where we are likely to find any room for debate in the matter of inter-level reduction. Indeed, I am inclined to feel that the proposed "upper level" is not at a distinct level unless at least most of the compositional redescriptions of upper-level phenomena in terms of lower-level entities meet condition (c), which would, in turn, guarantee that any inter-level reduction would be nontrivial.

Condition (c) gives explicitly the cost part of the cost-benefit condition, whereas the approximate equality in (b) guarantees that the benefits, if any, of using redescription D are small. Obviously, the deviation from strict equality in (b) and the cost-ratio in (c) required for the effective screening off relation to hold are interdependent, and are in turn both dependent upon outside factors that determine the importance of additional information and level of acceptable costs. These may vary with the purposes for which the theory is being used, and with any other factors (such as the current explosion in the development of computers and computational facilities) that may radically affect these costs or importances.

The situation where the approximate equality in (b) is in fact an inequality is by far the most interesting one, for *under these circumstances,* D *screens off* C (according to Salmon's definition) *but* C *effectively screens off* D (on my characterization). Thus, in this case, the two criteria would pick out different factors to include in an explanation of phenomenon B.

Condition (a) was also included for the same reason: it is the same as condition (i) in Salmon's definition of the screening-off relation, and thus points directly to a class of cases in which X screens off Y but Y effectively screens off X. Condition (a) would presumably be met in any case in which a successful and total theory reduction (along deductivist lines outlined by Nagel and Schaffner) holds between two theories, such that D is a description imbedded in the reducing theory and C is a description imbedded in the reduced theory. (I would guess that this should be provable as a theorem in the probability calculus from the characteristics of their model of reduction.)

I am not sure, however, how or even whether this result would be provable for reduction as I have characterized that relation. I rather suspect that it is not. Furthermore, in cases where no reduction or only a partial reduction has been accomplished, it would at least be true that condition (a) would not be known to be met for at least some descriptions C in the upper-level theory (and further, that on a subjectivist notion of probability, condition (a) would almost certainly *not* be met for these cases).

In fact, I see no reason why condition (a) should not be dropped for the effective screening-off relation, since conditions (b) and (c)—or (c')—seem to include all that is necessary; namely, the cost-benefit conditions. I have included it for the time being because it heightens the contrast between the screening-off and effective screening off relations, and because I think that substantial further work is necessary to see what if any other modifications and applications seem desirable in developing a cost-benefit model of explanation. The need for at least one further clarification should be immediately obvious: since Salmon (1971, p. 105) points out that his screening-off rule follows from his characterization of explanation, if I believe that the effective screening off relation says something fundamental about the notion of explanation (as I do), it is necessary for me to produce an appropriately modified concept of explanation. This is better left to some future date.

An important consequence of adopting the effective screening off relation rather than the screening off relation was assumed in the text: although upper-level descriptions meeting upper-level laws would effectively screen off lower-level redescriptions, upper-level anomalies—upper-level descriptions that failed to meet upper-level laws—would fail to effectively screen off lower-level redescriptions. This introduced an important asymmetry between cases that met upper-level laws (and which thus were acceptably explained at the upper level) and cases that were upper-level anomalies (and which thus had to be explained at the lower level). On Salmon's screening off relation, there is no asymmetry, since both cases that meet and cases that fail to meet upper-level laws are explained at the lower level, because lower-level variables screen off upper-level variables in either case.

This asymmetry arises in the following way for the effective screening off relation. Suppose as before that B* represents an upper-level description that is anomalous for upper-level theory. Presumably then:

(a) $P(B^*/A.C.D) = P(B^*/A.D)$

(b) $P(B^*/A.C.D) \not\approx P(B^*/A.C)$

The failure of condition (b) occurs because if B* is an anomaly, then P(B*/A.C) must either equal zero, or be very low, and much lower, for example, than the probability of states that are held to be explained by the upper-level theory under similar circumstances. On the other hand, if B* is explicable by an account in terms of lower-level variables, there must exist an appropriate description of B* such that P(B*/A.C) *is* appreciably greater than zero—and in general of the order that similar phenomena held to be explicable on the lower-level theory would exhibit. Thus, the failure of condition (b) means that the benefits of redescribing B* at a lower level are not negligible, and in general justify the greater costs implied by conditions (c) or (c').

CHAPTER TWELVE

Emergence as Non-Aggregativity and the Biases of Reductionisms

Reduction and Emergence

A traditional good reductionist might suppose that emergence is destined to be a thing of the past—with a reductionist explanation for a phenomenon, we have thereby demonstrated that it is not emergent. With the belief that reductionist science will continue to progress, claims that properties are emergent are nothing more than temporary confessions of ignorance. I grew up with this view—it was espoused by both Carl Hempel and Ernest Nagel (see, e.g., Nagel, 1961)—but it conflicts frequently with intuitions we have about when something is emergent, and makes emergence an epistemological, rather than an ontological question. Either this account is simply false, or else there is more than one concept of emergence. At least one of these—a very important one—is consistent with reductionism. An opposition between reduction and emergence forces people to take sides along an axis missing some of the most revealing cuts on the issue (Wimsatt, 1986b). One can be a reductionist and an emergentist too, with a proper understanding of these notions. Misunderstandings engender opposition to reductionism, and make emergence unnecessarily mysterious.

Objections to reductionism may be justified if directed at the right targets, but they often aren't. Claims involving emergent properties in discussions of non-linear dynamics, connectionist modeling, chaos, artificial life, and elsewhere give no support for traditional anti-reductionism or woolly-headed anti-scientism. But reductionists are not blameless either—they often misinterpret the consequences of their success. Emergent phenomena like those discussed here are often subject to surprising and revealing reductionist explanations.[1] But giving such an explanation does not deny their importance or make them any less emergent—quite the contrary: it explains why and how they are important, and ineliminably so.

Philosophers commonly suppose that emergent properties are irreducible, but some rather nice things fall out of a reductive account of emergence. This requires a broader view of reduction (and emergence) than is held by most philosophers, but one that self-proclaimed reductionist scientists would recognize.[2] *A reductive explanation of a behavior or a property of a system is one that shows it to be mechanistically explicable in terms of the properties of and interactions among the parts of the system* (see also Kauffman, 1971).[3] The relevant explanations are causal, but needn't be deductive or involve laws—contrary to conventional wisdom (Wimsatt, 1976b; Cartwright, 1983).

One might usefully classify concepts of emergence in terms of the kind of context-sensitivity they supposed for properties.[4] This account analyzes emergence in terms of dependence of a system property on the arrangements of the parts and, ultimately, on the context-sensitivity of relational parts' properties to intra-systemic conditions. Some accounts of emergence suppose extra-systemic context-sensitivity of system properties.[5] Neither need be anti-reductionist. The latter would require finding a larger embedding system, including the initially extra-systemic properties engaging the broader context-sensitivities. If an adequate mechanistic account could then be provided, a reductionist account in terms of the original system would have failed, but one at a higher level would have succeeded (including the original system together with more variables from its environment).[6]

Such level and context or scope switching—a well-hidden secret of reductionist approaches—can also be a useful strategy to remove model building biases resulting from "perceptual focus" on objects at a preferred level. These biases led to the invisibility of false simplifying assumptions made about the structure of groups in models of group selection (Wimsatt, 1980b). The models started by focusing on genes and individual organisms but in the process made standard simplifying assumptions appropriate for some questions at those levels, but inappropriate for almost any questions at higher levels of organization, *and equivalent to assuming that groups did not exist.* Not surprisingly with these assumptions, group selection was found not to have a significant effect! We must work back and forth between levels to check that features crucial to a phenomenon at an upper level are not simplified out of existence in modeling it at the lower level. On the analysis offered below, we have shown that these group properties are emergent at the higher level, although in a way consistent with a sensitive reductive analysis.

Some philosophers who defend "qualia"-type accounts of subjective experience suppose a third ("mystical") notion of emergence that—if coherent—would be anti-reductionist. I won't consider it here. Finally, many accounts appeal to multiple-realizability as a criterion for emergence, and link it with failures of reduction. I have argued elsewhere (Wimsatt, 1981a, 1994a) that multiple realizability is entailed by the existence of compositional levels of organization, is far broader than often supposed, and is neither mysterious nor contrary to reductionism.

Many cases were classically considered as involving emergence—cases motivating claims that "the whole is more than the sum of the parts"—like an electronic oscillator circuit. There's nothing anti-reductionist, mysterious, or inexplicable about being an oscillator. You can make one by hooking up an inductance, a capacitor, and a resistor in the right way with a voltage source. The system has the property of being an oscillator although none of its parts in isolation exhibit properties at all like this. And it is the way that these disparate parts are strung together that makes them an oscillator. (An oscillator must contain a closed circuit with these components.) A deductive theory relates properties of the parts to the frequency and amplitude of the oscillator. This is a reductionist theory even under the strong conditions of the formal model of reduction, and also one under the weaker characterization given here. This is intuitively a case of emergence, though it can't be if we tie emergence to non-reducibility.

More generally, *emergence of a system property relative to the properties of the parts of that system indicates its dependence on their mode of organization.* It thus presupposes the system's decomposition into parts and their properties, and its dependence is explicated via a mechanistic explanation. Fourteen cases from Table 12.1 fail one or more of the conditions for aggregativity, are consistent with totally reductionist accounts of the systemic phenomena in question, depend in some way on the mode of organization of the parts, and are *prima facie* examples of emergence. Not every counterexample kills an analysis, but too many that are too central do. These surely qualify. We need an analysis of emergence consistent with reductionism.

I now proceed indirectly by focusing on another question: When intuitively is a system "more than the sum of its parts"? This question has been classically connected with discussions of reduction and of emergence, and has stimulated a variety of erroneous but revealing views. How these views arise and may contribute to species of vulgar reductionism—"Nothing-but-isms"—through biases in reductionist problem-solving strategies is discussed at the end. Their pervasiveness is a backhanded tribute to the importance of approximations, idealizations, and limiting case arguments in science. These are powerful tools, but like any tool, they can be misused—we must learn how to use them properly.

Aggregativity

Emergence should involve some kind of organizational interdependence of diverse parts, but there are many possible forms of such organizational interaction, and it is hard to know how to classify them. One can't easily turn examples of the organizational interaction of diverse parts into a general analysis of emergence. It is easier to discuss *failures* of emergence, so I proceed in a backwards fashion (Wimsatt, 1986b) by figuring out what conditions should be met for the system property *not* to be emergent (i.e., for it to be a "mere aggregate" of its parts properties). This has a straightforward, revealing, and compact analysis. Forms of emergence can then be classified and analyzed more systematically by looking at how *these* conditions can *fail* to be met. Examples for different properties and decompositions of the system into parts are given in Table 12.1.

Four conditions seem separately necessary and jointly sufficient for *aggregativity*[7] or non-emergence. Aggregativity and emergence concern the relationship between a property of a system under study, and properties of its parts. For each condition, the system property must remain invariant or stable under modifications of the system in the specified way—a kind of independence of the property over changes in the mode of organization of the parts. This invariance indicates that the system property is not affected by wide variation in relationships between and among the parts and their properties. To be aggregative, the system property would have to depend upon the parts' properties in a very

Table 12.1 Examples of failures of aggregativity, as discussed in Wimsatt, 1986b (with indications of the decompositions assumed and which conditions are met)

	Invariance conditions R=*restricted or qualified (for some decompositions only)*				
Examples	IS	QS	RA	CI	Relevant parts or conditions
1. mass	yes	yes	yes	yes	none known. *Only case of true aggregativity in table.*
2. volume	R	yes	yes	no	not for solvent/solute. or packing diff. solids
3. critical mass of fissionable material. *(ignoring neutron absorbers in context)*	yes	no	yes	no	keep geometrically and density invariant
4. stability of rock pile *(yes for IS if same=exactly similar)*	no/yes	no	no	no	
5. oxygen binding in hemoglobin molecule	R	no	?	no	w/i alpha, beta chains
6. gamete frequency (classical model) *(these two cases differ in unit chosen)*	no	no	no	yes	units are genes
7. gamete frequency (classical model) (w/o disturbing mitosis)	yes	no	no	yes	units are chromosomes
8. gene effects (classical model) *(these two cases reflect a change in theory)*	yes	no	yes	yes	units are genes
9. position effect (molecular model)	no	no	no	?	units are genes
10. amplification ratio in multi-stage linear amplifier	yes	yes	yes	yes	idealization; not found in Nature (Figure 1)
11. amplification ratio in multi-stage non-linear amplifier	no	yes	no?	yes	analogous natural case (Figure 2)
12. solving a nearly decomposable problem	yes	yes	R	yes	units are subproblems; aggregativity usually is partial or qualified.
13. computer memory	yes	yes	R	yes	memory chips having same relevant characteristics

14. fitness (total additivity) *(QS is met if positive alleles are substituted for others at same locus without affecting genome size) (for different groups of genes in same organism)*	yes	no	R	yes	units are genes for a "quantitative trait"
15. fitness (partial additivity-classical)	no?	no	R	no	units are genes w. non-0 heritability of fitness
16. Levins' fine and coarse grained fitness functions	yes	yes	yes	yes	*approximations invalidate claim of aggregativity**

*Appearances are deceiving: conditions of derivation violate conditions *IS*, *RA*, *CI*, even though the composition functions for both the coarse-grained and fine-grained adaptive functions appear to meet all conditions. This situation should be generalizeable to many idealizations, or derivations involving approximations.

Notes:

1. In addition to the advantages of breaking down claims of emergence into these four criteria which can be met in various combinations, thus providing a finer classification of types of emergence, there is a substantial heuristic advantage in using these criteria in the analysis of complex systems. This arises because the criteria are met usually under restricted conditions on the parts decomposition involved, and sometimes with some additional restrictions. These criteria then become heuristic templates for uncovering these conditions, and for finding parts decompositions that realize these criteria to the greatest extent. These decompositions are prime candidates for specifying "natural kinds."

2. Example 16 shows that a purely formal account in terms of conditions on the composition function will not do; one needs also to look at the assumptions made in the derivation of the composition function.

3. See also subaggregativity, neighborhood aggregativity, modular aggregativity, semi-aggregativity, and near-aggregativity in Wimsatt, 1986b, for other kinds of qualified aggregativity.

strongly atomistic manner, under all physically possible decompositions—this last an almost impossibly strong demand. *It is rare indeed that all of these conditions are met* (see Table 12.1). Aggregativity is the complete antithesis of functional organization. Our reductionist science to date has focused disproportionately upon such properties, or properties that do somewhat less—meeting some of these conditions, approximately, some of the time—and studying them under conditions in which they are "well-behaved." Aggregative or even such pseudo-aggregative properties are treated as relatively fundamental (Martinez, 1992). In consequence, their import in the description of the natural world has been substantially exaggerated.

The conditions for aggregativity are: (1) a condition on the intersubstitution or rearrangement of parts; (2) a condition on size scaling (primarily, though not exclusively, for quantitative properties) with addition or subtraction of parts; (3) a condition on invariance under the decomposition and reaggregation of parts; and (4) a linearity condition that there be no cooperative or inhibitory interactions among parts in the production or realization of the system property. These conditions are not entirely independent of one another. In fact, there seem to be close connections between the first and third conditions and between the second and fourth conditions.

For a system property to be an aggregate with respect to a decomposition of the system into parts and their properties, the following four conditions must be met for properties in the following equation:

> Suppose $P(S_i) = F\{[p_1, p_2, \ldots, p_n(s_1)], [p_1, p_2, \ldots, p_n(s_2)], \ldots, [p_1, p_2, \ldots, p_n(s_m),]\}$ is a composition function for system property $P(S_i)$ in terms of parts' properties $p_1, p_2, \ldots, p_n$, of parts $s_1, s_2, \ldots, s_m$. The composition function is an equation, an inter-level synthetic identity, with the lower-level specification a realization or instantiation of the system property.[8]

1. *IS (InterSubstitution):* Invariance of the system property under operations rearranging the parts in the system or interchanging any number of parts with a corresponding numbers of parts from a relevant equivalence class of parts (cf. commutativity of composition function).
2. *QS (Size Scaling):* Qualitative similarity of the system property (identity, or if a quantitative property, differing only in value) under addition or subtraction of parts (cf. recursive generability of a class of composition functions).

3. *RA (Decomposition and ReAggregation):* Invariance of the system property under operations involving decomposition and reaggregation of parts (cf. associativity of composition function).
4. *CI (Linearity):* There are no Cooperative or *I*nhibitory interactions among the parts of the system that affect this property.

Note that conditions *IS* and *RA* are obviously relative to given parts decompositions, as are (less obviously) *QS* and *CI*. A system property may meet these conditions for some decompositions, but not for others. Table 12.1 presents different examples of aggregativity and its failure—species of emergence (many of these examples are discussed in Wimsatt, 1986b).

Figure 12.1 illustrates the first three conditions for the amplification ratio in an (idealized) multi-stage linear amplifier. The system property, total amplification ratio, is the product of the component amplification ratios (Figure 12.1a), a composition function that is commutative (Figure 12.1b, condition *IS*), and associative (Figure 12.1d, condition *RA*), and shows qualitative similarity when adding or subtracting parts (Figure 12.1c, condition *QS*). This example seems to violate the fourth condition (non-linearity), but we act as if it doesn't: geometric rates of increase are treated linearly here (and in other relevant cases) because it is the exponent that is theoretically significant for the properties we are interested in. Subjective volume grows linearly with the exponent (as our decibels scale reflects), and both of them grow linearly with addition of components to the chain. The fourth condition is violated for cooperative interactions in the hemoglobin molecule: the four subunits take up and release oxygen more efficiently by being organized into a tetramer than they would as four independent units. (A monomeric hemoglobin can be found for comparison in the lamprey.) The staged linear amplifiers are interesting because they show that aggregativity does not literally mean "additivity"—here multiplicative relations do equally well.[9] (Exponential growth is also much more common in biology than linear relations.)

The linear amplifier of figures 12.1a–d is example 10 of Table 12.1. To meet all criteria, we must assume that each subamplifier is exactly linear throughout the entire range—from the smallest input to the largest output—required of the entire system. No real-world amplifiers here: this is an idealization, which I will return to below.

Even this simple story has some important limits, however. Amplifiers are themselves integrated functional wholes with differentiated

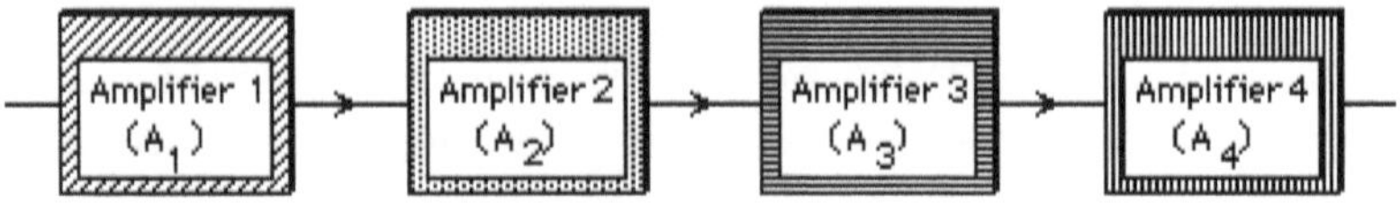

a: Total Amplification Ratio, At, is the product of the amplification ratios of the individual amplifiers: At = A1 × A2 × A3 × A4.

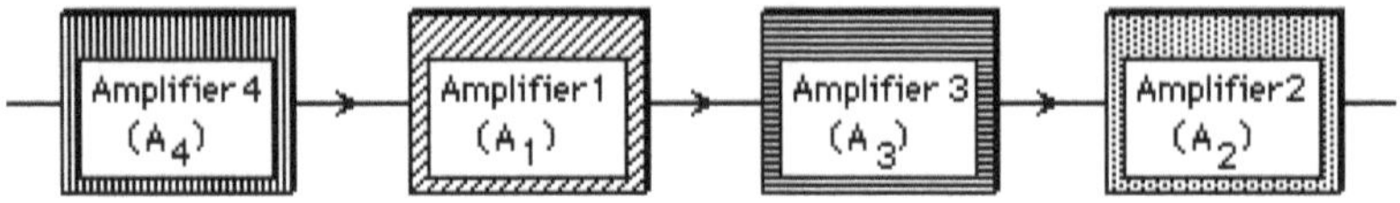

b: Total Amplification ratio, At = A4 × A1 × A3 × A2 remains unchanged over intersubstitutions changing the order of the amplifiers (or commutation of the A's in the composition function).

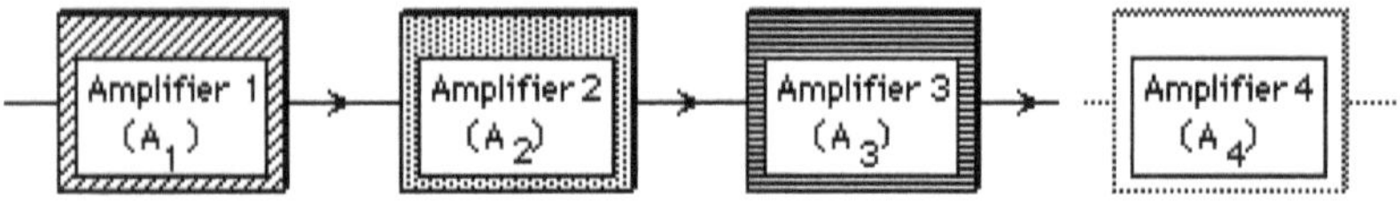

c: Total Amplification Ratio, At(n) = At(n − 1) × A(n), remains qualitatively similar when adding or subtracting parts.

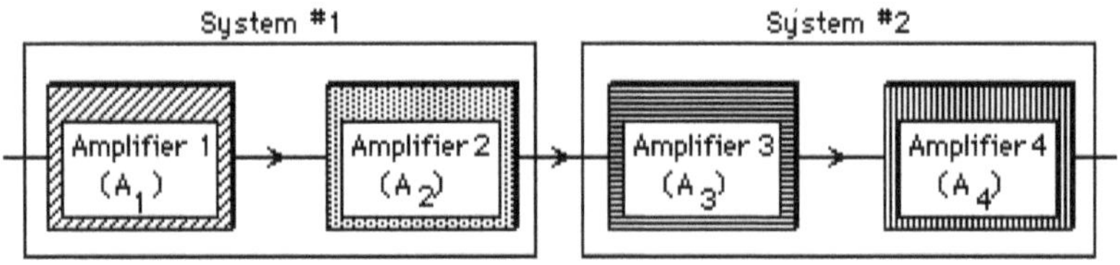

d: Total Amplification Ratio is invariant under subsystem aggregation—it is associative: A1 × A2 × A3 × A4 = (A1 × A2) × (A3 × A4).

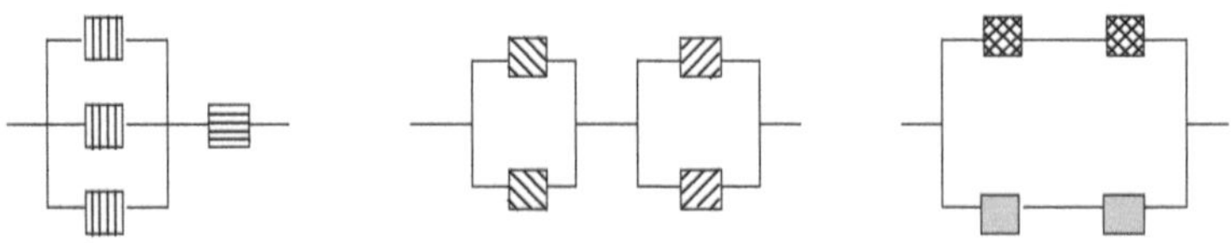

e: The intersubstitutions of a–d which all preserve a strict serial organization of the amplifiers hide the real organization dependence of the Total Amplification Ratio. This can be seen in the rearrangements of 4 components into series-parallel networks. Assume each box in each circuit has a different amplification ratio. Then to preserve the A.R. the boxes can be interchanged only within organizationally defined equivalence classes defined by crosshatch patterns. Interestingly, these classes can often be aggregated as larger components, as in these cases, where whole clusters with similar patterns can be permuted, as long as they are moved as a cluster. (See Wimsatt, 1986b.)

Figure 12.1. Conditions of aggregativity illustrated with idealized linear unbounded amplifiers.

parts—which cannot be permuted with impunity. (That is why we need circuit diagrams to assemble and to understand them—we cannot put them together in just any fashion!) Even the parts are integrated wholes. If you cut randomly through a resistor or capacitor, the pieces do not perform like the original. This is interesting: *testing these conditions against different ways of decomposing the system is revealing of its organization.* This suggests broader uses for the conditions and the analysis. I return to this in later sections.

We aren't done yet: notice that all of the examples so far (figures 12.1a–d) have the amplifiers arranged in series. This is an implicit organizational constraint on the whole system—readily accepted because our common uses of amplifiers connect them in this way. But we could also connect them differently, as in the three series-parallel networks diagrammed in Figure 12.1e, and then the invariances in total amplification ratio, ΣA, disappear. To calculate their amplification ratios, we make three (simplifying) assumptions: (1) branching parallel paths divide currents equally; (2) converging parallels add currents; (3) serial circuits (included aggregated subassemblies) multiply signal strengths (as before). In Figure 12.1e, starting with a signal of magnitude 1 unit and following these rules we get:

For the first circuit, $\Sigma A = [(A1 + A2 + A3)/3] \times A4$.
For the second circuit, $\Sigma A = [(A1 + A2)/2] \times [(A3 + A4)/2]$.
For the third circuit, $\Sigma A = [(A1 \times A3)/2] + [(A2 \times A4)/2]$.

If all component amplifiers, A1, A2, A3, A4 have equal amplification ratios, *a,* the ΣA's of these circuits are all equal. They each have two staged subassemblies, each with a net amplification ratio of *a,* so $\Sigma A = a \times a = a2$. (For these ideal linear amplifiers, parallating at any stage has no effect: the signal is just multiplied by the same amount along each path, so they collectively have the same effect as a single amplifier with the same amplification ratio.) But the decreased amplifying depth reduces ΣA from a^4 to a^2, so changing to any of these modes from a strictly serial circuit decreases the amplification ratio.

But there is worse to come. Suppose that the amplification ratios of the component amplifiers are not equal:

If $A1 = a$, $A2 = 2a$, $A3 = 4a$, and $A4 = 8a$, then
circuit $1 = [(7/3)\ a] \times 8a = 56/3\ a2 = 18.67a2$.
circuit $2 = (3/2)a \times 6a = 9a2$.
circuit $3 = (5/2)a2 + (10/2)a2 = 7.5a2$.

For comparison, all of them in series (as above) would give $64a4$. So how the amplifiers are connected together *does* make a difference. And the differences may be larger still. For example, the circuit on Figure 12.1e would have different amplification ratios along the two parallel arms of the circuit. As John Haugeland notes (personal communication), if this implies a voltage difference between them, it should give a reversed potential difference on the downstream side of the lower amplifying arm. Since amplifiers are not designed for current backflows, what would happen would depend (perhaps quite capriciously) on the hardware design of the amplifier and the magnitude of the voltage difference, and also on the load impedance of the rest of the circuit at that point, a contextual factor. So in yet another way, despite first appearances, amplification ratio is not an aggregative property, both because it is not invariant across decompositions internal to the amplifiers, or internal to their parts, and also because it is not invariant across aggregations of the amplifiers that are not serially organized. (This covers manipulations at three levels of organization: rearrangements of parts of the amplifier parts, rearrangements of parts of the amplifiers, and rearrangements of the amplifiers.)

Finally, we have assumed that each subamplifier is exactly linear throughout the entire range—from the smallest input to the largest output—required of the entire system. The amplifiers must multiply input signals of different frequencies and amplitudes by the same amount over this entire range. This is an idealization—example 10 of Table 12.1. Real-world amplifiers are *approximately* linear through given power and frequency ranges of input signals (see Figure 12.2a and example 11 of Table 12.1). (Frequency correction curves are published so that linearity can be restored by the user by "boosting" different frequencies by different amounts, but these curves are themselves functions of the amplitude of the input signal.) The amplifiers—not perfectly linear to begin with—become increasingly non-linear outside these ranges. They are most commonly limited on the low side by insensitivity to inputs below a certain value, and on the high side by not having enough power to keep the transformation linear. So with real amplifiers the order of the amplifiers *does* matter, even in the serial circuit.

Thirty years ago hi-fi systems had separate pre-amplifiers and amplifiers. Hooked up in the right order, they worked just fine. In the wrong order, the amplifier would be too insensitive to detect the signal from the phonograph cartridge, but small amounts of "white noise" it generated internally would be amplified enough to "fry" the downstream

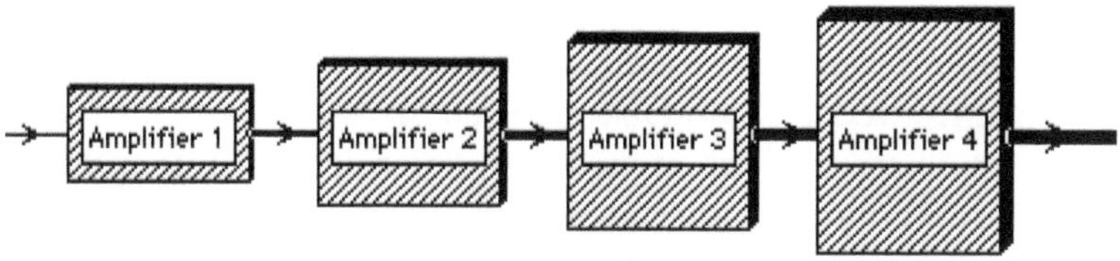

a: Amplifiers with minimum input and maximum output signal strengths, as indicated by their relative heights. Their order is not interchangeable without introducing non-linear distortion.

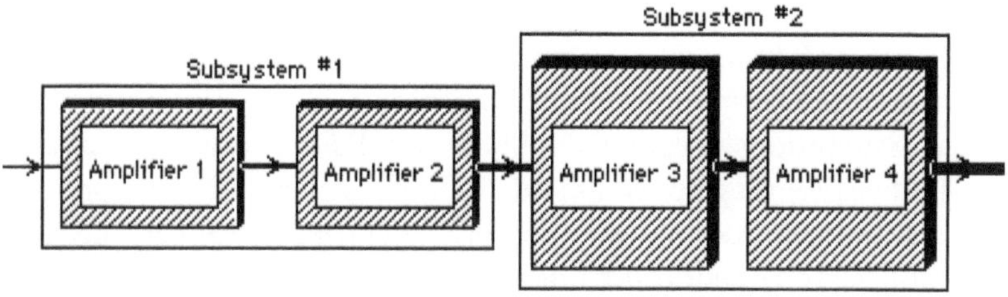

b: Partial or subaggregativity: The amplifiers in each of the two subsystems are interchangeable with each other without introducing deviations from non-linearity in the overall system, but intersubstitutions between subsystems introduces non-linear distortions.

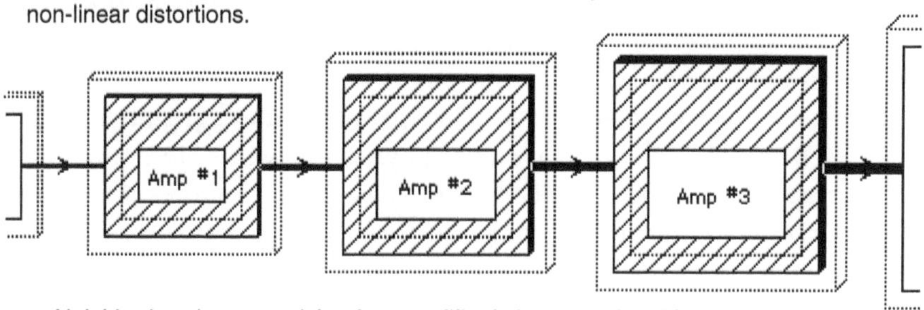

c: Neighborhood aggregativity: Any amplifier is intersubstitutable with its immediate neighbor to the left or to the right, but not further without introducing non-linear distortions. (Range of intersubstitution is indicated by dotted boxes inside and outside of each amplifier.)

Figure 12.2. Varieties of partial aggregativity illustrated with amplifiers that are linear only within a bounded range.

---

pre-amp. So-called linear amplifiers are really mean *approximately linear over a given range, within a specified tolerance,* ϵ. Different uses require different tolerances and ranges, so a system may be treated as linear for some purposes, but not for (more demanding) others (see below for more on the implications of tolerances).

The amplifiers may be linear through a large enough range that there is some adjustability—some restricted intersubstitutability among the components, yielding partial or subaggregativity (Figure 12.2b) or "neighborhood" aggregativity (Figure 12.2c). Restricted kinds of intersubstitutability (see also modular aggregativity, Wimsatt, 1986b) can also help characterize different modes or dimensions of organization in many kinds of complex systems. Similar qualifications for supposedly aggregative properties arise in the discussions below of additive fitness components, and fine versus coarse-grained patterns of environmental change. *Indeed, the appearance of common and unqualified aggregativity is a chimera, and is usually a product of uninspected assumed constancies, idealizations, and overlooked possible dimensions of variation.* The amplifier case shows us how known idealizations and uninspected assumptions (treating amplifiers and their parts as unbreakable modules because they are commonly treated as such in our theory and practice, and considering only serially organized amplifiers because these modes are common for functional reasons) can lead us to exaggerate the aggregativity of a property. *In the analysis of complex system properties, such kinds of errors are so easy to commit that they are almost the rule rather than the exception, contributing to design failures in engineering, modeling errors, and errors of experimental design in science, and conceptual errors in philosophy.*

But some properties at least seem like paradigmatic aggregative properties. The great conservation laws of physics—those of *mass* (example 1), *energy* (now replaced by the hybrid, mass-energy), *momentum,* and *net charge* (if we include its sign)—in effect indicate that these properties actually do fill the bill. They appear aggregative under any and all decompositions. Indeed, that's why there are conservation laws for them! Curiously, some properties you might have expected to don't measure up. Thus, *volume* (Table 12.1, example 2) isn't aggregative for solvent-solute interactions in chemistry. If you dissolve salt in water, the volume of the water+salt will be less than that of the water before the salt was added, and surely less than their volumes taken independently and separately. *(So sometimes the whole is less than the sum of its parts!)*

A system property may be aggregative for some decompositions but not for others, or, more generally, any of the conditions may be met for some decompositions, but not for others. (This is probably the most common situation.)

This fact has critical importance in theory construction, for these variations allow for and suggest feedback between these criteria and the choice of decompositions of a system for further analysis. We tend to look for invariances, and these conditions are treated as desiderata, so we tend in experimenting with alternative descriptions of and manipulations on the system to try to find ways to make them work—decomposing, cutting, pasting, and adjusting until these conditions are satisfied to the greatest degree possible. Furthermore, we will tend to regard decompositions meeting the aggregativity conditions as "natural," because they provide simpler and less context-dependent regularities, theory, and mathematical models involving these aspects of their behavior.[10] This is illustrated in the next section in considering chromosomes versus genes as units of analysis.

Aggregativity is not all that is relevant to "natural kinds." Other central grounds for regarding decompositions and the parts they produce as natural include "robustness" (Levins, 1966; Chapter 4 in this volume), and, more enigmatically, "generative entrenchment" (see Chapter 7). Other heuristics are also used with these conditions in constructing and validating decompositions—see the reductionist problem-solving strategies discussed in Wimsatt (1980b), Chapter 5, and Appendix B. These heuristics have systematic biases, which may give misleading results. One of the most systematic biases in discussions of supposedly aggregate behavior is to generate it under very special conditions or strong constraints on the system and its environment, and then forget these qualifications in subsequent discussions. Quite different kinds of examples of this are discussed in the next two sections.

Perspectival, Contextual, and Representational Complexities; or, "It Ain't Quite So Simple as That!" (An Example from the Genetics of Multi-Locus Systems)

The apparent formality of the four conditions for aggregativity might suggest a direct analysis of properties of theoretical equations. We can't do so without considering at least (1) the choice of a parts-decomposition, (2) idealizations and assumptions made in the *description* of a system, and (3) further idealizations and approximations made in the *derivation of equations* relating system-level and parts-level properties.[11] The first two considerations are discussed here, and the third in the next section. Examples 7 and 8 from Table 12.1 reveal how the apparent aggregativity of a system property depends upon the decomposition used.

Consider a multi-locus genetic system with the genes organized into chromosomes (example 8). A gamete is a haploid (or "half") genotype gotten by taking one or the other of each of the homologous chromosome pairs of its parental genotype. (Sperm and eggs are gametes contributed by males and females, which fuse at fertilization to form zygotes carrying whole genotypes.) The expected frequency of a randomly drawn gamete is assumed to be the product of the frequencies in the population of the different chromosome types that compose it. With random assortment at the level of whole chromosomes, meiotic processes reliably (usually, but not universally) produce gametes with exactly one of each chromosome-type. *This makes it look as if gamete frequencies* (with normal meiosis) *could be an aggregative property of chromosome frequencies.* Assume for now that they are. (They aren't always. I'll return to this below.)

We could also calculate gene frequencies and describe the genotype as a series of genes at each locus in each chromosome of the haploid genotype (example 7). But we *cannot* similarly get the frequency of a randomly drawn gamete as the product of gene frequencies in all of the chromosomes making up a haploid genotype. The linkage of genes into chromosomes means that unless genes are randomly partitioned to start with (they rarely are), they *won't* immediately distribute in this way. Ignoring selection, crossing over and recombination will gradually scramble separate and mix genetic combinations among mating members of the population over successive generations, but not immediately. (Without selection or other biasing forces, gametic frequencies exponentially asymptote to these multiplicative values—linkage equilibrium values—at rates proportional to their linkage distance, a function of their relative locations along the chromosome.)[12]

What does all of this mean? When genes come in chromosomes, *as they do in the real world,* one must take this into account when calculating gametic frequencies.[13] These lead to violations of condition *IS* that cannot be fully understood without recognizing multiple levels of organization: the gene or allele, and the structure of the chromosome, gamete, and haploid/diploid life cycle. In the real world even higher-level conditions on population structure are commonly relevant (Wade, 1996). Many applications of population genetic single-locus models implicitly assume an aggregativity of the gene in the population in making up gametes that simply is not there. Two populations having identical arrays of gene frequencies but different arrays of chromosome frequencies will produce different gamete frequencies (and thus different genotype frequencies and different results of selection) in ways determined most immediately by their chromosome frequencies. *For this reason, decompositions of the genotype into whole chromosomes are actually more aggregative than decompositions into genes.* Chromosomes are recognized *in the theory* as real natural objects via the structure of the relevant equations. These express gametic frequencies in a given generation as functions of recombination frequencies between genetic loci (products of their location in the linkage map) and whole gametic frequencies in the preceding generation. (Ernst Mayr's [1982] charge that population genetics was "beanbag genetics"—viewing organisms as a "bag of genes"—is false. On the more accurate multi-locus theory—which existed when he made this claim—organisms [or genomes] are more like *a can of worms* than a *bag of genes.*)

But this picture is still too simple. There is more structure in this case—more than is commonly recognized in population genetics texts. For some purposes, the gametes are kinds of meta-chromosomes: one inherits *gametic complements of chromosomes,* not chromosomes drawn independently at random from their sets of homologues in the population at large. These are assembled into diploid genotypes, not arbitrarily large polyploid assemblages of chromosomes, from which are drawn the haploid gametic complements of chromosomes and genes of the next generation. *Gametes and diploid genotypes are each assemblages structured via a systematic combination of constraints and random elements, and the results reflect both.* If we start out with non-random statistical associations between chromosomes in individuals (where gametic frequencies are not products of the frequencies of the component chromosomes), then, under random mating, independent assortment of chromosomes *will not* produce these frequencies in the next generation, but will go only half-way there in each successive generation. Failure to reach the maximally mixed state in the next generation (a generalized Hardy-Weinberg multi-locus equilibrium—misleadingly called "linkage equilibrium") reflects structural relationships among chromosomes, gametes, and genotypes—a more complex failure of condition *IS.*

This structure is best viewed in the two-factor Punnett square diagram of Figure 12.3 for two alternative alleles (A, a and B, b) at each of two loci (A and B). These come naturally packaged into the alternative gametic combinations A--B, A--b, a--B, and a--b, contributed by the male (alternatives across the top, invariant down the columns), and female (alternatives down the side, invariant across the rows). Resulting zygotic combinations are found in the squares at the intersection of their generating gametes.

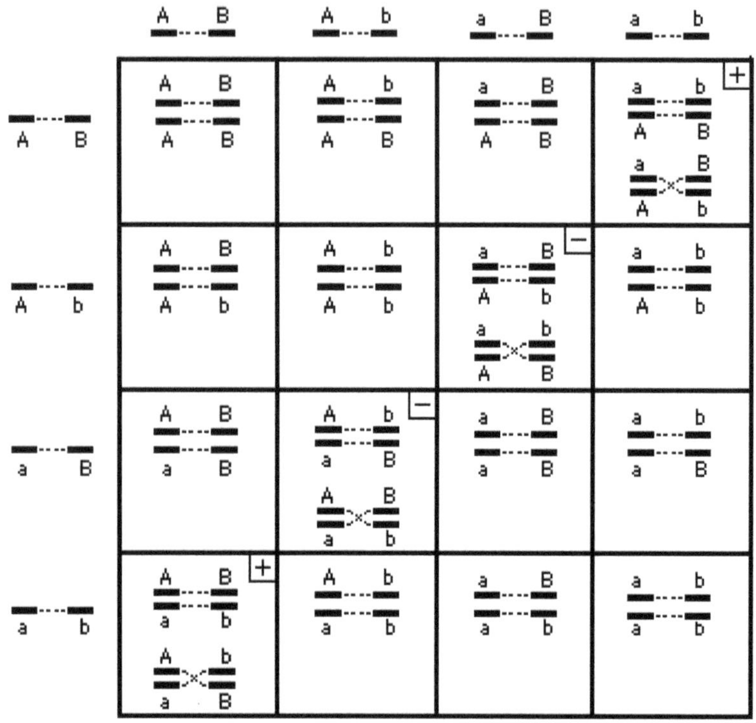

Figure 12.3. Composition of two alleles at two locus genotypes from gametes, and production of new gametes via independent assortment or recombination.

If close together on the same chromosome, genes will show a continuing slowly decaying statistical association, given by the linkage distance r, ranging from 0 to .5 as a function of how frequently recombination events separate them—a property of their relative locations on the chromosome. But if they are either very far apart on a long chromosome, *or are on entirely different chromosomes,* they will still show an identical linkage effect of .5. *The identity of this effect (when r is .5) with the genes far apart on the same chromosome or on different chromosomes shows that this association is not—or not just—a product of chromosomal organization.* It is a property of the genotypic-gametic life cycle, as can also be seen in Figure 12.3.[14]

Consider recombinations first. For this assume that each gamete is a single chromosome, with the A and B loci separated by a dotted region of the chromosome in which crossing over and recombination occurs. Assume equal recombinations (chromosomes line up so that no loci are gained or lost in any reassortments), and occur with equal frequency, r, for each of the 16 possible pairings of gametes. In each square along the reverse diagonal we see the chromosomes before recombination above, and the results after recombination below. Recombinations happen in all squares *but only in the squares of the reverse diagonal do the recombination products differ from what went before.*[15] In all others, either both chromosomes are identical (as on the forward diagonal), or they differ only at one locus (as for the rest). In these cases, recombination will not produce new combinations. So new recombination products—when post- and pre-recombination chromosomes differ—can occur in only 4 out of 16 squares.[16]

But this story can be replayed with the same diagram for independent assortment! Now A and B are on *different* chromosomes and the dotted line—indicating direct physical connection when they were on the same chromosome—now indicates that these chromosomes come into the union packaged in gametes. The "crossing over" now indicates free or random interchange due to independent assortment of different chromosomes and the genes they contain in the production of new gametic combinations. Independent assortment yields an equal probability that two chromosomes that came in together in the same gamete will stay together in outgoing gametes—reflected in an r of .5. But here we have the same situation again! Independent assortment happens in all squares *but only in the squares along the reverse diagonal do the products of independent assortment differ from what went before,* so new products of independent assortment may occur in only 4 out of the 16 squares. The similar results in these two cases, as compared with the single-locus case, reflect the structural constraints of the diploid-haploid life cycle.

Compare the single-locus case, again using the same diagram. Consider the B-locus, with B and b as alternative alleles in the four contiguous squares of the upper left quadrant (this pattern for the B-locus is repeated in all four quadrants, so we can pick one with no loss in generality). These four squares are a 2×2 Punnett square for a single factor cross among heterozygotes at the B-locus—the one-locus analogue of the 4×4 two-locus Punnett square of the whole figure. Heterozygotes are formed in two out of the four squares (again along *its* reverse diagonal).[17] This different proportion of squares in which new arrangements of elements can occur—2/4 versus 4/16—has consequences. It means that total mixing (and Hardy-Weinberg equilibrium) can be achieved instantaneously in one generation for the single-locus case, rather than asymptotically over many generations as for two or more loci. The asymptotic rather than single-generation approach to equilibrium reflects the presence of a structural condition, a higher-level "segregation analogue," retarding the rate of loss of variance among larger genetic structures. A structural relation among parts with consequences in the equations for gamete production, it should not be surprising that it produces a failure of aggregativity in which the arrangement of the parts matters. Indeed, this is but one of several segregation analogues, reflecting different higher-level aspects of population structure (Wimsatt, 1981b), and providing in effect a kind of "external genetics."

To illuminate the structural relationship producing gametic linkage in another way, consider what would be required to negate its effect; that is, to make the two-locus case come to equilibrium in one generation as the single-locus case does. Suppose a population starting with equal numbers of AABB and aabb homozygous genotypes. Only gametes A––B and a––b would be produced, so the middle two rows and columns would be empty. Only the four corner squares would count, producing a "reduced" 2×2 table. Equilibrium in this population in one generation requires equal numbers of the four gametic types in the next generation. One of two counterfactual modifications of how chromosomes assort or recombine would do it: (1) if recombination or assortment were no longer random but obligate, so that if reassortment could happen, it did. That is, if r = 1, then the two occupied squares in the reverse diagonal produce enough A––b and a––B gametes to balance the A––B and a––b gametes produced in the upper-left and lower-right squares. (2) As an alternative, we could leave recombination and independent assortment alone, and have obligate dissimilar matings rather than random matings. Then all of the matings are in the upper-right and lower-left squares, and an r of .5 will produce equal numbers of the four gametic types in the next generation. These counterfactual thought experiments indicate structural aspects of the cycle producing gametes and genotypes that violate aggregativity assumptions. The second is a special case of assortative mating indicating super-individual population structure. Assortative mating is common in nature, and has important evolutionary consequences, indicating yet another failure of condition *IS*! The fact that it is commonly ignored reflects more about our entrenched idealizations than anything else.

To summarize: the mode of structural description of the genome (with genic versus chromosomal partitions) affects the apparent aggregativity of the properties in question—the frequency of gametes and genotypes produced. And chromosomal decompositions are more aggregative than genic ones because they reflect intra-chromosomal linkage, but even chromosomal decompositions ignore another factor that I have called gametic linkage. All of these partitions—gene, locus, chromosome, gamete, and genotype—are needed in different combinations for different problems, as are super-organismal assemblages in other cases and conditions: mating pairs, families, groups, and demes. (I list only genetic assemblages. As Brandon, 1982 argues, these won't suffice for all questions of evolutionary dynamics—we need other properties of phenotypic units.) For sufficiently constrained problems and conditions, the smaller simpler partitions may appear to be aggregative, but this is usually misleading. And we are not yet done with this case: there are other idealizations hiding in the wings.

Gametic composition is still not aggregative at the level of whole chromosomes or gametes: there are also problems with conditions *QS* and *RA*. These problems are partially hidden by assumptions of standard models of the recombination process, and are occasionally violated (showing non-aggregativity of the relevant properties). But even more interesting is the fact that they *are* so regularly met is a special product of design features of the meiotic process. *Thus meiosis operates so as to increase the apparent aggregativity of processes for producing gametes, thereby increasing both the average fitness and the heritability of traits and fitness in offspring.* This apparent but highly conditional aggregativity arises because of a "special hookup" of processes and parts—a special and quite complicated adaptation of the hereditary machinery.[18]

Models of recombination and linkage commonly used in population genetics suppose that the number and arrangement of loci in chromosomes is invariant.[19] This (false) assumption prevents variation in genome size in ways that test *QS*, and don't allow reaggregation except through recombination of homologous segments that preserve their orientations. This assumption is commonly true, but, *crucially,* not always. Rarer but evolutionarily important translocations, deletions, duplications, and inversions violate this idealization. *These standard models thus exaggerate the aggregativity of the actual physical processes.* These larger changes produced by inversions and other more arcane reconstructions of the genome often cause major further changes both in fitness and in the types of gametes produced in ways often characteristic of speciation events.

Existing theories also can't show whether gamete frequencies are aggregates of gene or chromosome frequencies for another reason—*they are not conservative.* Current theories give equations for the *characteristic* products of these processes, not for *all* of the products, or perhaps more accurately, not for their products under all circumstances, because not all products would be classified as gametes. The lack of conservation in these theories is hidden by dealing with gamete frequencies, rather than numbers, so there is never a full balancing of the equations (as in chemical reactions) for producing gametes from parental genotypes. Aggregativity would require the equations to be conservative.

Consider a hypothetical example. (I combine elements from two different scenarios, but inversions produce scenarios qualitatively like that sketched here.) Suppose unequal recombination plus disturbances of meiosis in a female *Drosophila melanogaster* (having a haploid chromosome number of 4), produces a fragmented piece of X chromosome, a gamete with the other three chromosomes and the other fragment of the X, and three normal gametes with all four normal chromosomes. The first two are complementary fragments—one small and one nearly complete—of a normal haploid genotype.[20] Different criteria for what counts as a gamete might count just the three normal gametes, or add the fourth gamete with the part of its X chromosome missing (if the deletion was small), but the X-fragment would never be counted as a gamete, on grounds that it could not combine with any normal gamete to make a viable and or reproductively competent fly.[21] (This alone, or any selection, makes surviving gamete frequency a non-conservative function of gene or chromosome frequency.) One might also question counting the partial X-deletion as a gamete, on the same grounds. The harder line of saying that any changes in genome size and the arrangement of loci should disallow the result as a gamete violates conservation (with the ignored products subtracted) even more severely and causes other problems: mutations with such changes have survived and played a role in speciation events. Ignoring them cuts micro-evolutionary models off from important kinds of macro-evolutionary events. *Our idealizations partition the world for us along lines convenient for some purposes, leading us to ignore connections that are essential for others.*

Descendant organisms need to inherit (at least very nearly) a full complement of genes to have any chance of surviving and reproducing. Thus fitness definitely fails to meet both conditions *QS* and *RA* for contributions of genes. But how did we get from gametic frequencies to fitness? Evolutionists want to deal only with organisms that survive long enough to leave descendants. And gametes are generally produced in such excess that it is only too easy to consider only those that are survivable. So population geneticists often implicitly add conditions that are normally met and whose violation would virtually ensure inviability or reproductive incompetence—such as failure to inherit a significant fraction of the genome. Most models of mating assume conservation of genome size—surely also a design feature of meiosis and fertilization, as experimental or rare natural conditions readily demonstrate. The consequences of tetrad formation in inversion heterozygotes show that changes in genome size or arrangement—simple reversals in the order of some of the genes in a chromosome—can have significant fitness consequences, and must be controlled for. Boveri's classic (1902) experiments showed that dispermic fertilization, where disturbances of mitosis make genome size and composition a random variable—had disastrous consequences to be avoided. *That one has to impose constancy of size, composition, and arrangement as side conditions in population genetic models of genome formation reveals that we are not dealing with aggregative properties.* (These conditions are so taken for granted that they are rarely stated—or studied.) *With these conditions and theories, things may appear aggregative that are not.* Instead these things are highly sensitive to the arrangement of mechanical parts necessary to produce—and have the function of producing—this apparently aggregative behavior. (A more familiar analogy might help: that computers can do sums accurately doesn't make them "mere aggregates" either—even if we limit the case to special-purpose machines that can *only* do sums. If we cut them in half, they won't do sums half as fast, or half as big.)

This fictional or quasi-aggregativity is particularly pronounced for quantitative genetic multi-locus models of additive traits, where it is supposed that each of a number of genes contribute additively to that trait. The expression of any of the genes and the additivity of their contributions usually depends on phenotypic conditions produced by many other genes that are a normally presupposed part of the genetic background. (To worry about the intensity or additivity of eye pigment, you need eyes!) The assumption of additive fitness contributions of genes has played a central if contested and ambiguous role in discussions of higher-level units of selection (Wimsatt, 1980b, 1981b; Sober, 1981, 1984b; Brandon 1982; Griesemer and Wade, 1988; Lloyd, 1988, 1989; Sarkar, 1994; Wade, 1996).[22] Following Lewontin (1978) on "quasi-independence," I have argued (Wimsatt, 1981b, 1986b) that the additivity of fitness components that exists is local and context-dependent (though it can *appear* to be context-independent for small and limited changes). This local additivity does not show that fitness is an aggregative property of genes.

Adaptation to Fine- and Coarse-Grained Environments: Derivational Paradoxes for a Formal Account of Aggregativity

An even more paradoxical case is that of the relation between Levins' fine and coarse-grained adaptive functions (Levins, 1968, pp. 17–18; Wimsatt, 1980a, 1986b). These mathematical functions model the fitness of an organism in a sequence of environments in terms of its fitnesses in the individual environments. Both appear to be aggregative, but the relationships they are taken to describe cannot be, because they aggregate in different ways, and are not equivalent.

Levins' fine-grained adaptive function is given by a sum of products:

(1) $$W_f = \sum_{i=1}^{n} p_i W_i$$

In this equation, W_f is the net fitness of the organism in a mixture of temporal subenvironments $E_1, \ldots, E_n$, in relative proportions $p_1, \ldots, p_n$, in which it has fitnesses $W_1, \ldots, W_n$. Levins supposes that organisms "experience" the composite environment as an "average" of component environments—thus their linear contributions to its fitness. The specified component fitnesses for subenvironments are those that would be realized by that organism in a pure environment of the corresponding type for the entire interval. The form of the equation is like that given in decision theory for "expected utility," with W_i's as utilities, and p_i's as probabilities.[23] Note that the fitness specified by the equation depends upon the relative frequencies or proportions of the subenvironments, *but not on their order.*

His coarse-grained adaptive function, using the same variables is given by:

(2) $$W_c = \prod_{i=1}^{n} W_i p^i$$

This fitness function is also independent of the order of the subenvironments. It suggests the multiplicative law for combination of probabilities, and conjures up an image of the organism jumping through a series of hoops, with its chance of getting through each being *independent* of whether it has passed through any of the others.[24]

Each of these two "adaptive functions" has a mathematical form meeting all conditions for aggregativity (the first is additive, the second multiplicative), so fitness in both cases seems to be an aggregative property of the component fitnesses and the frequencies of their environments. But this would lead to a direct contradiction. Both functions start from the same general expression for fitness, and transform that expression in different ways making different mathematical assumptions. They are not equivalent, and produce different answers except under very special limiting cases (Strobeck, 1975). This is common enough for approximations, but that doesn't remove the paradox. *If fitness were really aggregative as each equation—taken separately—suggests, then one should be able to transform situations meeting either equation into situations meeting the other by simply reordering the subenvironments!* So one or the other or both of the equations must be false, and—despite what the equations say—the fitness of an organism cannot be an aggregative function of its fitnesses in the subenvironments. (Cases 1 and 2 of Figure 12.4 depict "fine-" and "coarse-grained" environments for organisms with temporal integration ranges intermediate between 1 and 12, *and not too close to either.*[25])

The coarse-grained adaptive function is derived by twice making an approximation that is literally false:[26] (1) The fitness of an organism in a subenvironment is assumed to be a function only of that (sub)environment—there are no "historicity" effects. (Equivalently) (2) later, it is assumed that the fitness function does not change over the discrete period of integration used, even if this period (Δt in equation 2.2 of Levins 1968, p. 18) is allowed to be quite long. But if true, this would allow unlimited applications of conditions *IS* and *RA* (intersubstitution and reaggregation) to reorder the subenvironments, changing the "small" and "well mixed" subenvironments of the fine-grained adaptive function into the "large" and "well separated" subenvironments of the coarse-grained adaptive function. Suppose that all of the micro-environments of one type were lumped together, and followed by all of the micro-environments of the other type (*transforming* case 1 into case 2). But the adaptive functions for these two cases are different mathematical functions, and yield different answers when given the same fitnesses and frequencies for their components, so they can't both be right at the same time. What has to give is unrestricted application of *IS* and *RA*. But with this goes the claim that either of the adaptive functions is truly aggregative.

Consider three idealized environments, which are checkerboards, all

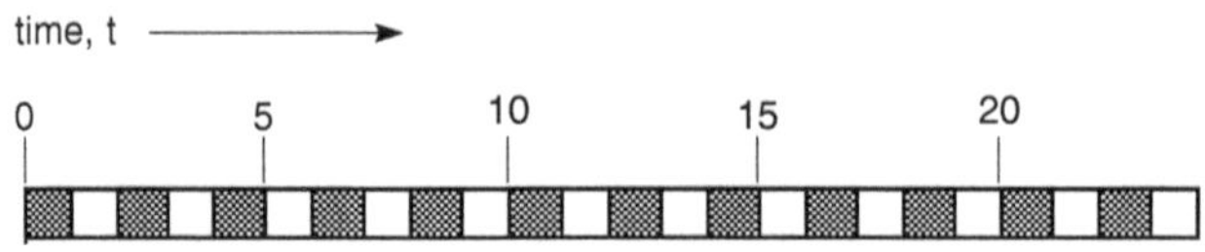

Case 1: p(E1) = p(E2) = .5; regular repeat of subenvironments with unit length; fine grained (for both environments) for an organism with tolerance, integration range, or threshold 1.

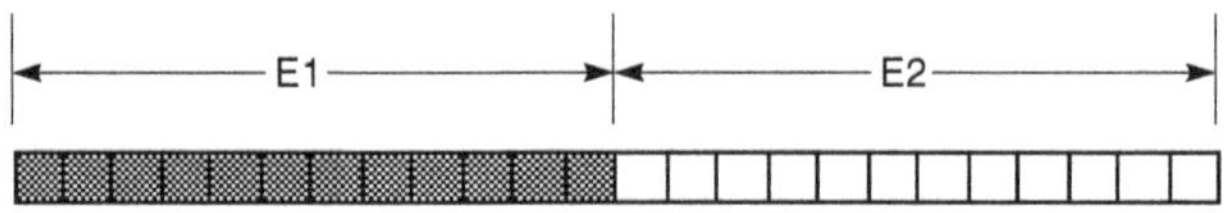

Case 2: same, but with variation on a different scale gotten by re-ordering subenvironments; coarse grain for organism with thresholds (for both environments) of < 12.

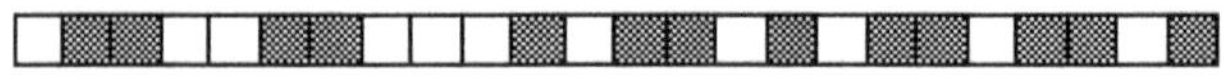

Case 3: p(E1) = p(E2) = .5; Sample trial random variable, with 13 E1, 11 E2. Fine grain for organism with thresholds 3 for white, 2 for "gray". (In calling this "gray" we are treating a regular 2-D repeat of black and white pixels within the squares as a (perceptually) fine-grained property! To a "LaPlacean demon" which calculates exactly and does no averaging, there is no gray—only arrays of black and white pixels showing a variety of "homogeneous" regularities on different size scales.) Also note (compare case 1) that random noise coarsens the grain.

Figure 12.4. Patterns of environmental grain.

with equal proportions of black-and-white squares; in the air above the temperature is 50°C and 0°C, respectively (this mimics solar heating effects). They differ only in the size of the squares, which are 10 mm, 10 meters, and 10 kilometers on a side. We will compare the fates of two organisms in them—a water buffalo and a *Drosophila* (or fruit fly), the universal test instrument of classical genetics. (The grains of their environments are compared for two different niche dimensions across a range of size scales in Table 12.2.) Assume that 0°C is too cold, and 50°C to hot for either organism to stand for long periods of time, and 25°C is about optimum for each. At roughly 3 mm and 3 meters in length, the fly and water buffalo have a length ratio of about 1:1000. Thus they bear about the same relationships to neighboring checkerboard scales. Each is about a third of the length of the smallest and middle scales, respectively, and about 1/3000 of the length of the middle and largest scales. How would they do when moving through these different scale checkerboards? Because their nervous systems are tuned to detecting things (including temperature differences) on their own size scales, each would detect those variations. Detection on that scale is relevant to locomotory decisions over that or much larger distances—avoiding local hot or cold spots or going up or down larger temperature gradients. But they also buffer physiologically on the scale to which they are perceptually sensitive: they have enough thermal mass (and low enough thermal conductivities) to be unaffected by air temperature variations for that combination of size and temperature ranges. (They wouldn't notice variations in patches the next scale down—10 microns and 10 mm, respectively—but would perceive only their comfortable "average" 25°C.) But they would be in real trouble the next scale up (squares 10 meters and 10 kilometers on a side), dying or being sorely stressed before they could get to the other side. The smallest scale for each organism (10 microns and 10 mm) is both perceptually and physiologically fine grained (for thermoregulation). Their respective middle scales are perceptually coarse-grained, but physiologically fine-grained for each. And their relative large scale is coarse-grained for both organisms in both respects. But, save for extremely large and small grains, the *Drosophila* and the water buffalo will experience any given environment differently because of their different scales relative to the grain of that environment. Figure 12.4 actually illustrates both perceptual and physiological graining. The scale of the "gray" denoting type E1 subenvironments was chosen to be fine enough to be conventionally treated as an average gray, but coarse enough still to be discriminable as an array of black and white dots (at least to some of us, or when I'm wearing my glasses!).

Levins' two adaptive functions are designed for different kinds of limiting cases, which make different approximations appropriate. Ignoring this can lead to explicit contradiction. He suggests that real cases will fall somewhere on a continuum between them (1968, pp. 18–19). The simplest mathematical assumption of unlimited re-orderability (through application of *IS* and *RA*) is too strong—stronger than Levins actually needs for deriving his fine-grained adaptive function. (The derivation only requires the weaker condition that *the subenvironments can be arbitrarily re-ordered as long as that leaves a representative sample of the environments of the whole in any sequence of a length range important to determining fitness.*) This length range is the tolerance, integration range, or threshold of Figure 12.4. This constraint on the representativeness of any appropriate subsequence of the sequence allows some reordering of the subenvironments, but prevents unlimited reorderability. This prevents transforming a fine-grained environment

Table 12.2 Environmental grain for different niche dimensions, organisms, and size scales

Environment (Grain Type for Organism)	Scale: 10 microns	(×1000) → 10 mm.	10 meters	10 kilometers
TEMPERATURE: (Perceptual Grain) Water Buffalo	*fine*	*fine*	COARSE	COARSE
(Perceptual Grain) *Drosophila*	*fine*	COARSE	COARSE	COARSE
TEMPERATURE: (Stress/Mortality) Water Buffalo	*fine*	*fine*	*fine*	*transit*IONAL
(Stress/Mortality) *Drosophila*	*fine*	*fine*	COARSE	COAR

into a coarse-grained one but also shows that the fine-grained adaptive function is not a true aggregative property of the fitnesses of the subenvironments. Indeed, stochastic fluctuations from the average (illustrated in case 3 of Figure 12.4) require higher thresholds or tolerances than the regular periodicity of otherwise comparable case 1, and the variance in length of "white" and "gray" subenvironments would be reflected in the larger biological thresholds we would find in nature.

The subsequences of the environment important in determining fitness may differ in length and composition for different adaptive problems (e.g., thermoregulation, mating, or predation) that determine an environmental scale. This is a temporal or spatial size scale for determining relevant changes in the environment as a function of properties of the organism—it could thus equally be thought of as an organismal scale, or most accurately as a scale relating organismal and environmental properties. (This was set up deliberately with the water buffalo and the fly.) As we have already seen, the relevant scale differs for different organisms as a function of their properties and capabilities, for different functional subsystems of the organism, for the environmental variables in question, and for how far these variables deviate from their ideal values for those organisms. (The tolerable size for checkerboard squares for each organism would have been smaller—possibly much smaller—if the temperatures had been –50°C and 100°C, rather than 0°C and 50°C—even with the same mean temperature.) Threshold or transition regions occur where environmental variations much less than that are averaged (added), and those much larger than that are treated as independent obstacles (in effect a multiplicative sieve) that must all be gotten through. *These different scales or limiting cases provide motivations for the two distinct adaptive functions.*

Thresholds are common, and inconsistent with the unrestricted application of either or both of conditions *QS* and *CI*. How far do temperatures fluctuate from my current (preferred) body temperature, how rapidly do they change it (a measure of my thermal mass, surface area, and surface conductivity), and how big is their spatial extent relative to my rate of travel through them? Or how great is the distance between prey captures, and how much net energy do I get per capture relative to how far I can go between captures? Is it large and concentrated enough to be worth claiming or defending?[27] For many adaptive problems there are couplings between size and time scales in terms of the rate and frequency of various energy flows, and inequalities that must be satisfied for the organism to survive. These different thresholds are rarely totally independent of one another. Larger mammals usually better survive the extended cold temperatures of the arctic than smaller ones, and for longer, but this depends on how well fed they are.

So we learn from this example that *one cannot simply look at the form of an expression relating system and parts' properties to tell whether a system property is aggregative.* We must look also at the assumptions made in deriving the "composition function" for the system property, and make sure that all of the assumptions are empirically adequate for the case in question. This fact places important limitations on a formalistic account of aggregativity, for it isn't enough to look at the form of equations in the finished empirically adequate theories, you also have to know how you got there, *because the approximations you made along the way cannot be forgotten in evaluating aggregativity.*

Aggregativity and Dimensionality

Aggregative or near aggregative relations reduce dimensionality of equations and necessary theory—producing simpler theories in obvious ways. This is nowhere better illustrated than in Lewontin's famous table (Table 12.3) of the dimensionality of population genetic theories under different assumptions about gene interactions (Lewontin, 1974, p. 283; Wimsatt, 1980b). As one might expect, the fewer the kinds of

Table 12.3 Sufficient dimensionality required for prediction of evolution at a single locus with *a* alleles when there are *n* segregating loci in the system

Level of Description		Zygotic classes	Gametic classes	Allele frequencies	Allele frequencies
Dimensionality:		$a^n(a^n+1)/2-1$	a^n-1	$n(a-1)$	$(a-1)$
Simplifying Assumptions:		none	1	1, 2	1, 2, 3
# loci, *n*:	# alleles, *a*:				
2	2	9	3	2	1
3	2	35	7	3	1
3	3	377	26	6	2
5	2	527	31	5	1
10	2	524799	1023	10	1
32	2	9.22×10^{18}	4.29×10^{9}	32	1

Source: Table is modified and extended from Lewontin, 1974, p. 283, table 56.

Simplifying Assumptions:
1. Random union of gametes (no sex-linkage, no assortative mating).
2. Random statistical association of genes at different loci (linkage equilibrium).
3. No epistatic interaction (inter-locus effects are totally additive).

relevant interactions between alleles and loci involved in the relevant equations for fitness of evolutionary units and the determination of evolutionary trajectories, the more aggregative the phenomena appear.[28]

The following modification of Lewontin's table is from Wimsatt (1980b).[29]

With either assortative mating or sex-linkage, the frequency of genotypes is required to determine the frequency of matings of different types, and from them the frequencies of offspring genotypes. But even if different genotypes pair randomly, genes may be clustered in a non-random fashion in gametes.[30] As we saw above, this will result in both non-random production of gametes, and of genotypes, when individuals producing these gametes mate. With different genotypes having different fitnesses, different genotype frequencies will produce different net effects on gene frequencies, so these higher-level units—frequencies of genotypes and of gametes—are required to predict their outcomes of selection correctly. And finally, if genes contribute additively to fitness, this contribution is statistically independent of genetic context—what genes are found at other loci. The dimensionality specified in Table 12.3 is the number of independent equations that must be solved simultaneously (actually, iteratively) in order to predict the outcome. At each locus, the gene frequencies must sum to 1, so if $n-1$ of the alleles are specified, that determines the frequency of the last. Similarly the gametic and genotypic classes in the second and first columns must sum to 1, so the frequencies of the last gametic and genotypic types can be calculated as 1 minus the sum of the others.

The number of independent equations is reduced as the simplifying assumptions are made—but so also is the complexity of each equation. Multi-locus gametes and genotypes can arise in a larger number of different ways from other multi-locus gametes and genotypes than genes can (they are more complex and therefore can be assembled in more different ways). These ways of origination must be reflected as terms in the equations for how the frequencies of gametes and genotypes change in successive generations.

The net effect, building the more complex cases, is that as the relationships of the lowest-level parts in the larger structures become causally relevant, changes in these relationships change the outcomes, and the complexity of a dynamical theory grows rapidly. Thus, in search of the simplest workable theory of a system, it is natural to start with one assuming the minimal number of causally relevant relational properties. In the logical extreme, this is to assume aggregative behavior. If a property is aggregative, then the value of that property for the whole system is all that matters in its dynamical behavior. If the property of the system is invariant over aggregative operations on its parts, then it is independent of variations in these changes, which is to say that their individual values do not matter as long as the value for the whole remains invariant. If that is not possible, structures are assumed that preserve the largest possible degree of invariance of system properties on organizational rearrangements of parts' properties.

Aggregativity as a Heuristic for Evaluating Decompositions, and Our Concepts of Natural Kinds

The examples covered in this chapter have yielded many interesting features of claims of aggregativity or partial aggregativity. Seeing them together suggests interactions among the development of theory, methods of decomposition, and experimental design, and what we make of what we have found. Together they give a somewhat different picture of the nature and uses of aggregativity, and have further implications for the assessment (and biases) of reductionist methodologies.

1. Table 12.1 shows that very few system properties are aggregative

functions of parts' properties, so emergence—as failure of aggregativity—is extremely common. It is the rule, rather than the exception. The conservation laws of physics pick out aggregative properties, but little else does. So this could, perhaps fairly, be criticized as yielding a very weak notion of emergence. But it accords with intuitions of most scientists I know, who are not willing to give up either their reductionism or their emergence, and who agree with its classification of particular cases.

2. So, then, why is the temptation for "Nothing but-*ism*"—the ontological war cry of what Dennett (1995) calls "greedy reductionism"—so strong? We see statements quite regularly in science like "Genes are the only units of selection," "Organisms are nothing but bags of genes," "The mind is nothing but neural activity," "Social behavior is reducible to or nothing more than the behavior of individuals." If total aggregativity is so rare, why are claims like these so common? While true aggregativity requires invariance of the system property under *all* decompositions and reaggregations, I suggest that we often (fallaciously) think of behavior at higher levels as being aggregates of the behavior of parts for *particular* decompositions that *do* show this invariance—or show it only partially (for some of the criteria) or approximately (invariant within an ϵ for the criterion in question—see below). Such properties *look* aggregative for some decompositions, but reveal themselves as emergent or organization-dependent for other decompositions or conditions. We saw this in the discussion of multi-locus genetic systems. Analyzing such practices naturally changes our focus from ontological to methodological questions: from how to specify relations between system and parts' properties to looking at the reasons for, process of, and idealizations made in choosing and performing a decomposition, and the broader effects of those choices.

3. Properties may be aggregative for some decompositions and not for others, or more so for some than for others. The degree of aggregativity may then be used—consciously or unconsciously, but in any case quite rationally—as a criterion for choosing among decompositions: *we will tend to see aggregative or more aggregative decompositions as natural decompositions, and their parts as instances of natural kinds, because these decompositions provide simpler and less context-dependent regularities, theory, and mathematical models for the behavior they capture.*[31]

That simplifications arise from reductions in context-dependence is clearly demonstrated with the dimensionality reductions accompanying standard simplifying assumptions in classical multi-locus population genetics. These are summarized above in Table 12.3. These decompositions may be particularly revealing cuts on nature, but we must not over-generalize their import. (They may be the right cuts for the wrong question.) *We will tend to see these parts as special, and to make "nothing but" style reductionist claims for them. This is a particularly pervasive kind of functional localization fallacy—a move from the claim that a decomposition is particularly powerful or revealing to the claim that the entities and forces it yields are all that matters* (Chapter 9 in this volume; see also Bechtel and Richardson, 1993). Such "nothing-but" claims are false or methodologically misleading if taken to suggest that one shouldn't bother to construct models or theories of the system at levels or with methods other than that of the parts in question, or that these preferred entities are the only "real" ones (see Chapter 10), or that questions one can pursue with such decompositions are more important.

4. When partial aggregativity leads to greater physical or functional modularity of the parts (likely in evolving systems—see Lewontin, 1978; Wimsatt, 1981b, pp. 141–142; Schank and Wimsatt, 2000; Brandon, 1999), it may promote (evolutionarily) a consilience or robustness of parts' boundaries individuated using different properties of the system. System boundaries in one property may act as symmetry-breaking factors leading to accumulation of differences and boundaries in other properties along the same dividing lines (Platt, 1969). *Greater robustness* (see Chapter 4) *of parts under that decomposition—a standard criterion of objecthood—will also strongly contribute to the judgment that this is a decomposition into natural or real parts.* Such claims are *prima facie* reasonable. (Robustness is a degree-property, so justifying these claims is not an all-or-nothing affair.) These judgments are all context-sensitive, so they still don't support "nothing-but" style claims.

5. The four conditions all specify invariance of the system property under operations on the parts. For quantitative properties one can easily produce a family of criteria for approximate or local aggregativity, in which variation of the system property within $\pm \epsilon$ is tolerated for various values of ϵ (Wimsatt, 1986b); I used this strategy in Chapter 9 to describe different degrees of near-decomposability or modularity in systems. Tolerances are useful theoretical tools because we use quantitative or formal qualitative frameworks as templates that nature may meet in varying degrees. With a particularly adaptable framework that

can be fitted to nature in various places in different possible ways, we may try many such mappings, looking for "best fits." (The "coefficient of determination" or r-value of an equation in a linear regression is a particularly simple example.) Using "tolerances" for key qualitative concepts is a particularly useful strategy in a messy, inexact, and approximate world that has many regularities and stable patterns, but few exceptionless generalizations. Given the "noise" present in experimental situations, we need tolerances anyway, but this provides an additional important reason.

6. Since approximations are frequently used to produce equations of aggregative form, we must investigate the accuracy of the approximation under relevant conditions. This may take two forms, with different consequences:

a. Emphasizing *conditions* for accuracy focuses attention upon the context-dependence of system properties, with failure of aggregativity often leading to using or studying the system under different conditions, hoping to find conditions under which its behavior is more aggregative, or other conditions where it may be qualitatively different. (Here the search simplifies conceptualization and analysis of data, and will also likely generate better experimental control—indicating conditions one should attempt to realize or avoid.) If conditions for aggregative behavior are found, we will be tempted (see (3) above) to regard that decomposition, and the conditions that produced it as reflecting the *real* nature of the system, however unjustified that may ultimately be.

b. Emphasizing *required* accuracy highlights our purposes and demands of our applications, adjusting demands or methods as required—using methods suited to the study of non-aggregative behavior (for greater accuracy) or weaker methods treating it as aggregative (for less). Awareness of our role in classifying the property as aggregative or not may make us more self-critical of our idealizations, and of possible biases in our problem-solving heuristics (Wimsatt, 1980b, and chapters 5 and 6). Such awareness is important, but fixation on it may yield complementary errors: the view that classification of the system property is *merely* instrumental, or—even less accurately—is just conventional or socially determined. (*Determines* is commonly misused in such contexts—equivocating between "plays a role in determining" and "by itself, is sufficient to uniquely determine.")

7. One can also assess aggregativity in past and present theories of the relation of system and parts' properties. In doing so, one must look not only at the final derived equations, but also at their derivation, to see if the idealizations and approximations used *assume* one or more of the conditions of aggregativity, and whether they are legitimate for the conditions at hand. One may thereby derive an aggregative model, and also come to understand its conditions of applicability. As the genetic and environmental examples of this chapter show, apparent units of aggregation can be very misleading, and the respects in which they can be treated as aggregates quite limited.

8. One can track systematic changes in our view of the relation between system and parts' properties, both analyzing their status in historically important disputes, and comparing their changing status in successively better theories of a phenomenon. Does this raise Hempel or Nagel's worries (Wimsatt, 1986b) that aggregativity and emergence are really about our *knowledge* of the world, not about the world itself? But this information is *just* what a sophisticated realist needs—to see how the world changes as viewed through our theories, when our theories change. *Discussions of theory-dependence always somehow suppose that the effects of this theory-dependence are impossibly confounded with those of the world. But there is no reason to suppose that this is so. As long as you can tell one from the other—object of study from tool of access—you're okay, and to do this you need to be able to vary both to assess and separate these effects.* Criteria for aggregativity can be viewed either as statements about ontological relations in the world, or as tools for constructing and characterizing theories. It's just that these are *different* uses. If one make their aims clear, there are no inescapable threats to objectivity.

9. Earlier reductionist theories of the behavior of a system tend to have more simplifying assumptions, controlled variables, and assumed constancies, and predicates treated as monadic or of reduced order than later ones (Wimsatt, 1980b, 1985). More realistic models are often suggested by the failure modes of these simpler ones. Higher-order relational properties and more complex interactions between parts are recognized through increasingly detailed specification and analysis of the internal structure and environmental relations of the system. These kinds of progress should increase the degree and kinds of emergence postulated of system properties. *This is just the opposite of that predicted on the classical positivist model of emergence, which saw emergence disappearing with the progress of science, but appears to be just*

what is actually happening with recent increased interest in the study of complex systems. Consider, for example, the increased talk of holism and emergence accompanying the rising interest in non-linear dynamics—systems that violate (at least) the fourth condition for aggregativity. This move is clearly a confirmation of the strategy of analysis for emergence advanced here.

Reductionisms and Biases Revisited

I will now use this analysis to address some broader questions concerning reductionism, particularly how it is used and perceived in the context of an incomplete analysis; that is, our usual situation! The four conditions of aggregativity represent a powerful approximate and adjustable framework. In any given case one could evaluate how well each condition is met across different decompositions of a system into parts. The better a decomposition meets these conditions, the more easily we can treat it as factoring the system into a set of modular parts having monadic, intrinsic, or context-independent properties. With particularly simple and theoretically productive decompositions, we will tend to view these parts and properties as instances of natural kinds, as robust, and to regard the system as "nothing more" than the collection of its parts. *We have here turned an architectonic distinction between kinds of properties into a search heuristic for finding preferred, simple, "maximally reductionist" decompositions of systems into parts*—decompositions that lead readily to extremes of "nothing but" talk and disciplinary imperialism.

A reductive explanation of a system property or behavior is one showing it to be mechanistically explicable in terms of properties of and interactions among the parts of the system. What does this kind of explanation have to do with "nothing but" style reductionism? In principle, nothing—but in practice they are temptingly connected and easily confused. With total knowledge of a system, the two species of reduction are clearly distinguishable, but we don't have total knowledge, and with increasing degrees of ignorance about a system they come to look more and more alike! *A closer look at what we do in conditions of partial ignorance is especially important for fields and explanatory tasks whose major questions are still "in process," and just the kinds of judgments we should seek for limited, fallible, and error-prone scientists.*

A system that is aggregative for a given decomposition is almost trivially mechanistically explicable: the parts all have the property in question, and enter into the explanation of how the system has it in the same simple way. Relationships with other parts are usually either monadic (i.e., nonexistent) or of relatively low order, and would tend to meet strong conditions of symmetry and homogeneity. The system would be relatively uninteresting: its parts would show no functional differentiation. *But none of these conditions follow from saying that the properties of a system are mechanistically explicable.* So to say that the behavior of a system is totally explicable in terms of the behavior of its parts is *not* to say that it is an aggregative function of the parts. (The inference *does* go the other way. That could contribute to the confusion, but given the rarity of true aggregativity, even this fallacious inference doesn't explain much: there are too few occasions to invoke it.)

Now let's consider the human genome project as an example. Suppose that early in the investigation of a system (say, an organism) we think we know a good set of parts (e.g., its genes). If we don't yet know the diversity of ways in which these genes may interact with each other and with the physical conditions in the organism (on all of the relevant size and time scales), we may tend to treat their interactions as all alike. We may do so *either* in a first-order simplified model of the system, which we simulate or analyze to explore its behavior, *or* in the "out of sight out of mind" blissful ignorance that often accompanies our view of a complex task before we really get into it. In either case we are likely to overestimate how aggregative the system is, how simple it will be to understand its behavior, and to make the most simple-minded reductionist claims about what can be learned from studying it at the lowest (or indeed *only* at the lowest!) level. (Wimsatt, 1979, 1980b, and 1997a provide examples of how reductionist approaches often involve these assumptions, and how they emerge and can cause trouble in seemingly benign applications of commonly effective reductionist problem-solving heuristics.)

This would explain (indeed, it predicts!) (1) the characteristic oversimplifications in early claims made for the human genome project, and (2) the subsequent (necessary) broadening of the project to make it viable. (This included adding parallel comparative studies of a diversity of genomes of other species at different phylogenetic distances to determine what varies across those species groups, and to get some idea of its significance; and developmental and physiological studies at a variety of levels of organization of the expression of the genetic traits of interest.) It also explains (and predicts) (3) the increasing moderation of claims for what we will learn from it. And if I am right, (4) explanations that come out of it will also be far more contextual and qualified (and may involve the discovery of qualitatively new kinds of mechanisms and interactions). *This pattern—these four changes in the character of the program and the claims made for it—are not only explicable after the fact, but predictable in advance, given normally applicable reductionist research strategies and their biases* (see Wimsatt, 1980b, 1997a, Chapter 5, and Appendix B). They apply not only to the human genome project, but chart the expected trajectory of any successful reductionist research program in the empirical sciences.

I neglect here the obvious political purposes served by exaggerating how much could be achieved with how little. They aren't needed to explain the phenomena, which could arise simply from cognitive bias, but phenomena are also often overdetermined, and reductionist claims often *do* serve political ends, inside and outside of academe. (Most of the offending claimants do—or should, if they are honest—start with the disclaimer, "In principle, . . ." Yes, the road to hell *is* paved with good intentions![32]) To assess these claims fairly, we *must* recognize the limitations of our knowledge, the heuristic character of our tools, and specific biases likely to result from their application. Knowledge earlier of how we tend to construct models and theories in contexts of partial ignorance might well have produced a better project, at a more reasonable pace, with more realistic ends, and at lower cost to the other fields that we will have to support and develop in any case in order to decode the texts we find in "the book of life."

But this chastening skepticism about reductionism appears to run counter to the facts. If this is true, why should reductionist methodologies have appeared to be so successful? A crucial property of heuristics (one of six—see Appendix A) is that a heuristic principle succeeds in part by transforming a problem into a different but related problem that is easier to solve. But if it does so very effectively, there will be a strong tendency to identify the new problem as the old one—saying, "Now that we've clarified the problem so that it can be solved, . . ." or some such thing! In this way quite substantial changes in a paradigm can be hidden—particularly a cumulative string of such changes, each too small to be regarded as "fundamental." I think that this kind of ex post facto reification is central to the exaggeratedly high opinion we have of reductionist methodologies, and more generally to the largely mistaken belief that work elaborating a paradigm is merely playing out options that are already given.

But wait a minute, hasn't the reduction succeeded? Well, it has and it hasn't: its successes are often genuine, but quite misleading—it may succeed *via* a series of subtle shifts in the question such that important aspects of the original question have not been answered. And by shifting them outside of the bounds of the new science, there will be a natural tendency to downgrade their importance. We work on problems that yield to our methods, and we all have a tendency to overestimate their power, and the importance and centrality of our own field. (It's what we know best after all!) In this respect, scientists share with others a cognitive bias that is quite general across fields and contexts. So questions that can't be addressed using our own very successful methods must not be very scientific, not very important, or both! This sounds like it's worth a good laugh, but it may be even more dangerous than failing to solve a problem—we may now even fail to recognize that it exists!

Abbott (1995) provides a revealing analysis of perceptual distortions leading individuals toward the extremes of an income distribution to see themselves as closer to the middle than they really are. He sees this as a generalizable phenomenon. In this context, if we exaggerate differences among cognitive positions near our own, and contract differences among positions further away, systematic and predictable metrical distortions will exaggerate the centrality and importance of our own position. *This is a quite general property of perspectives and of any perspectival view of the world.* If, for example, we tend to assume unconsciously that the importance of our specialty is proportional to the fraction of *our* knowledge of the world that it represents, we would both explain our systematic overestimates of the importance of our own areas, and predict that this distortion should be more extreme as we get more narrowly specialized. (This should apply both across individuals, and for an individual over time.)

The whole of embryology was differentiated out of the study of heredity with the rise of classical transmission genetics between 1900 and 1926, even though developmental phenomena had figured centrally there earlier, and were regarded through much of this period as important constraints on the form of acceptable theories of heredity.[33] Morgan was originally skeptical of Mendelism for just that reason—until the spectacular successes of his research group led them to simply ignore its quite paradoxical inadequacies on that score. Development again achieved the promise of a resurgence in genetics in the early 1960s with the discovery of the *lac operon*—the first account of how a gene could perform a complex and conditional control function in the expression of a trait, but a simplistic and incorrect promise it turned out to be. The genetics of elephants (or any eucaryote) was *not* like the genetics of bacteria after all—especially not the developmental genetics of metazoa. Development is now again at center stage (accelerating since the late 1970s or early 1980s) for various reasons. Important aspects of it can now be studied molecularly, and we have discovered some extremely powerful and general large-scale gene complexes (the related *HOM, HOX* and *DHOX* families) that give us a handle on many more (but far from all) developmental phenomena. Is this a triumph for reductionism? In part, but it has succeeded in this by entraining and using a successively broader diversity of kinds of data and theories from other sources, and recognizing a whole new cast of causal entities. This very diversity of major players makes it much less reductionist. The methodology (and even more, the rhetoric!) are still quite reductionist (reflecting the character of the heuristics used), but neither the ontology nor the epistemology are anywhere close!

PART FOUR

Engineering an Evolutionary View of Science

This closing essay was designed as an introduction, but took a different turn. It traces my path from a simplistic reductionism as a hopeful engineering physicist to a fascination with heuristic methods for dealing with complexity, especially among the surprisingly intelligible systems of the evolutionary sciences—in effect the engineering sciences of the natural world. Complementary experiences in theoretical applied physics and engineering design work stimulated my interests in methodologies for complex systems. These experiences, thoroughly at odds with contemporary philosophical opinion on how "high" theoretical science and mundane "applied" science are practiced, are recounted and explored here.

Chapter 13: Epilogue: On the Softening of the Hard Sciences

We like to say that philosophy is a reflexive discipline: we investigate problems, and pride ourselves in reflecting on our methods. One might suppose from this that other disciplines don't. Wrong—many do. Scientists design, calibrate, and debug theories, instruments, and methods. Sociologists study questionnaire design and statistical methodology, and historians take courses on historiography. Intelligent practitioners are commonly critically self-reflective. (And philosophers aren't always. We have largely uninspected assumptions too—different ones for different times. And our self-reflectiveness is at times more preached than practiced.) Engineering has particularly strong claims that its practice is part of its subject matter. Engineering design bridges both concerns: making, studying, designing, maintaining, manufacturing, and repairing objects and processes that show adaptive complexity, and also studying and creating methods for accomplishing these tasks. These methods are also engineered objects, thus part of the subject matter. And our bodies and minds are naturally engineered objects—evolved cognitive capabilities and all. Reason and rationality, traditional domains of the philosopher, are no less engineered than the rest. As Dennett suggests (1995, p. 187): "Biology is not *like* engineering—biology *is* engineering."

Herbert Simon's "sciences of the artificial" (1996) embraced evolution, and this broader view of engineering.[1] Donald Campbell urged that not all evolving systems are biological, and his pioneering work on evolutionary epistemology from 1956 on drove a search for a more general science of evolving systems (Campbell, 1974a, is an influential statement of his views).[2] The legacies of Simon, Campbell, and others from the 1950s have waxed and waned since the late 1960s, but have burst on the scene full-born in the last decade.

Campbell saw selection processes as the only methodologically acceptable explanations for fit between a system and its environment—embracing all areas of our cognitive, social, and cultural experience. He advocated what were in effect "memes," but with more structure, in 1965—long before Dawkins named (and claimed) them. Diverse activities developed as secondary selection processes (perception, various forms of learning, social imitation, and the organized selection systems of language and science). These "vicarious selectors" began as adaptations serving natural selection or other already established vicarious selection processes. (Note the inductive/recursive definition!) Most could develop a decoupled dynamics generating runaway positive feedbacks elaborating, redirecting, or even opposing other selection processes, thereby substantially increasing the complexity of cultural change. Campbell didn't emphasize material technology, though technological selection and evolution were implicit in his views, which influenced both historians of technology (e.g., Constant, 1980; Vincenti, 1990) and theorists of cultural evolution (Boyd and Richerson, 1985; Sperber, 1996).

Engineering is full of heuristic methods applied to solve and rationalize the most complex construction problems we know. (Evolutionary constructions are still more complex, but we lack any design plans. Instead we practice "reverse engineering"—trying to figure how organisms work, and why they are designed that way. We commonly get a simplified and partial account. And because these "designs" are layered kluges, we can't expect to find a "unitary" design plan "from the ground up.") Engineering shows—writ large—the robust pragmatic realism (Chapter 4) and other heuristic elements permeating methodology *as practiced in all sciences* (chapters 5 through 8) but often obscured in their more formal statements.

If engineering has a separate theoretical component, it is the science of design—many of whose elements are laid out in Simon's essay of that title (in Simon, 1996). Design makes evolutionary complexities intelligible, and—following Dobzhansky—nothing makes sense in biology except in the light of evolution. But in both engineering and biology intelligibility requires a sense of history. The architecture of engineered objects is deeply heuristic: cost-effective means-end solutions using whatever comes easily to hand. Newly engineered objects become both environment for other objects and ancestors of new ones. This fundamental insight is heightened through the ideas of "exaptation" (Gould and Vrba, 1982), and "generative entrenchment" (see Chapter 7 in this volume). Iterating this process gives all things a cumulative character and guarantees the importance of history—or, as newly liberated economists now like to say: "path dependence."[3]

Engineering methodology advertises results over arguments of principle, but engineers embrace principles—fundamental and heuristic—that serve their ends. (Philosophical instrumentalisms or empiricisms are too far removed from practice to derive the benefits and good sense flowing from such a richly articulated heuristic and realistic view.) Principles of good design and engineering practice inform and occasionally transcend engineering specialties (Simon, 1996; Vincenti, 1990; Alexander, 1964). Specialized techniques (and engineering theories) of broader epistemological and methodological import include ideas of near-decomposability and modularity, reliability theory, robustness, and sensitivity analysis, much of what used to be called "operations research," and the theory and practice of modeling and simulation that have transformed all traditional scientific activities.

We are constituted by, develop in, and work on and among designed systems molded by natural selection and its emergent congeners. A striking number of philosophers—many of whom profess a great respect for history—are refugees from engineering. This yields an arresting paradox: most seem eager to shed any association with that prior experience while seeking to analyze problems in the three great areas of design—body, mind, and culture, without pondering whether they have learned anything they can use! Engineering can show us much about scientific methodology and these key areas of design.

Engineering needn't be arcane. My experience with heuristic modeling practices before computers used nothing more than elementary geometry, mechanics, drafting, and some intelligent thought. The remarkable tasks performed with these simple methods recall Hutchins' (1995) discussions of navigation practice—done as a distributed computation and representation task by a navigation team of enlisted men on a navy carrier as it enters port. My design task, like theirs or like any large scientific or engineering project, is a "distributed computation," but our ideology focuses on the individual at the same time as many of our activities and the norms of science require public and shared activity. Both individual design projects and their articulation into larger complexes are accomplished with methods that are heuristic throughout (at individual and social levels), so it is puzzling that hypothetico-deductivism is the only model we have found for the construction of large conceptual structures. But we were looking in the wrong place. These practices are pages from the working manual of the backwoods mechanic, and are rich in materials for further study. They permeate all of the sciences. It is in them that contingency is to be found—but also the roots of our confidence in our results.

These ubiquitous heuristic practices should temper those who approach scientific methodology with *in principle* arguments only. They are common throughout pure and applied science, and link both methodologically to engineering. Fundamental sciences are rather special cases, but most philosophical attentions are directed to them—though few still believe in the foundationalisms that motivated their study. *Non*-fundamental and applied sciences are more representative and revealing of the practice, aims, and commitments of science. Methods of heuristic inference need further study in all disciplines (Gigerenzer, Todd, and the ABC Research Group, 1999). As design methodology and systematized practice, engineering deserves study in its own right. Talk of "meta-engineering" (Dennett, 1995) is increasingly common in discussions of adaptive design, artificial life, and visualization.[4] Simon's *Sciences of the Artificial* (1996) was a prescient text of meta-engineering; probably the first with an appropriately broad conception of its scope, which extended from evolution to economics, with the whole of the human and design sciences in between.

This closing chapter seeks to redirect you like a (theoretically and methodologically inclined) engineer through what went before, and reflects my methodological origins in the fertile period of the 1950s and 1960s that produced these works—one more dominated by an awareness of the dimensions of adaptive control than the hope of a universal computation. These perspectives have once again become cutting-edge in the new systems biology (Wimsatt, 2007a) and the newer synthesis of evolutionary developmental biology (Wimsatt, 2007b).

CHAPTER THIRTEEN

Epilogue

On the Softening of the Hard Sciences

This title is a deliberate anomaly. Shouldn't one talk more naturally about the hardening of the softer sciences, indeed of all the sciences, with the rise of the computer? Mathematical modeling, automated high-volume execution of experimental procedures, high-tech data and database manipulation, statistical analysis, and visualization—all on personal desktop computers more powerful (and better supplied with software) than the university mainframes of 30 years ago have transformed and automated every corner of our science. Isn't this what any good positivist would have wanted?[1] This is surely the trend visible from reading ads in any recent issue of *Science* or *Nature.* But along with this, for most philosophers, historians, and sociologists of science, there has been a softening of our *conception* of the sciences as we move away from logical positivism in our various often orthogonal directions. For many, this has lead to a generalized rejection of objectivity and espousal of a deeper valuational (and usually socialized) conceptual relativity. This kind of global facile rejection of scientific objectivity is a serious mistake: it ignores the strengths and successes of science—in effect by finding yet another way to see human nature as prior to nature rather than as a part of it.

In the post-Kuhnian period, we spend more time looking at how scientists actually practice science through the "laboratory studies" of sociologists of science, and increasing "participant observation" by philosophers of science. Scientists often don't behave as they are "supposed" to, but their practice still seems rational and well adapted to their circumstances. They do far better than with the simple-minded, context-free rules from our overly idealized normative theories of scientific behavior.[2] If this is softening, then so be it. A better-founded softening of our visions of science arises throughout, from realistic exposure to what scientists actually do, and this applies also to their access to increasingly powerful computational tools.

Increased computational power has made us less satisfied with our prior *in principle* claims about how scientists are supposed to do science. Since we can, individually, *actually* do things that were almost unthinkable before, we must reconceptualize the boundary between what is doable in practice (even if not quite yet) and what is achievable only in principle. The latter is not a distant though visible destination point (Chapter 11). It says more about the mathematics of computability than about the range of our science, though that fact is well hidden in our idealizations of scientific method. We have become more aware, practically speaking, of what we *really* can and cannot do. We have increasingly rich theory of how easy or hard it is to do various things in practice. The theory of computational complexity allows us to classify problems (usually under various representations) as computable in polynomial time (and of what degree!) versus undergoing combinatorial explosion with increases in the number of underlying variables. We try to stay away from these last problems, or if possible reformulate them so that they don't blow up, or do so more slowly.

So are we limited to a blind and uncritical descriptivism? We still need generalizations—including normative ones, though ones more sensitive to the finer details of context than in the past. Our normative claims should more often be rooted in heuristics of effective and efficient problem solving and scientific practice, or demands of the situation, than in general logical or deontological claims. In Kantian language, this is the domain of the hypothetical rather than the categorical imperative. And good hypothetical imperatives should give effective guidance. *I seek to elaborate a philosophy of science that can be practiced (not in principle, but in fact) by real human beings with the real instruments we can bring to bear, now and in the future.* But might this just anchor us to a temporally and technologically dated framework? It need not! For example, we get new and interesting conclusions just by assuming that our reasoning processes must be boundedly rational, in Simon's sense (1955, 1996/1969). This assumption will never become obsolete. Bounded rationality suggests preferred ways of doing things (and thus is normative), and predicts how and when our tools—our reasoning heuristics—should work well and when they should break down (see the appendixes). Knowing the limits of our tools should improve our scientific inferences and productions. We should be as much therapists or consultants as theoreticians, though the boundary between them is less clearly marked in the New Naturalism than it once was.

This is far from the mainstream views common in philosophy when I was a student in the 1960s, and still quite different from mainstream views today. It is more common in the areas of science I have been attracted to—areas that have only recently ceased to suffer from what used to be called "physics envy"—including, paradoxically, much of physics. We should learn from such areas to base our epistemology, metaphysics, and pragmatics on more representative regions of science. Much of what they have to teach is "knowing how" rather than "knowing that," and the acquisition of changed attitudes and values—a new way of seeing, rather than (just) the collection of a new set of facts.[3]

Classical physics envy[4] was eliminative and full of talk about what was reducible or *computable in principle.* It was austere and deductive. Errors were to be weeded out up front by avoiding methods that would allow them to creep in. Physicist Richard Feynman[5] called this a Euclidean methodology. This vision more accurately reflects a stereotype of early logical positivism (perhaps closer to the logical atomism of Bertrand Russell) than the view of any scientist. The new sciences, the softer hard sciences, have rich phenomenological strains, use numerical simulations as much as analytical results, are compositional rather than eliminative, and *computational in practice.* They celebrate emergent phenomena (Chapter 11), and all kinds of higher-level entities and properties. Approximations and heuristics are "up front" in attempts to get the answers they want, as are course corrections in eliminating errors that infect our efforts at every stage. To a good Euclidean, heuristics would look downright sloppy—elevating ad hoc-ery to a methodological principle. Nonetheless, it is what real scientists do regularly and do well. It has much to recommend it in philosophy and in science. I want to contribute to an understanding of this approach, and to accelerate the transition to their new way of seeing and doing.

Several friends have suggested to me that some might enjoy and be helped by hearing how an unrepentant proto-positivist came to stray so far from the beaten path. So I'll relate some experiences with how scientific problem solving is *actually* practiced that led me—then a worshiper of simplicity and theoretical elegance—to develop a special taste for nearly intractable messiness. By now, *pace* Quine (high priest of Ockham's eraser), I hope I have traced some surprisingly well-marked paths through the ontology of the tropical rainforest inhabited by primitive "folk-methodologists" of the biological and social sciences. I have tried to provide a beginning and partial manual for the arts of the inexact sciences. Denizens of these "ontological slums" seemed always on the verge of eviction, as philosopher-kings plotted how to practice urban renewal on their neighborhoods. But in the last three decades, increasing numbers of "hard scientists"—mathematicians, physicists, economists, and some biologists, with interests in fractals, chaos, computer simulation, visualization, and non-linear dynamics, are moving into the area and proclaiming themselves "holists." New developments of theory in these areas have helped to make messiness and ontological profligacy fascinating, attractive, and even respectable. But not any mess will do. There are standards, some with roots in traditional methods and others transplanted from scientific practice, for how one should do it. And many exemplars of "the new messiness" do not fare well.

From Straw-Man Reductionist to Lover of Complexity

I've been messy from the start—more of an engineer than a theoretician, and seldom able to get interested in an important classical philosophical problem until I found it blocking the way on some path I wished to travel. I grew up liking big flashy equipment with lots of dials and controls—the kind in 1930s "mad scientist" movies. I entered Cornell in 1959 as a freshman intent on doing a degree in engineering physics on the way to becoming an aeronautical engineer and running in the space race.[6] My biologist father was a classical histologist and embryologist who worked also on the physiology of reproduction and hibernation. He was, by avocation, a naturalist, a falconer, and a woodsman. I grew up playing around his lab at Cornell, going with him on field trips, and building various mechanical (and sometimes explosive) things in our well-equipped basement shop. We had more unusual pets than any collection of kids on the block: little brown bats (that's a species, *Myotis lucifigus,* as well as a description), boa constrictors and king snakes, a raccoon, a tarantula, a sparrow hawk, a goshawk, and more. The more exotic and dangerous things (dad's colony of vampire bats and a South American pit viper for which the closest anti-venom was 250 miles away in New York) usually stayed over in the lab—

though we were welcome to bring friends in to see them and be impressed. I remember a demonstration of rattlesnake locomotion on our living room rug. Spelunking trips to collect bats and rock-climbing trips to collect pregnant female rattlesnakes for lab work were great fun, though that work (and its products) seemed too messy to be *real* science. This perception wavered only temporarily when (in eighth grade) my father let me help him with some pioneering tracer work using Iodine-131, to figure out how the thyroid worked. (*That* after all, or at least the radioisotopes, was almost physics!) The staccato relay race of decaying iodine flashed diodes on the scintillation counter (and counts were coded in binary) in the lab at Brookhaven where we moved in the fall of 1954 to learn the procedures. I still feel the excitement of the cold walk out from the lab to the car late in the evening after a "run"—an evening spent counting radiation in bat thyroids, with the ever-present steam whistling out of the red-and-white checkerboard of the nuclear reactor stack dominating the scene a block away. *That* was science, just like in the science-fiction movies, and I lived daydreams of portentous discoveries!

Through this period and most of college, I was an engineer at heart, but an engineer with a theoretician's values—values in unstable equilibrium with my own practice and abilities. My high school career was populated by frequent but irregular dates and infrequent but regular science fairs. I built a smoke tunnel for investigating flow around airfoils, and a (barely) working ramjet with a rig for static testing. (It was far too heavy and unreliable to fly!) I designed a rocket test stand, and did an abortive mathematical project on gravitational equipotentials for the sun-earth-moon system, which allowed me to make some neat 3-D drawings. By the time I entered Cornell, I was—apparently spontaneously and naturally—a hard-core reductionist who worshipped the adamantine clarity, precision, and deductive power of classical mechanics and similar disciplines, which all sciences should try to imitate or deserved to wither. I expected my college education to show me *how* all of the important phenomena reduced to Newtonian mechanics, or its descendants—I already *knew* that it was possible. I was a walking breathing straw-man. On the other hand, I soon discovered that I wasn't great at doing problem sets, particularly when they weren't interesting and the instructor already knew the answers. (It was different if the problem was interesting: I remember discovering to my delight that you would be able to see further around the backside of the moon [it, in effect, visually rotated] the faster it traveled past you—and quite a ways around, if it was travelling at relativistic velocities.) Like many physics students, I also "discovered" tachyons, but quickly dismissed them as a physically meaningless solution to the equations, about five years before Nobelist Steven Weinberg took them seriously. ("Imaginary" mass?—you've got to be kidding! But "imaginary" numbers are useful for something . . . ?)

I took two logic courses in my freshman year, mistook that for philosophy, and wanted more. In my sophomore year, I took introduction to philosophy, in which we read the Platonic dialogues (fun and thought provoking), Berkeley (what kind of nut would wonder whether a tree was there if no one was looking at it?), Mill (I hadn't yet warmed to ethics), and Russell (real "scientific" philosophy—like classical mechanics!). (Since corrected, these youthful first reactions probably still sometimes tint my perceptions.)

In my junior year I transferred from Engineering Physics to Physics—thinking I would get more theory, but found instead only more math. (I later found philosophy of science to be the kind of "theory" I was seeking when I discovered Reichenbach's *Philosophy of Space and Time* [1958].) By then I was taking an overload of physics and math courses but discovering lots of people who did problem sets a lot better and faster than I could. I had put aside philosophy, bought a motorcycle, and fell in love four times. I was doing a physics lab (in which—surprise—nothing ever worked out the way it was "supposed" to, and the error analyses were both more creative and more interesting than what we were supposed to establish). A year-long course in electromagnetic theory was very hard—about half of the students were new physics grad students. There was still more applied math (ditto!), and I was finally taking classical mechanics (much easier—though the class was taught—in Italian, I think—by an Italian physicist). I decided to read the book rather than go to lecture, and instead audited a course taught at the same time in the graduate school of aerospace engineering by Ed Resler (Edwin L. Resler Jr). Resler had delighted all of us EPs with a beautiful course in statistical mechanics the year before. He was (and is) a theoretician among engineers—later chairman of Cornell's applied math program—and a marvelous teacher who communicated with a wry wonder the excitement he felt about his subject matter.

Messiness in State-of-the-Art Theoretical Physics

This was 1962, still early in the space race, and "we" were interested in "re-entry" problems. Resler's research course was on magnetoaerodynamics—a specialized sub-area of magnetohydrodynamics, though hardly one that founders of the parent discipline could have anticipated. (Or: What happens to and around a Mercury space capsule—which generates a lot of drag and a little lift—when you dip it into the upper atmosphere going fast enough to ionize anything that isn't ionized already?) Mathematically speaking, I was hanging on by my fingernails, but so was almost everyone, and it didn't matter much: theory was still under construction and one paid more attention to the qualitative behavior of equations than exact solutions. Besides, the area was full of fascinating phenomena. Do you know that if you shoot a high velocity rifle bullet through an evacuated tube filled with a low density plasma, there will be *forward* facing shock waves—a wake that gets further ahead of the bullet the farther away you get from its axis of motion?!? (They are called Alfven waves, after the imaginative physicist Hannes Alfven, who won a Nobel prize a decade later for that and a variety of other things.) Now widely recognized, they appear (though in more regular configurations) in such mundane places as the bow-wave of the earth as it plows through the solar wind, but they were then well-kept secrets from undergraduates. At first blush, to intuitions honed on the behavior of waves around moving bodies in water and air, forward facing shock waves seemed to violate causality—for the same reason as a forward facing light cone would in special relativity. But there was a sensible explanation: the bullet carries with it an entrained or frozen electromagnetic force field that extends further ahead as you got farther away from the axis. And you could get that out of the equations.

The course fascinated and scandalized me, and marked my first move away from an enthralled worship of formalism, simplicity, and elegance. Resler spent the first lecture on methodology. He covered three boards listing the equations relevant to solving problems in magnetoaerodynamics—22 simultaneous equations starting with Newton's laws of motion, but rapidly progressing to non-linear partial differential equations for hydrodynamic flow, compressible aerodynamics, the electromagnetic field equations, diffusion equations, equations for ionization kinetics, and the like. (I found that the father of one of my childhood friends, Nobel laureate Peter J. W. DeBye, had a length [and an equation] named for him—the mean free path of an ion before it encounters an oppositely charged particle. And I had thought he was a chemist! What was *he* doing in a physics course? Didn't physicists *teach* chemists all they had to know?) Resler had written down about every physical equation we had studied so far, and a few we hadn't. "These are the state equations for our system," he said, "How many unknowns?" (Count them!—which he did just before the bell rang, as the suspense built.) "Note that there are also 22 unknowns" he said with an air of discovery as if he'd never counted them before, "Therefore they are solvable in principle!"

There was a pause. I waited expectantly for him to produce the analytical solution in closed form—or to say he'd do it in the next class. I started to draw the kind of box you do in your notes for closed form solutions—but none were forthcoming. I didn't know it yet, but this was the first time I had encountered a set of equations that we weren't going to be expected to learn how to solve. He continued: "But you can't *really* solve them, of course. What we do is lump a few variables to make dimensionless parameters, let 18 of the variables go to zero or infinity, make a few more approximations, pick interesting values for the other 4, and put it on the computer for the weekend.[7] [Remember, it was 1962.] Then, if the computer hasn't crashed or the run hasn't bombed, we collect the box of printout, and try to make sense of it and decide what other values to try." This was supposed to be theory, which I thought would be tightly deductive. But it all sounded very tentative, experimental, and full of approximations—I was deliciously scandalized. This was state of the art theoretical physics, but it seemed more like crudely inductive and exploratory experimental science—only with equations, not organisms. The art, it seems, lay not in constructing a general solution, but in finding sneaky ways to produce situations in which you could ignore most of the complexity and generality you had gone to so much trouble to build into the model in the first place. (In fact, the art was to find a *different* and revealing way—or ways—of simplifying it so you weren't just removing the complexity you had put in, but finding new simplicities that permitted you a decomposition along different fault lines.)

I've always been fascinated with taking limits of mathematical expressions. It's almost as if the original equation has multiple personalities: if you tweak it the right way, it behaves like something else and another equation is produced. Of course the potential for that behavior is in the original equation—with the right initial and boundary conditions, a master equation is like a machine that has other machines hiding in it—which you get to see when you take limits in the right way. Taking limits also helps to conceptualize the effects of heuristics, which solve problems by simplifying them: limiting approximations are prime examples of such heuristics in the "exact sciences." Limits also provide a way in which macroscopic order and simplicity can emerge from microscopic confusion and complexity (Monte Carlo simulations are another.) These are all physically based metaphors for how the same system can appear differently to observers who interact with it differently—anticipating the levels and "perspectives" of Chapter 10.[8]

This seemed more like Kurasawa's famous cinema tale, *Rashamon,* than my conception of canonical physics. The arguments were pretty far from first principles. There was much more room for intuition and exploration than I had supposed, and deductive arguments didn't convey the certainty I expected. (In fact, they were usually called "derivations," not "deductions," and because of the use of limits or approximations, they were *not* truth-preserving, something I remembered [and used] when I moved to philosophy and was surrounded by people who thought that theories were "deductive systems.") It appeared that no one knew exactly what results would follow, and when, and which of them to trust—not Resler, not the authors of the high powered applied physics monographs we were reading, nor the writers of the current research papers, who made an awful lot of unrealistic assumptions and approximations, and surely, least of all, us. I was impressed and intrigued. It looked like physics wasn't at all like the physics I (thought I) knew. This was extraordinary physics, with a world of new phenomena and equations (most derivative, but new nonetheless), and I found out that work at the frontier wasn't at all like I though it was. But I had another surprise coming, for I was soon to find out that quite ordinary physics wasn't quite as advertised either.

Hidden Elegance and Revelations in Run-of-the-Mill Applied Science

That revelation came the next year, which I spent working in the engineering department of the adding machine division of NCR,[9] which was conveniently located in Ithaca. (Cornell had perceived that my dedication to doing problem sets was less than total—indeed much less than appropriate for a physics major—and suggested that I take a year off to think about it.) Ithaca was a backwater division of the company, an environment in which you didn't expect state-of-the-art problem solutions. Apparently I knew more mathematics and theory than anyone else in the department, so I was often given problems that hadn't yielded to the standard "cut and try" methods. My immediate boss, Bernard Hogben (Cornell EE 1935), was one of the few people there with a full-fledged engineering degree. The scuttlebutt was that Oscar, the head engineer, didn't believe in electricity. He had come "up from the line" (promoted from the assembly line to the engineering department), and always preferred complex mechanical designs to simpler, cheaper, and more reliable electrical ones. This wasn't an argument about twentieth-century science; we hadn't yet successfully navigated nineteenth-century physics. And he certainly didn't believe in theory. Bernie Hogben, who had some faith in theory, had to give me these projects in secret. To Oscar, I was just an unusually good draftsman with too much education.

This story is full of seat-of-the pants heuristic design principles (not only for how to do things, but how to break the problem into parts [near-decomposability], for when something could be ignored, or for when it had to be taken into account). Many of these are left to the footnotes, which are unusually long for this section, so look to them not only for context-setting details, but also for evidence in support of the claim that "It's heuristics all the way down."

We had received an order from a Swiss bank for 3,000 special adding machines that could imprint magnetic numbers on the bottom of checks that lacked them—as most checks did in those days. It was a big order, and an important one—imagine a manufacturing order now with a price tag that would buy 1,500 new cars. For each check, a cam had to advance 3 inches of fresh carbon ribbon tape to imprint the number. The ribbon cam designed by another engineer was breaking the tape. (He had just carved out what looked to him like a plausible shape—a strategy that works surprisingly often because large safety margins are built into normal engineering practice.) Fresh with theory and confidence, I was certain that the problem was just to find the cam profile that would advance the ribbon the necessary distance in the time required while minimizing the maximum value of acceleration (and thus the force on the tape). This I set out to do.

The machine was a complex thicket of mechanical linkages driven from an electrically powered driveshaft. It was basically a 1919 "full keyboard" cash register mechanism (initially crank-driven) with 43 years of layered added mechanical kluges increasingly crammed into the existing case to accomplish new functions or modulate existing ones. Our computing facilities consisted of a single *Frieden* scientific mechanical calculator that cost $1,200 (back when a new car cost less than twice that much) and did less than a modern pocket calculator—less reliably, more noisily, and much more slowly. (Taking square roots required calculating and reentering successive approximations using

Newton's method, and took 2–3 minutes if you didn't do it all the time. And anything trigonometric was out of the question. You solved it graphically, or used tables or a slide rule.) Given that the solution had to be mechanical, I had to design a set of linkages to transmit power from the driveshaft to the tape advance mechanism with a velocity profile that didn't break the tape. The technique used for this kind of problem involved an ingenious kind of analog modeling that they called "layout work." I learned it from those who knew it as I went along. I've never seen it described before, and it is a fascinating addition to the history of computational and simulation methods.[10]

The first step in this kind of modeling was to make very accurate drawings of the proposed mechanical parts in the linkage, enlarged by a factor of 4 or 8 (limited by the size of the largest drafting table you could find). Enlarged drawings gave more accurate results. One then cut out paper drawings of the parts corresponding to the various links (four or five if I remember), and pinned them down at appropriate locations in a master drawing that located all the major attachment points and components of the machine. Starting at the driveshaft (the causal "input variable"), you pin and rotate the link coming off the rotating driveshaft through small angular increments (2 degrees), plotting the location of the attachment point for the linkage at the other end. That also gives the motion of one end of the next link as a function of the motion of the driveshaft. The process would then be repeated for the next link in the series, moving it as driven by the first link, while constrained by its other connections. This process was followed through the linkages until one got the angular rotation of the cam arm as a function of the rotation of the driveshaft. *One in effect modeled the constraints and motions in the machine by cutting out and playing with paper dollies!* I no longer have my drawings from this project, but the construction of the limb-motions in a high jumper (using Marey's [1895] pioneering light-tracing studies to capture motion) gives a sense of the kind of articulated motions produced and plotted (see Figure 13.1).

This procedure only worked for motions in the same or parallel planes—and fortunately the machine architecture was remarkably planar. (We only do what we can!) But remarkably, planar design also allows for tight packing of components that had to be capable of independent motion without interfering with one another—a fortuitous consequence of design practices that led people to think primarily in two dimensions while building machines in three!

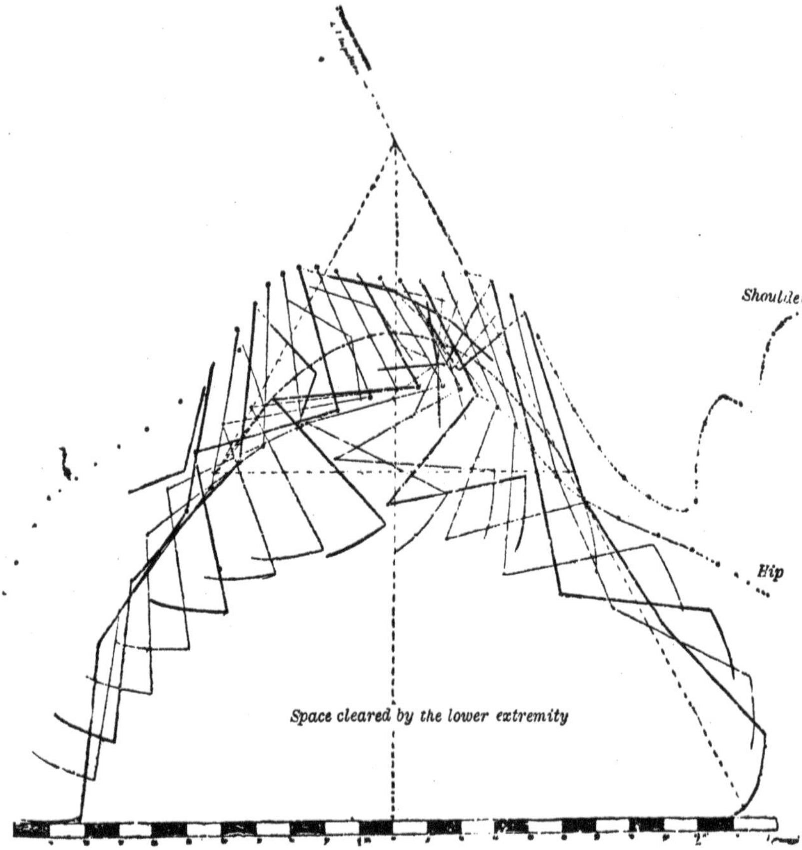

Figure 13.1. Geometrical chromophotograph of the movements executed in taking a high-jump (from Marey, 1895, p. 155, fig. 105). This figure suggests, with its articulated arcs, the phenomenology of the constructed velocity profiles leading to the construction of the cam profile as described in the text. Both depict the motions in an articulated structure in terms of a series of successive displacements.

In principle, one could have modeled and solved the transformation induced by the linkage as a nested series of trigonometric transformations yielding the desired input-output relation. Clearly out for the *Frieden!* Doing it with trig tables would have been both slower and more likely to introduce major errors (by misreading the table). Paper dollies would have less precision than the tables, but the errors introduced were comparable to the "slop" in the mechanism,[11] and could thus be ignored. These tradeoffs—lower accuracy in general, but less chance of large errors and more rapid computation—were characteristic differences between analog and digital techniques and computers. This method was very fast: you got a new point on the trajectory of a link as fast as you could repin its base at a new location and swing an arc—6 to 8 points per minute, or half that if you wanted to be extra careful. No one could calculate nearly that fast.

The next stage of the problem was relatively simple, and also involved a geometrical analog solution to a computational problem. Given the angular displacement per unit time of the cam arm and the desired acceleration curve for the required tape displacement, one can generate the desired cam shape through a geometrical construction in which points on the cam surface are generated from intersecting arcs swung from points on the trajectory of the cam follower (the link driving the cam) and the desired velocity trajectory of the tape. Theory came into this, but only in determining and solving for the appropriate velocity profile for the tape, and that never by solving an equation. Once that was solved, it was plotted onto the drawing to give the desired displacement as a function of time. This was one of two necessary inputs to the geometrical construction technique determining the cam profile. That technique, and the other input from the driveshaft linkage train, and the modeling technique—all of the complicated stuff—fell out from normal engineering practice. But there was more.

So far this is all applied geometry and Newtonian mechanics, but the problem isn't solved yet. I had shown how a constant angular velocity at the driveshaft end would be transformed into a changing velocity at the cam-linkage end. But this is a solution to the wrong problem: the shaft *doesn't* rotate at constant velocity because it has variable loads on it at different parts in its cycle. It is driving all of the other parts of the machine through similar linkage arrangements, each with varying resistance at different parts of its particular cycle. Even if the real rate of rotation *could* be calculated *in principle* (not likely!), no one in their right mind would do so, even today.[12] They would do what we did: get high-speed motion pictures of the shaft of a normal machine to determine its actual rate of rotation, measured from successive movie frames, as the machine went through its cycle.[13]

The movie gave me a velocity profile at the driveshaft end. It yielded the transformation to run through the linkage to convert the "constant velocity" rotation curve into the "real" angular rotation of the cam arm per unit time. (Since I already had a scale of 2 degree "tic" marks propagated through the linkages by the construction, laying out the variable transformation only required "reading" their projections at each linkage and placing the transformed marks appropriately.) That the "ideal" and "real" rotations could be solved independently and then used together in this way makes the problem "nearly decomposeable," an important property first noted by Simon, and elaborated in several of the chapters here.

Empirically based as this solution was, it still made at least two idealizations (in fact, there was a quite revealing third).[14] *First,* it ignores variations and slop in the linkage—for example, variations in the exact size and location of bolt holes—which would affect when and with what exact trajectory the next link in the chain would begin to move. Generally, there was no way to correct for it. The amount of slop would have varied from machine to machine—though upper bounds for allowable slop could be specified in parts' tolerances—calculated in terms of allowable failure rates, which would have been a cost-benefit process. (Indeed, since this slop was comparable to that found in the modeling with paper dollies, the modeling process "calculated" the answer to the "right" number of significant figures automatically.) *Second,* it ignored the additional load on the driveshaft induced by the new mechanism, which would slow it down some while the load was applied. *This* was ignored because there was no way of correcting for it without knowing the cam profile already, since the cam profile would determine what load was necessary to drive the tape advance mechanism. If it hadn't worked it would have been easier and more accurate just to try again using the prototype-designed cam profile as a "first approximation" and, if necessary, retaking high speed pictures of the main driveshaft with the prototype linkage in place, re-iterating the design process with the new "approximate" solution. This kind of iterative feedback-correction process is ubiquitous and essential with tuning complex mechanisms and their redesign processes. With luck, the additional load would be "negligible," and the first approximation would work. Ultimately (some time later) I was lucky—it did.

Now I thought I was finished. I generated the cam profile and completed the drawings for the necessary parts. I was done several days early, but I had a nagging suspicion that it had all been too simple. Cam design was an elective fifth-year engineering course, so there might be additional complications I didn't know about. I headed off to the engineering library at Cornell over the weekend to read up on cam design. I'm glad I did. The first sentence of the first chapter of the first book I opened read as follows: "*The first law of cam design is to minimize the jerk.*" It was in the second book too.

Was this a joke? I'd never even heard of the "jerk." (How many of *you* have before now?) Reading on, I discovered that this was the derivative of acceleration—the third derivative of displacement! The book nowhere said why you had to pay any attention to the third derivative, and neither did any of the others. Was Newtonian mechanics wrong? Didn't **F** = m**a**? And wasn't force what broke things? I mean—we all knew Newtonian mechanics was *really* wrong, but *here?* We weren't near either the relativistic or quantum limits of classical mechanics. (Our adding machines were neither that fast nor that small!) Then I realized that the problem was not with Newtonian mechanics, but with the implicit Newtonian theory of materials: all of our problems had always been worked with perfectly rigid bodies or perfectly elastically deformable ones. Real materials are neither—they can take a surprisingly large force without breaking if the force is slowly applied—but apply it quickly, with a *jerk,* and CRACK! With this new "law" about the jerk derivative in hand, I generated the new displacement curve, and from it the new cam profile, and everything worked out because the new cam didn't break the tape. (Substituting the third for the second derivative in the calculations is also a nearly decomposable part of the problem—now at the *other* end of the calculation from determining the velocity profile of the driveshaft. It only changed the desired trajectory, which was used with the [idealized] "actual" displacement gotten by solving other parts of the problem, to generate the cam profile.)

My puzzlement only grew though: Why had I never heard of the jerk derivative? I had taken some potent applied math, physics, and applied physics courses, as well as some *very* applied engineering courses (drafting, metal-casting, machine shop, circuit design, . . .), and lots of things in between. Nowhere had anyone thought it important to mention the jerk, which could have been taught in a high-school physics course. It was an intuitively motivating example for elementary differential calculus, with potential applications everywhere.

My puzzlement got deeper, and my curiosity led me farther afield. I learned that this phenomenological law was exceedingly general and applies to the failure of all kinds of materials. It is important for crash safety in automobiles twice over—because it described the tendency for structures (including bones) to break, and also because in moving or stabilizing ourselves, we counterbalance or adjust to impressed forces (read: accelerations) and detect sudden changes in force (read: jerks).

The lag time for our nervous systems to detect and respond to these changes produces potential stress and chaos-inducing deviations from our optimal set points. By the time we respond to a strong jerk, usually the damage has already occurred, even if we could have handled a more slowly applied stress. If theoretical physicists didn't have to learn about the jerk, why at least weren't future engineers taught about it?

I made a practice of asking anyone who might possibly know anything about the jerk. Barely anyone had ever heard of it. Those who did had no explanations. It might as well have been a brute fact of nature. I first got a move toward an answer 16 years later on a cross-country air trip. A neighbor pulled out a powerful programmable calculator—just like mine![15] He was a high-powered materials science engineer, steeped in solid-state physics and quantum mechanics. Yes, of course, he knew about the "jerk." He said that it was exceedingly general but no one knew exactly *why* it worked. Solid-state physics models had not yet explained it, and it was probably true for different reasons in different materials. (Supervenience in materials science!) He thought it was avoided in early courses because people were embarrassed that they couldn't explain it. Another faculty member at a major engineering school opined wryly that to fail to explain such a simple law would give budding engineers the wrong message. Nascent scientists are seldom treated any better! As modern critics of science education are quick to point out, that would have been just the *right* message to give—more intriguing and seductive than papering over what you don't know to present a speciously seamless body of knowledge. (Philosophers also sometimes paper over holes in arguments with specious *in principle* claims—see Chapter 11.) Part of what made Ed Resler's course so delightful to me was that he rarely papered *anything* over, and when he did, it was well marked.

I thought the mystery of the jerk was solved—as a problem in social psychology. But not quite! I found out the likely truth in the summer of 1996 when I met Richard Dunlap—a versatile engineer and former head of the Navy's research division for undersea technology. He suggested that the "law" of the "jerk" is not a law (despite descriptions), but a "curve fitting" correction used to improve predictions for real materials when the second derivative didn't work adequately (personal communication, August 1996). So going to the third derivative was like taking the next term in a Taylor series expansion or using a higher order polynomial to fit a set of data. The "jerk" that breaks could just as well be some function of all higher derivatives and not just the third.

There likely was *NO* theoretical basis for a minimization principle with the third derivative of displacement. Using it (as the next term) just gives a tolerably low error in predictions. Indeed what kind of theoretical basis *could* such a law have without some kind of derivation?

Nor, without specific attempts to find and to rigorously test it, are we likely to falsify it in engineering situations. Petroski (1994) relates how Galileo's *incorrect* analysis of deflection in beams was missed for 200 years. (It was dimensionally correct, so it scaled appropriately to different size structures, but it was off by a factor of 2.) Galileo's prestige, the engineering practice of using a safety factor of about 3, and other possible explanations for any failures that did occur—local inhomogeneities in materials, crack formation with aging, incorrect design estimates of the actual beam loads—all make this explicable. Absent mechanisms in solid-state physics motivating a law with the third derivative, or other mechanisms ruling it out, the answer is indeterminate. Perhaps this richer story should be taught in high-school physics: it could motivate future condensed-matter physicists. Dunlap attributes his idea to lectures he heard at MIT in the 1930s.

My experience with asking about the "jerk" derivative, and the blank stares I got, suggests that avoiding such questions may be self-amplifying. If the teachers of too many physicists were embarrassed about being unable to explain it, too many of them would never encounter it—and go on to teach others without ever knowing their error. And (if it was discussed) it was only engineering after all—right? Let the "applied boys" worry about it. We do basic research around here! I've met quite a few such people in science faculties around the country. Almost none of them had ever heard of the jerk derivative.

"Pure" versus Applied Science, and What Difference Should It Make?

This scandalized me again, and I wondered how much of the theoretical physics we learned had been deliberately or inadvertently "laundered" in this way. Idealizations and abstractions are okay in science; it's the name of the game in model-building (Cartwright, 1983). But you must be able to recognize *where* it is idealized, and *when* it is overextended (see Chapter 6). How many of the roots of the physical sciences in applied problems had been hidden as a result? This was not a new question for me and Cornell was a good place to ask it. Cornell was full of basic science being pursued in applied places. Theoretical solid-state and plasma physics were being done in engineering departments. There was nothing particularly applied about them—they were just areas of basic science being pursued by people who ultimately needed answers for real problems.

Thus, plasma physics was studied in electrical engineering, where EEs wanted to know about reflection of radio waves off ionized layers (plasmas) in the upper atmosphere; geophysics (then part of geology), where people wanted to know about the Aurora Borealis (plasma again!); aerospace engineering (us!); astronomy, which included astrophysicists who studied Van Allen belts, stellar structure, and other plasmas; and others (cosmologists) who in those days seemed to talk philosophy and other creation myths in high physics dress. The shock tube—a key tool in plasma research—originated in the 1950s in aerospace engineering as a way of generating and studying hypersonic plasma flows in the range of Mach 20 (20 times the speed of sound) and higher. It then migrated over to chemistry, where it gave a way to study reaction kinetics in very fast reactions, with the temporal course of the reaction spread out in space behind the shock. Lasers (which can make plasmas) came to aerospace engineering a year later, and Tokomaks (magnetic plasma confinement vessels for fusion reactions) entered nuclear engineering a few years after that. Plasmas developed such a large and varied experimental and theoretical following that one could no longer pretend it was just applied physics, rather than the physics of gaseous matter under a range of relatively high-energy conditions. Foundational, no—real, basic, and important, yes!

The "pure" versus "applied" science dichotomy is specious in another way. *An experimentalist in any science (pure or not) has to be part engineer—to know what sorts of instrumentation are possible, and how to design, build, calibrate, and maintain it, how to tell when it is not working, or biased, how to improve it, and how plausibly to redesign, extend, or adapt equipment and experimental designs to other purposes.* Engineering knowledge is so central to experimentation that it often won't do just to hire an engineer.[16] Good experimentalists need to know a fair amount of how to do it themselves—in part because they need to know when not to trust the results too much, a skill also crucial in heuristic reasoning. Nor does the theoretician/experimentalist dichotomy work as a barrier between engineering and basic science: unless theoreticians limit themselves to generating analytic solutions (forgoing simulations or use of numerical methods), the structure and limitations of the tools and methods used will inevitably introduce an element of engineering into their work.

Another applied science at Cornell—genetics—was then (in 1962) wholly in the agriculture school. Why? Because as it grew at Cornell after 1900, the agriculture school had money and relevant research problems and the "pure" School of Arts and Sciences had neither.[17] What were students to think if they had to go up to the "Ag" school to learn about it? To "real" scientists, Ag was even lower status and more applied than engineering. After World War II engineering had the romance of a new scientific discipline. But I remember the consternation and disbelief of pre-meds 40 years ago: first, that they had to take an Ag course at all, and second, that it should turn out to be so difficult. But times change: both sciences have since become "pure," and genetics has even become "foundational." So much for the distinction between the pure and applied sciences! Making the distinction is not as easy as you think. The most important lesson to be learned here is that *the flow of knowledge is not a one-way stream from theory to application* (as most philosophers and historians of the basic sciences appear to think).

A delicious irony of this all-to-common scorn for application among basic scientists emerged recently when Leo Kadanoff lectured on model building in physics to our seminar series. Kadanoff is a leading (almost legendary) theoretician in the now sexy land of non-linear dynamics and chaos. He asked if we could do the seminar at his place where he had the audiovisual and computer equipment to show off his stuff (it clearly isn't true as philosophers suppose that the words are all that matters). His talk was on the application of lattice-gas modeling methods to a variety of problems in fluid dynamics. Lattice-gas methods idealize particles to move and interact in particularly constrained ways such that their motions and collisions can be represented in two dimensions in a (usually triangular) lattice according to simple, well-defined rules. Such models are readily simulated in parallel, and speed and simplify computations by allowing them to be done with integers rather than real numbers.

Halfway through the talk, he turned the meeting over to a visitor, Dan Rothmann of physical sciences at MIT, with the comment that Rothmann was doing really interesting stuff that went farther than his own. Rothmann took the floor, showing videos of supercomputer simulations of the behavior of non-miscible fluids under various conditions. In his last example, two fluids were modeled as interacting under high pressure in a porous solid. Suddenly I realized that what we were seeing was petroleum geology—oil and water in a porous rock matrix! Sure enough, over dinner, Rothmann told me that he had come to MIT from Schlumberger (the French petroleum giant), and that had motivated how he framed his physics problem.[18] (Rothmann was in geophysics, not physics: interactions with meteorologist Edward Lorenz, credited as one of the modern discoverers of chaos, had interested him in chaos and non-linear dynamics.) Fair enough—so theoretical physicists now do some applied problems! But tout the work of geophysicists?

I mentioned that this was ironic and reminded me of a faculty resident in our undergraduate hall at Cornell who had done the strangest things in the early 1960s. Walter Graff, professor of civil engineering, modified the Navier-Stokes equation for incompressible fluid flow from hydrodynamics to allow for the inclusion of solid masses of varying sizes and densities. He wanted to develop a theory of the motion of solid matter (rocks and earth) in irrigation ditches. This occasioned patronizing sniggers from some physics friends about "trying to make what he was doing look scientific." If only they could have seen the Kadanoff-Rothmann talk! Applied work that was beneath contempt has become "High Church science."

Considering all of this, I am led to an only partly tongue-in-cheek proposal: *could applied science be, at least in part, anything that is too difficult for the current generation of theoreticians to solve exactly?* Until recently, and still for the most part, numerical analysis and simulation has lower status than derivations and closed form solutions. This remains true even when you can't visualize the latter, or have to make so many idealizing assumptions to get them that they are no longer solutions for—or even useful guides to—the same problem.[19]

Abstraction is often a useful heuristic in problem solving, but there can be too much of a good thing, and sometimes one learns far more by a close analysis of a real case in all of its detail. Philosophers should worry about this more often. This is one dimension, not the only one, of the relation between pure and applied science, and, a step further, between applied science and engineering. There are salutary practices here, which I have already talked about: engineering has much to contribute both to traditional and to new topics in philosophy of science, and this perspective is almost untouched. A sense of the importance of having application to real problems is one obvious and important starter, but we can go on from there.

Theory construction and testing have more than a superficial resemblance to or connection with fault localization and error analysis in integrated circuits and software,[20] reliability theory and reliability testing (Barlow and Proschan, 1975), prototype construction and testing, medical diagnosis, and the diagnostic part of auto mechanics. It's for sure that experimental design does (Schank, 1991). I have throughout this book urged that we take a close look at the functions of scientific practices, and this calls for the use of intuitions from engineering design also. Careful study of these areas could provide much more.

Hortatory Closure

A new realistic, mechanism-based, application-centered paradigm for philosophy of science could provide metaphors more productive for philosophy and useful to real scientists than the recurrent debates between "scientific realists" and "neo-empiricists" or "neo-instrumentalists," whose views somehow manage to use all of these words without exemplifying any of them. Realism is not primarily a topic for a branch of philosophy of language, or even of metaphysics: it has deep implications for scientific problem-solving strategies. Engineering design is not irrelevant elsewhere. Thus it seems a natural and desirable part of the curriculum for evolutionary biologists, who might better appreciate adaptive design in nature if they come to appreciate the multi-faceted constraints on their designs (see also Simon, 1996; Wimsatt, 2006a).

Above all, realistic models of the scientist as problem solver and decision maker are essential. We are beings with *small* finite powers (a "dinky error-prone computer") rather than with the denumerable but impossible ones favored by mathematicians and most philosophers. Heuristics are powerful and permeate all of our activities (Gigerenzer, Todd, and the ABC Research Group, 1999). We have many talents: these talents should be applied effectively—in sickness and in health—across the disciplines of a science, and in its laboratories, rather than exclusively to its foundations. This perspective carries with it natural recommendations for how to proceed:

1. Science and technology are the most productive systematic sources of knowledge production we know. Despite the discovery that the sciences (and scientists) have clay feet (something surprising only because we have believed too much in our too simple idealizations of scientific method and practice), they are the best we have. Look to how science is really done for new methodological insights—not just for selective confirmation of current philosophical theories. When testing philosophical theories look for the tough cases; ones capable of producing deep and rich counterexamples we can learn from (Maull, 1976; Garber, 1986). Participant observation in science is a good idea.

2. Closer experience with scientific practice suggests that we should embed our more traditional concerns of justification and discovery in a theory of practice. (Historians and sociologists of science properly make failure to do so a complaint about much philosophy of science.) Consider an analogy: interest in embodied theories of consciousness issue from beliefs that the logical structure of the central nervous system does not exhaust the biological contributions to consciousness, and that the physical implementation of the being-in-context has much to tell about the nature of cognition. Do we need analogous moves to analyze the nature of our scientific activities? We have pursued a disembodied theory of science—which still requires human agents. (Notice how often *theories* are described as if they were agents—producing theorems and applying themselves—a process sociologist Elihu Gerson calls "deleting the work.") One assumes perfect and omnipotent agents more easily when they are depersonalized.

In *our* world, objectivity, reproduceability, control, and systematic investigation are all secured via regularized procedures that are finely tuned and contextually adapted on the fly to our abilities and the task at hand—modulated, not mechanized. They are embedded within a hierarchically nested set of social interactions that regulate the flow of information and our attitudes toward it while managing people and resources. We need to recognize these structures, and learn how to use them in theorizing about science.

3. This search for how science is done leads one naturally to close study of an area or subarea of science. Detailed knowledge of a case enriches our knowledge of generalizations, and improves our sense of their boundaries and exceptions. But one can go too far. Some have proclaimed this the heyday of the "special sciences" and see the "disunity of method" as following directly from their (too precipitous and total) rejection of the aims of the unity of science movement. Absolute generality is not the only worthwhile aim. Restricted generalities are often valuable, both for their own sake and as ways of leveraging greater generality. Science is a complex system and most truths about complex systems are richly filigreed with understandable qualifications that are often avenues to further systematic knowledge (Wimsatt, 1992). Not all counterexamples are alike, and superficial ones will not threaten robust theories. But we need better accounts of how we do this and when it is productive to do so to help us decide when a counterexample *is* superficial.[21]

4. Look at contexts of discovery, creation, and invention—in engi-

neering and in science, and not only to contexts of justification. But don't let contexts of discovery absorb the whole range of interesting methodological problems, as justification once did. Justification is not suddenly irrelevant, but it does have to be seen in a new light (see chapters 7 and 9). And look for new contexts: one whose significance is still significantly underestimated is calibration (but see Galison, 1997).

5. And while we're at it, let's not forget application. *Applying* theory is not a trivial matter: it requires lots of imaginative work—something positivists never really took seriously.

6. Don't accept self-trivializing, extreme, brittle, and fashionable theories that prove too much too easily; example, that the results of scientific investigations are (just) socially determined. Beware of all varieties of "nothing-but-isms" (Chapter 11), just as we should be suspicious of "single-factor" theories of anything. If the only value of a view is as a *reductio,* its target is probably an unusually hollow straw-man, and the problem has probably been misformulated. Extreme views can serve a function in science (and in philosophy), but remember how they do so (Chapter 6), and don't mistake them for true, plausible, or even (very) interesting theories. Curiosities, yes—theories, no!

7. Show practical sense about the limitations of a line of argument: reliability considerations (Chapter 4) suggest you should start inferences relatively close to your chosen subject matter. Thus, if a theory of meaning is relevant to scientific practice, it is going to be that part of scientific practice that deals with meaning—aspects of linguistics and neighboring areas of cognitive science and anthropology. It is not likely to reach any further—*pace* Kripke. *Thus, for example, rigid designation has absolutely nothing interesting to tell us about scientific theories of mind.* Kripke's work could only be relevant to philosophy of science through a long and torturous path. With many steps in the argument, each large and eminently contestable, the inference is risky. Why should we even be interested in such long-range inferences except under the rarest of conditions? Would you want to fight a fire in *your* house with a hose from a hydrant half a mile away? Not if there were one on your block!

8. You have a much greater chance of having something useful to say to scientists, or in getting them to pay any attention to you at all, if they can recognize the salience of the issues you raise. And that requires starting close, unless they already know and trust the intervening terrain very well.

9. Finally, don't accept any philosophical theory that has as a conse-

quence that most scientists don't know what they are doing, or are doing it wrong. Science (and its technological ancestors and extensions) has worked too well for too long for that to be true. Expect to be pleasantly surprised when you look more closely.

I now close this variegated exploration of grassroots scientific and engineering practice and their implications for philosophy. May your opportunities for exploration be as real, as varied, and as productive. But above all, may they be as enjoyable.

Appendixes

Notes

Bibliography

Credits

Index

Important Properties of Heuristics

I use the term *heuristics* more broadly than is common in the artificial intelligence (AI) literature, but closer to Herbert Simon's original (less formal) use when he got the term from Georg Polya's *Patterns of Plausible Inference* (1954). In AI, heuristics are formal procedures or inference rules. But it is a natural extension, using the properties described below, to see adaptations like the extended waggling "bait"-like tongue in the open mouth of the anglerfish as a heuristic procedure for attracting smaller fish, which become prey. Similarly, to a migrating animal the presence in a strange environment of a conspecific is an indicator that the environment is suitable: organisms characteristically stay longer in places that are fit, and quickly move on in hostile environments. Here a heuristic is viewed as a regularity for action wherein a kind of action (behavior) is characteristically undertaken under specifiable kinds of circumstances to achieve an end, or as part of a larger plan that is designed to do so. These "action-patterns" have important characteristics that explain why heuristics are adopted, calibrated, combined in larger methodologies to correct for biases and increase robustness, and have such rich characteristics and wide applications. These properties are shared with adaptations, and so heuristics are plausibly seen as problem-solving specializations of a broader class of adaptive tools. Items 1–3 appeared in Wimsatt (1980b, discussed with item 4 in Chapter 5), item 5 in Griesemer and Wimsatt (1989), and item 6 here. I make no claims to closure.

1. By comparison with truth-preserving algorithms or other proce-

dures for which they might be substituted, heuristics make *no guarantees* that they will produce a solution or the correct solution to a problem. A truth-preserving algorithm correctly applied to true premises *must* produce a correct conclusion. A heuristic need not.

2. By comparison with procedures for which they are substituted, heuristics are *cost-effective* in terms of demands on memory, computation, or other limited resources. (This is why they are used instead of methods offering stronger guarantees.)

3. Errors produced by using a heuristic are not random, but *systematically biased*: (a) The heuristic will tend to break down in certain classes of cases and not others, but not at random. (b) With an understanding of how it works, it should be possible to *predict* the conditions under which it will fail. (c) Where it is meaningful to speak of a *direction of error*, heuristics will tend to cause errors in a certain direction, which is again a function of the heuristic and of the kinds of problems to which it is applied.

4. Application of a heuristic to a problem yields a *transformation* of the problem into a non-equivalent but intuitively related problem. Answers to the transformed problem may not be answers to the original problem, though various cognitive biases operative in learning and science may lead us to ignore this.

5. Heuristics are useful *for* something: they are *purpose relative.* Tools that are effective for one purpose may be bad for another and increases in performance in one area are commonly accompanied by decreases elsewhere (Levins, 1968). This may help to identify or predict their biases: one expects a tool to be less biased for applications it was designed for than for others it is co-opted for.

6. Heuristics are commonly *descended from other heuristics,* and often modified to work better in a different environment. Thus they commonly come in evolutionarily related families, which may be drawn upon for other resources or tools appropriate for similar tasks. On different scales of resolution, a family of heuristics may look like a single heuristic, or conversely. (Thus Lenat, 1981, exhibits some 60 heuristic rules for his theorem prover that are equally sensibly seen as 60 instanciations of the same heuristic with slightly different antecedent conditions.)

Chapter 5 provides the context for the heuristics that follow. There I show how the assumption of limited powers plus a minimal interest in reductionism (seeking to explain system behavior in terms of interactions between parts of the system) biases the problem-solver against recognizing environmental causes of system behavior.

Common Reductionistic Heuristics

Most scientists suppose that a reductionistic analysis offers a lower-level mechanistic account of a higher-level phenomenon, entity, or regularity. To do so, one commonly decomposes a complex system into its parts, analyzes them in isolation, and then re-synthesizes these parts and the explanations of their behavior into a composite explanation of some aspect of the behavior of the system. (*Decomposition* and *recomposition* [Bechtel and Richardson, 1993] is a "near-decomposeability" meta-heuristic for reductionistic problem-solving.) With this approach, we use heuristic strategies with systematic biases that ignore or downplay the context-sensitivity of the results and the importance of the environment.

Biases of Conceptualization

1. ***Descriptive localization.*** Describe a relational property as if it were monadic, or a lower-order relational property. Thus, e.g., describe fitness as if it were a property of phenotypes or genes, ignoring the fact that it is a relation between organism and environment. (This strategy may be justified/facilitated [and its strong assumptions hidden] by fixing the environment, thus making it artificially disappear as a variable—see complementary heuristics 6, 8, and 9 below.)

2. ***Meaning reductionism.*** Assume that lower-level redescriptions change the meanings of terms, but higher-level redescriptions do not. This reflects a kind of atomistic essentialism. Thus we suppose that the

meaning of *gene* is changed when we discover the structure of DNA, but that *iron* is not when we discover that it occurs as a crucial chelating ion in hemoglobin. The result: philosophers (who view themselves as concerned with meaning relations) are inclined to a reductionistic bias.

3. ***Interface determinism.*** Assume that all that counts in analyzing the nature and behavior of a system is what comes or goes across the system-environment interface. This has two complementary versions: (a) *black-box behaviorism*—all that matters about a system is how it responds to given inputs; and (b) *black-world perspectivalism*—all that matters about the environment is what comes in across system boundaries and how the environment responds to system outputs (e.g., Fodor's "methodological solipsism" or Searle's Chinese room). Either can introduce reductionistic biases when conjoined with the assumption of "white box" analysis—that the order of study is from a system with its input-output relations to its subsystems with theirs, and so on. The analysis of functional properties, in particular, is rendered incoherent and impossible by these assumptions.

4. ***Entificational anchoring.*** Assume that all descriptions and processes are to be referred to entities at a given level, which are particularly robust, salient, or whatever. This is the ontological equivalent of assuming that there is a single cause for a phenomenon, or single level at which causation can act. Thus the tendency to regard individual organisms as primary, and more important than entities at either higher or lower levels. (Cf. methodological individualism for rational decision theorists and other social scientists. Similarly for genes for some reductionist neo-Darwinians.) See also *perceptual focus* (#19 below) and *multi-level reductionistic modeling* (Wimsatt, 1980b).

Biases of Model-Building and Theory Construction

5. ***Modeling localization.*** Look for an intrasystematic mechanism to explain a systemic property rather than an intersystemic one. **Corollary 5a:** *Structural* properties are regarded as more important than *functional* ones (since functional ones require reference to embedding systems).

6. ***Contextual simplification.*** In reductionistic model building, simplify environment before simplifying system. Thus the environment may be treated as homogeneous or constant (in space or in time), regular in some other way, or random. This strategy often legislates

higher-level systems out of existence, (see the "migrant pool assumption" in models of group selection, Wimsatt, 1980b) or leaves no way of describing systemic phenomena appropriately.

7. ***Generalization.*** When starting out to improve a simple model of the system in relation to its environment, focus on generalizing or elaborating the internal structure, at the cost of ignoring generalizations or elaborations of the existing structure. **Corollary 7a:** If a model doesn't work, it must be because of simplifications in the description of internal structure, not because of simplified descriptions of external structure.

Observation and Experimental Design

8. ***Focussed observation.*** Reductionists will tend not to monitor environmental variables, and thus will often tend not to record data necessary to detect interactional or larger scale patterns.

9. ***Environmental control.*** Reductionists will tend to keep environmental variables constant, and will thus tend to miss dependencies of system variables on them. (*Ceteris paribus* is regarded as a qualifier on environmental variables.) Mill's (1843) methods applied with this heuristic (vary the system variables one at a time while keeping all others—always including the environmental variables—constant) will yield as a systematic bias apparent independence of system variables from environmental variables, though the right experiments won't have been done to establish this.

10. ***Locality of testing.*** Test a theory only for local perturbations, or only under laboratory conditions, rather than testing it in natural environments, or doing appropriate robustness or sensitivity analyses to suggest what are important environmental variables or parameter ranges.

11. ***Abstractive reification.*** Observe or model only those things that are common to all cases; don't record individuating circumstances. Dangers of this approach: (1) lose sense of natural (or populational) variation, and increase danger of typological or stereotypical thinking; (2) lose detail necessary to explain variability in behavior, or exploit in experimental design. (Raff, 1996, notes that evolutionary geneticists focus on intraspecific variability, while developmental geneticists focus only on genes that are invariant within the species. This produces problems both of methodology and of focus when trying to relate micro-evolution and macro-evolution or evolution and development. Similarly, cognitive developmental psychologists tend to look only for

invariant features in cognition, or major dysfunctions, rather than populational variation.)

12. ***Articulation-of-Parts (AP) coherence*** (Kauffman/Taylor/Schank). Assuming that studies done with parts studied under different (and often inconsistent) conditions are *context-independent,* and thus still valid when put together to give an explanation of the behavior of the whole. (Schank, 1991: Checking this gives a non-trivial use for computer simulation.)

13. ***Behavioral regularity*** (Schank/Wimsatt). The search for systems whose behavior is relatively regular and controllable will result in selection of systems that may be uncharacteristically stable because they are relatively insensitive to environmental variations (Schank: regular 4-day cyclers among Sprague-Dawley rats are insensitive to conspecific pheromones; Wimsatt: Mendel's selection of 7 out of 22 characters that are relatively constant and insensitive to the environment probably resulted in unconscious selection against epistatic traits, which [happily] made his model ignoring them less problematic.)

Functional Localization Fallacies

14. ***Deficit reification.*** Assuming that the function of a part is to produce whatever the system fails to do when that part is absent (e.g., spark plugs as "sputter suppressors"), or produced when that part is activated or stimulated. More generally, the error involves *reifying added or subtracted behaviors* of the system *as functional properties* of the manipulated unit. Gregory (1958, 1962) notes that the things not done with lesion or deletion experiments may simply be the most obviously affected (rather than the most important). The part could have more importance to functions that are strongly canalized, or may have deficits not revealed under the testing conditions. More generally, if a part does realize a function, it does so usually only against a background of activities by other interacting components. **Corollary 14b:** Judgments of modularity are often insufficiently justified.

15. ***Assuming simple 1–1 mappings between recognizable parts and functions.*** This can lead to problems in two ways: (1) ignoring pleiotropy: stopping search for functions of a part when you find one (e.g., the newly discovered region of hemoglobin implicated in NO+ transport, because it was assumed that *the* function of hemoglobin was oxygen transport); (2) ignored division of labor (when a part's necessity

is shown through deletion studies, etc.), thus missing other parts' roles in the hypothesized function because they are part of the constant context, and always there to provide it. See item 9 above.

16. ***Ignoring interventive effects and damage due to experimental manipulation.*** First noticed in neurophysiological studies, it occurs also in many other places (e.g., marking specimens in mark-recapture studies may affect their fitness).

17. ***Mistaking lower-level functions for higher-level ends,*** or misidentifying system that is benefited. This is common in units of selection controversies—either of the apocalyptic variety as with Dawkins (1976), who denies all units of selection at higher levels than the gene, or for eliminative reductionists, who want to deny the existence or significance of large domains of cognitive function. There are legitimate concerns of level in both disputes, but the extremists are commonly seriously wrong.

18. ***Imposition of incorrect set of functional categories.*** Common in philosophy of psychology when it neglects ethology, ecology, and evolutionary biology.

Other Important Biases

(Items 10, 11, and those following can generate either reductionistic or holistic biases in different contexts.)

19. ***Extra-perspectival blindness or perceptual focus.*** Assuming that a system can be exhaustively described and explained from a given perspective because it has been very successfully and powerfully so described. (Not all problems of biology are problems of genetics, or of molecular biology, physiology, or anatomy [to cite other past excesses] and [as we can now see from a safe distance], not all problems of psychology are problems of behavior.) Perceptual focus can artificially inflate the number of properties attributed to a level of organization. Thus, the individual psyche—though perfectly real—has attracted social properties through improper (reductionistic) functional localization fallacies, and other phenomena (e.g., Mach bands) better explained at lower neurological levels. This bias interacts with #11 to give *extra-level blindness,* which can be counteracted by doing *multi-level reductionistic modeling,* in which a process is modeled at several levels with results that are then cross-checked.

20. ***Tool-binding.*** Becoming sufficiently bound to a specific (usually

very powerful) tool that one chooses problems for it, rather than conversely ("The right job for the organism," rather than "The right organism for the job"!) This applies to theoretical models and skills as well as to material tools and model organisms. It is an efficient division of labor if mastery of the tool is very demanding, and problematic only when it facilitates errors 11 or 16.

APPENDIX C

Glossary of Key Concepts and Assumptions

(With references to relevant articles; primary reference is in **bold.**)

AGGREGATIVITY. The condition of a system property in which it can legitimately be said that it is "nothing more" than the properties of its parts, justifying **nothing-but-ism.** For this to be true, roughly, the system property must not depend upon the mode of organization of the system's parts. (One productive way of defining a property as emergent is to say that it does depend upon the mode of organization of the parts, so aggregativity can be regarded as the opposite of emergence.) Being aggregative requires meeting four conditions that are almost never satisfied, but these conditions provide useful tools to determine how a system property depends upon the organization of the parts. Reduction is sometimes mistaken for aggregativity, which is a much stronger condition. That is, Reductionism is *not* "nothing-but-ism." Our tendencies to make this confusion are explicable in terms of the use of reductionistic problem-solving heuristics, and are discussed in **Chapter 12.**

BOUNDARIES. Robustness for objects involves coincidence of boundaries. This brings boundaries into focus. Strong gradients in values of many properties can cause other properties to develop gradients in the same place. This reinforcement of boundaries can lead to "spontaneous" emergence of robust systems as objects, natural and biological

(Platt, 1969) and social (Abbott, 1995). But systems may have robust boundaries while their parts do not, leading to a situation that naturally promotes functional localization fallacies and failures of near-decomposability (**chapters** 9, 10, and 12). Systems with multiple partially overlapping boundaries have richer possibilities for interpenetrating interactions, but are for that reason harder to individuate. These constitute a new class of *complex objects* found much more commonly in the biotic and social realms. The multiple boundary crossings yield causal thickets and disciplinary conflicts (Chapter 10).

ENGINEERING PERSPECTIVE. A cluster of theses derived from the assumption that theory has much to learn from practice and application. Teleological: Design is design for an end. View scientific activities as functional, and evaluate their designs for that supposed end (Wimsatt, 1979; Chapter 10). Relation to practice: Focus not only on theory and *in principle* arguments, but on the practical implications of a view of science, how to apply it, and how it must be adjusted or qualified to do so. The central role of heuristics as fallible inferential tools, rather than sources of certainty. Applied not only to our theories and methods as instruments, but also to our mental capabilities and inferences. Most engineering is re-engineering, recognizing that we rarely start from scratch, but will use what comes readily to hand, as quicker, cheaper, more convenient. This has two consequences: (1) history matters; to understand our methods we must understand where they came from and how. The genetic fallacy is not a fallacy. (2) There is no "perfect adaptation" *ex nihilo:* adaptation commonly co-opts something else to a new role, so exaptation is common. This view is profoundly instrumental, but denies any necessary tension between instrumental usefulness and truth or realism.

ERROR. The axiomatic (or "Euclidean") worldview, and its descendant computational worldview, are specialized adaptations for dealing with structures that don't need to be changed. Ideally (or "in principle") they have no errors. In the real world, their implementations have small error rates and relatively easily localized errors, when applied to simpler problems. This is also an uncommon and profoundly non-biological model of organization. Optimal performance in such networks requires quite different strategies than are appropriate for structures with higher error rates. Evolution and evolutionary ecology are laboratories for the design of "error tolerant" (and even "error metabolizing") systems. We have invested lots of effort in investigating the potential of Euclidean systems, but almost nothing in theoretically characterizing organic/robust systems (**chapters 1–4**) or the *metabolism of error*, the deliberate use of error (in model building) to increase knowledge (Chapter 6).

EXAPTATION. When something is used to serve a function that it wasn't originally designed (or selected) for, it is an exaptation rather than an adaptation (Gould and Vrba, 1982). Gould championed the idea that most elaboration of functional organization is through exaptation rather than adaptation (design for that function). Thus the reasons for a given functional design are inextricably historical and complex, and rarely transparent to someone looking only at current function. So in evolutionary history, organisms are not so much engineered as re-engineered through a succession of kluges interpolated with retunings to make them work better together.

FUNCTIONAL LOCALIZATION FALLACIES. A functional localization fallacy occurs when a function is attributed or localized incorrectly in a system using a reductionistic heuristic. With reductionist methodologies, the most common kind of mislocalization is to attribute a function properly attributed to a whole system to just a part of that system. This error is particularly easy when one part is particularly central to that function, as e.g., the brain is to consciousness. This bias may misattribute many social properties as individual psychological ones. Related errors occur in treating a relational property as monadic, or in underestimating the degree of a relational property. For different kinds of functional localization fallacy, see **Appendix B** and (in passing) Chapter 12. More fully discussed in Bechtel and Richardson (1993).

GENERATIVE ENTRENCHMENT (GE). A measure of how many things depend upon an element and thus likely to change if it changes. (In an abstract network, if means of access [robustness] are paths *to* a node, then GE is the reach of paths *from* a node. Thus robustness and GE are complementary measures of local order in a complex system. What is derived from what depends upon the manipulations or inferences involved, so whether a given relation involves one or the other may depend on the specified operations.) Things with higher GE are more evolutionarily conservative because the chance that random changes in them will be adaptive declines exponentially with increasing GE. They also generate more massive changes when they do change. Things that stay around long enough get entrenched and more resistant to change because they have more things depending on them and depending on them to greater degrees. Evolution is an ongoing dialectic of local adaptation to an (internally and externally) changing environment that is also partially a biotic product. This correlates with the layered cyclic process through which adaptations to some things are co-opted as exaptations, and through modification, become adaptations to others. This is the major way through which the evolutionary history of an adaptive system becomes relevant to its current form (**Chapter** 7 and Wimsatt, 1981b, **1986a,** 1997b, **2001;** Wimsatt and Schank, 1988; Schank and Wimsatt, 1988).

HEURISTICS. Our adaptations meet the defining characteristics of heuristics—or to put it in the phylogenetic order: our heuristic problem-solving principles are specialized cognitive adaptations, and are still marked by six important characteristics of this origin (**Chapter** 5, Appendixes A and B, and Wimsatt, 1980b).

KLUGE. Originally a programmers term; an unpretty but conditionally effective fix for a program or design failure or "bug." It may be inelegant or unpretty by violating common principles of design (and thus prone to other failures), by using something not intended for that purpose (see "exaptation"), or by doing the fix in a way that is not efficient or robust. In conscious design processes, a kluge is more likely if the cause of the failure it "fixes" cannot be determined, leading one to look for other ways to block the failure than the preferred way of redesigning to remove the cause. Since a kluge (because inelegant) is itself harder to understand (especially if undocumented), use of kluges increases the probability of having to use more. Inelegant solutions that are products of past historical commitments are kinds of kluges that have analogues in evolution. Mutations have diverse effects dispersed throughout the phenotype. They are "selected" when their net benefit exceeds net loss, but may fail anywhere due to sampling processes (genetic and ecological "drift"). The standard of selection is more likely met for changes with greater positive excess, but applies only locally and incrementally: there is no design process directing the accumulation of successive kluged increments, so they generally are not systematic. Thus they are incorporated not for ultimate optimality or efficiency, but on a first-come basis, leading naturally to generative entrenchment of the earlier modifications.

LAWS. The compositional multi-level materialism I favor takes mechanisms as more fundamental than laws at all levels of organization above that of the physically most fundamental. Whether expressed as principles (Hardy-Weinberg principle) or law (Mendel's law of independent assortment) these are mechanism-based expressions of the operation of causal factors under specifiable standard or idealized conditions (discussed here in **Chapter 11,** but more fully in Glennan, 1996, 1997). Laws are likely to be "sloppy, gappy" generalizations (Wimsatt, 1990).

LEVELS. Natural systems are commonly found in compositional levels of organization, with the entities at neighboring levels related by part-whole relations. *Size scale* is a pivotal indicator of the effectiveness of whole families of causal processes (because of the robustness of most objects at levels), and also largely determines (given the forces at work) a correlative characteristic time scale or "relaxation time"—a relatively narrow range of rates over which most processes at that level happen or go to equilibrium. As a result, our theories come in levels (language tracks nature rather than conversely), because it is in the relations among robust objects that you can get the "biggest bang for a buck" in theory construction, and most of what "happens" between them takes characteristic times within that range. The aim of science is the articulation and coordination of different entities and phenomena at different levels—the discovery of mechanisms (static mode) or processes (dynamic mode) relating phenomena and regularities at different levels rather than eliminative "nothing-but"-ism (**Chapter 10;** Wimsatt, 1976a).

MULTI-PERSPECTIVAL REALISM. Realism is connected with robustness (detectability via multiple independent means). A particularly important variant of this occurs when the different means derive from different perspectives. When a conceptual scheme claims exclusiveness or exhaustiveness, or treats other conceptual schemes as competitors when it cannot establish its primacy, this leads to relativism. But if these schemes are recognized as perspectives (severally incomplete, mutually complementary, and possibly co-calibrating), the consequence is a realism recognizing their respective co-referring objects of study as robust multi-dimensional trans-perspectival objects. See chapters 9 and 10 for analysis and examples.

NEAR-DECOMPOSABILITY. The ability to break structures into parts, and then reassemble them to solve or engineer problems is an impressively powerful heuristic (Simon, 1996). When this is possible, the system is said to be "nearly decomposable" (**Chapter 9; Wimsatt, 1986a, 1986b**). Reductionism in science, structured programming, engineering, mass production from stable subassemblies, and tinkertoys all use (and teach) this strategy. It is so powerful and so endemic to Western industrial society that we find it hard to see when it breaks down, but it often does. Reductionistic problem-solving heuristics, their strengths, limits, and systematic biases are covered in **chapters 5** (heuristics) and **12** (aggregativity). (See also "aggregativity" and "functional localization fallacies"; Wimsatt, 1980b.)

PERSPECTIVES. In complex systems, *perspectives* give organized approaches to a cluster of problems and techniques, often span levels, cross-cutting levels and each other, and give knowingly incomplete descriptions of the systems to which they are applied. (Levels can be viewed as special cases of perspectives ordered by hierarchical part-whole composition relations.) Their relation to each other, how they partition the systems to which they are applied, and the degree to which problems for a given system are bounded within individual perspectives on it are crucial to characterizing the complexity of a system. This kind of analysis is also required to understand the behavior of robust systems that must be simultaneously described using multiple boundaries and decompositions (**chapters 9 and 10;** Wimsatt, 1976a). Reality is multi-perspectival and robust. Some systems get so complex (the causal interactions among their variables are sufficiently disordered) that levels and perspectives break down, failing to have the partial dynamical and explanatory closure characteristic of both. Problems can (and must) be approached from a variety of directions, any one of which makes competing methodological claims with the others. In this situation, all we have is *causal thickets* in which methodological disputes are rife between researchers from different levels and perspectives over the common territory that they claim, and functional localization fallacies are hard to avoid.

REDUCTIONISM. There are three types of "reduction" or "reductionism" that are often confused: *successional* reduction (which applies either to same level theories or to universal theories) is a similarity relation (and is thus *intransitive*) in which one seeks to localize similarities and differences between a new theory and an older one in order to further the development of the newer theory and to delimit the useful range of the older one. This is plausibly thought of as a kind of theory-reduction (Nickles, 1973; Wimsatt, 1976a). *Inter-level* reduction involves an attempt to explain upper-level phenomena in terms of an articulated account of the operation of lower-level *mechanisms* (**Chapter 11**; Kauffman, 1971; Bechtel and Richardson, 1993). It commonly involves identification of descriptions from two or more levels of organization as referring to the same thing or, more commonly, weaker species of identity relation (qualified context-dependent identification, instanciation, realization, or localization). These forms of identifications are assumed to be *transitive* as one moves up and down levels of organization. Nonetheless, neither the successional relation nor the identification can be regarded as a truth-preserving deductive derivation because of the use of limiting case approximations (in both cases), and truncation of relevant contextual details (in the second). In the compositional sciences, inter-level reduction proceeds by developing mechanistic explanations through an iterative feedback process relating the levels (**chapters 10 and 11**; Wimsatt, 1976a). A common mistake of critics and advocates alike is to suppose that reduction supports *nothing-but-ism* (as in "the whole is nothing more than the sum of the parts"), but this would require the much stronger conditions of *aggregativity*, which are rarely met **(Chapter 12)**.

ROBUSTNESS. *Robustness* is the existence of multiple means of determination or access to an entity, property, process, result, or theorem that are at least partially independent. This allows even less reliable individual components to generate higher reliability and adaptability of the overall structure. (This is the fundamental organic design principle, not genetic determination, of phenotype.) Robustness is the ultimate methodological criterion for reality *and is so in all sciences*. It also characterizes the appropriate kind of stability for ecosystems (**chapters 4 and 10**; Wimsatt, 1974, 1976a, 1980a, 1991).

SELECTION. We are creatures (and cultures) engineered by *selection* processes at a variety of intersecting—sometimes coupled, sometimes dynamically independent—levels (**Chapter 10**; Wimsatt, 1976a). Biological and cultural evolution use different transmission channels and somewhat different principles, but show significant theoretical similarities (and also differences forcing different theoretical approaches to many of their problems). Both are still covered by Darwin's Principles (Lewontin, 1970); however, for virtually all significant evolutionary processes, Darwin's Principles must be supplemented by two additional principles that yield a developmental structure of the phenotype and a life cycle in terms of generative entrenchment (Wimsatt, 1999a, **2001**; **Chapter 7**). These additional principles and the developmental organization they bring to bear are even more important for cultural evolution than for biological evolution. I see self-organization as continuous with selection, though it is often useful to treat them as separate processes (Wimsatt, **1971, 1986a** [last section]).

A Panoply of LaPlacean and Leibnizian Demons

Nineteenth-century physicists invented "demons" to go along with the various perpetual motion machines forbidden by newly discovered fundamental laws. Philosophers have contributed to this lore, without noticing its negative lesson. Here are some of my favorites.

Heuristic for finding demons: Look for *in principle* claims that are treated as if they could be delivered if anyone were sufficiently interested—but aren't and can't. (Given how important they are to the argument, if they can, why aren't they?) There is bound to be a demon lurking in the background (see Chapter 11). Computability variants of this are useful in mathematics, but when applied to the real world they require impossible amounts of computation, degrees of knowledge, and—often ignored—a systematic way of ordering the states or alternatives so that they can be exhaustively specified. Dennett's (1995) versions of Borges' labyrinth contain several ingenious examples.

1. LaPlace's original demon and the definition of determinism. A deterministic system is one in which, given the equations of motion of the system, and a complete description of it at any point in time, one can predict or retrodict its state arbitrarily far into the future or arbitrarily back into the past. Notice that genetic determinism cannot be formulated as true, subject to this constraint: LaPlace's demon would require non-genetic information since *even if one were to accept that all organic products are genetic products,* what happens at any given time re-

quires the various levels of biological organization (e.g., the cell) and thus requires the action of the genes over a temporally extended period of time (in a context of non-genetic elements).

2. The rational demon. The definition of *expected utility* requires (1) mutually exclusive and (2) exhaustive specification of alternative actions, (3) mutually exclusive and (4) exhaustive specification of outcomes and (5) their utilities, application of scientific theories and initial conditions to compute (6) complete specification of probabilistic arrays mapping actions to outcomes, (7) leading (then almost trivially) to computation of expected utilities of all actions and choice of maximal one. (Note that this, in effect, requires a state-space representation of the system.)

3. Descartes' evil demon and the skeptical regress (and modern brain-in-the-tank descendants—cf. Fodor's "methodological skepticism" and Searle's (reductio) "Chinese room").

4. Epistemological demons: knowledge-and-belief structures that are closed under entailment (and globally consistent).

5. Physical (or biological, or neurophysiological, or econo-psychological) reductionist demon, able to provide the *in principle* reductionist translations obviating the need for the relevant upper-level theories and descriptions of phenomena.

6. Lewontin's (1966) probabilistic demons (their estimates of means converge, and the means are sufficient for prediction of optimal behavior) verses the capriciousness of nature from decay of information on different time scales and pathological distributions (Pareto distribution) for some kinds of events.

7. Darwinian instrumentalist demon. Like Darwin's view of natural selection, infinitely sensitive to detail (but only to functionally relevant properties), and selects always for the best. Although apparently weaker than the LaPlacean demon, it is not clear that this demon can get by with knowing any less (because of the causal structure of the world and intertwining of functional, neutral, and dysfunctional consequences).

8. Chaos has led to the denial of a "chaos demon," a LaPlacean demon who knows all parameters of the universe with the infinite precision necessary to accomplish LaPlace's task in a chaotic world. (A temporally bounded version of this demon would only need to know what is going on within the bounded neighborhood necessary for prediction within the chosen time interval.)

9. The "meaning daemon"—able to keep track of all the semantic

linkages (and to keep them updated in the light of new connections) in order to realize the maximally connected network required by holistic functional theories of meaning. Meaning holists don't distinguish importantly different cases here. Must everything be directly connected to everything (n^2 linkages for n elements, assuming pairwise connections only), or will it do just for everything to be reachable from every other element in the set (then as few as $n-1$ connections for n elements)? Also, the topology of connections should matter as (in any more realistic models) their directionality.

10. The Gricean demon—able to keep track of the recursively enumerated nested intentions necessary to unpack speakers' meaning on Grice's account, and to update the changing purposes of the parties of the conversation (42 times in 7 minutes, according to Nancy Stein's analysis of a recent conversation) sufficiently rapidly to do this. (The Gricean demon—though respectable—is actually a dinky finite-state automaton by comparison with the others.)

11. The Turing/Church daemon: able to complete any computable task. Led to "Turing machine functionalism," and—with Kleene's proofs (in Shannon and McCarthy, 1956) that mapped the powers of networks of McCulloch-Pitts neurons onto various classes of Turing machines—various curve-fitting but radically biologically unrealistic solutions to the mind-body problem. Useful against someone who claims that there is a specifiable behavior that is *in principle* not realizable by a machine, but otherwise irrelevant. Continued as plausible through behavioristic arguments that only behavior could matter to an analysis of mind, since only that was accessible to us as we learn language. (Ignores the problem of how—i.e., with what hardware, and consequent limitations and failure modes—*we* accomplished it.) The dual of the brain in the tank or "Chinese room" puzzles.

12. The Craigean elimination daemon. Use Craig's theorem and Ramsified sentences to eliminate the theoretical entities of any theory, therefore rendering it more "observationally justifiable" though essentially unusable.

13. Dawkins/Kitcher/Sterelny "bookkeeping daemon": Keep track of all contexts of all genes (including their genetic contexts to adjust for epistatic interactions) in all organisms so as to be able to calculate and update as necessary each generation (asynchronously, for different generation times of different organisms) the net selection coefficients of all genes so as to plug into the "bottom up" genic theory of natural selection required for Dawkins' reductionistic vision.

Notes

1. Myths of LaPlacean Omniscience

1. Simon's legacy here is double edged: he advocated formal and foundational methods but sought to build a theory of human problem solving on satisficing and heuristics. I adopt the second edge, but not the first. I met Simon in 1965, and used his work in my dissertation (Wimsatt, 1971a) and here explicitly in chapters 5 (Wimsatt, 1985), 8 (Wimsatt, 1974), and 9 (Wimsatt, 1976b). In several other essays (Wimsatt, 1979, 1980b) I make central use of satisficing, heuristics, and computational limitations. See also Nickles (1980b, 1981, 2002), Giere (1988), Darden (1991), Bechtel and Richardson (1993), Nozick (1993), McClamrock (1995), and Tyson (1994). I take pride in recruiting many of these to "bounded rationality," and encouraging others in our increasingly common heresy.
2. Michael Williams' (1996) epistemological views naturally resonate here—see chapters 4 and 9 in this volume.
3. Such views are coalescing around "Developmental Systems Theory," or DST. See Oyama, Gray, and Griffiths (2001).
4. *Generative entrenchment* is discussed in Chapter 7. See also Wimsatt (2001).
5. Nozick's (1993) liberating book, *The Nature of Rationality,* is an important ally for many of these points. He adopts evolutionary readings of rationality, and draws many of the same conclusions.
6. McClamrock's (1995) arguments for the embodied nature of consciousness, Hutchins' (1995) explorations of the socially and culturally distributed nature of cognition, and Clark's (1997) explorations of bootstrapping are pivotally important. See related moves in Chapter 8 (Wimsatt, 1974), the ancestor of Chapter 9 (Wimsatt, 1976a), and my analysis of the innate-acquired distinction (Wimsatt, 1986a). I focus more on individual or

generic problem-oriented heuristics, but this approach also applies to social heuristics. Resonances with McClamrock's date from his time at Chicago. Those with Hutchins and Clark are unanticipated gifts.

7. Nozick (1993) urges philosophers to use heuristics from Simon's (1973) "On the Structure of Ill-Structured Problems" in theory construction. His recommendation is in the spirit of those made here.
8. I'm not against *in principle* claims, or even these ones—just these ones used as they have been to rule out other approaches or considering other factors. See Appendix D.
9. See appendixes A and B for properties of heuristics, how biases arise from the reductionist methodology, and a list of many reductionist heuristics and their biases.
10. My inspiration here is the protagonist in Douglas Harper's evocative study, *Working Knowledge: Skill and Community in a Small Shop* (1987). Another moving articulate and timely expositor of related themes is Wes Jackson (1994). A researcher-activist for sustainable agriculture and lifestyles, Jackson urges an "ignorance-based philosophy," explicitly criticizes Cartesian assumptions and LaPlacean demons hiding in our faith in "technological fixes," and laments the loss of local and contextual knowledge of the land and conditions for living with it accompanying the loss of rooted local communities.
11. This was a central message of Lakatos' classic *Proofs and Refutations* (1976), traced through the elaboration of much of nineteenth-century mathematics. He also drew heavily on Polya's (1954) work on "plausible inference" in mathematics.

2. Normative Idealizations versus the Metabolism of Error

1. I use the term *model* as scientists commonly do, not in the model-theoretic sense familiar to most philosophers (contrast van Fraasen, 1990, with Downes, 1992). Scientists' models are tentative idealizations of phenomena or relationships in nature, true or false in various ways for how accurately they describe their targets in those respects. They are often *deliberately* configured to be false in known respects in service of the aims of modeling. See Chapter 6.
2. Gigerenzer and Goldstein (1995) show that an important family of plausible heuristic choice strategies ("probabilistic mental models") perform as well or better in realistic environments as "traditionally optimal" computationally greedy strategies, and do so faster with less computation and memory. They call this striking result the "less is more" effect. See Gigerenzer, Todd, and the ABC Research Group (1999).
3. Darwin's theory explained the generally adaptive character of organic structure and behavior, but too much perfection was for him an embarrassment—an invitation to Natural Theology. Arbitrary and apparently poorly designed features, such as opportunistic adaptations of things designed for other pur-

poses, were evidence of the fortuitous cumulative improvement of design through natural selection.

4. In fact, for studies after the fact, this indicates a source of bias in claimed confirmations of maximization behavior!
5. Utilities are indices of preference, not preferences, but if this is used to keep them constant at any cost while preferences change systematically, we risk making them operationally useless and theoretically problematic. Another explanation for this phenomenon provided by Hoffrage and Hertwig (chapter 9 on "hindsight bias" in Gigerenzer, Todd, and the ABC Research Group, 1999) seems complementary rather than conflicting.
6. Other cases where interaction of "rational" (self-interested) decision processes with "irrational" and "biasing" emotions (keyed to social situations) can yield higher optima than the first acting alone are discussed by Frank (1988). He argues that such cases plausibly explain the evolution and maintenance of emotions in humans and other species.
7. For more on model building see Wimsatt (1980a, 1980b) and Wimsatt and Schank (1993). On approximations, see Ramsey (1992), Sarkar (1998), Batterman (2002), and chapters 11 and 12.
8. Defining expected utility requires specifying a set of exhaustive and *mutually exclusive* alternative actions for the agent. (I've not seen a non-trivial decision for which this can be done.) Then we must determine for each action the array of probabilities with which it yields a set of (again exhaustive and mutually exclusive) outcomes, whose utilities to the agent are known and determinate (meeting a set of conditions allowing their measurement on a linear metric). Knowing these alternatives, outcomes, and probabilities supposes a kind of complete knowledge of the world, its laws, and its conditions which we don't and couldn't possess, and doing infinitely precise computations with them that are manifestly impossible. LaPlace's demon was a stand-in for God, but a poor approximation either for economists or their subjects (see Wimsatt, 2006b, 2006c).
9. Howard Stein notes (in conversation) that approximations used with care needn't lead to contradictions. Agreed, but the care is often lacking: approximations are often used as if they *are* true—not approximately true. Pi may be set equal to 22/7 in a computation, but that doesn't make them equal. Taking the equality for granted leads to contradiction. Approximations often aren't announced either. Numeric routines used in every computer for trigonometric and transcendental functions are all approximations, usually truncated infinite series expansions. They produce radically deviant values if used too close to singularities—for example, tan ϑ for ϑ too close to 90°. This is rarely checked unless problems arise, and we simply assume we'll detect any serious problems. But we often don't (Peterson, 1995).
10. Petroski (1994) considers various paradigms of failure and of good design so one can understand how errors are likely to arise, how they can be detected, and eliminated. He urges the study of case histories to hone these skills and intuitions in the education of engineers. Similar themes are echoed periodi-

cally in most professions. Only a neo-Cartesian philosopher would be tempted to think that you could eliminate all errors "up front."

11. See Schank's (1991) groundbreaking work on parallels between model-building, programming, and experimental design.

3. Toward a Philosophy for Limited Beings

1. Just as *know-how* is not (or not just) a species of propositional knowledge, neither is *know-why*. To value their practices as a native is at least partly to identify with them, an involvement that makes total detachment impossible. This was once a sure admission of methodological guilt. But it shouldn't be, as long as we can converse and negotiate with those who have other values.
2. At this level of abstraction, the recommendation looks Wittgensteinian. It has some things in common with later Wittgenstein, but except as indicated in the text, had other sources.
3. Nozick (1993) urges use of heuristics from Simon's (1973) "On the Structure of Ill-Structured Problems" by philosophers in theory construction. I heartily recommend both.
4. Old boundaries between cognitive, conative, affective, perceptual, and motoric dimensions of our socialized embodied selves are breaking down. But these domains need reconceptualization to see how this is possible (Griffiths and Gray, 1996).
5. The logical empiricists were interested in multiple problems of demarcation. But this was usually with the supposition that a distinction existed yielding a crisp boundary, and no recognition of complications arising from multiple relevant overlapping boundaries, or boundaries that changed over time. Two worthwhile papers on this topic are Platt (1969), and Abbott (2001).

II. Problem-Solving Strategies for Complex Systems

1. This criterion is more fully developed than Hacking's (1983) similar views, which are limited to experimental manipulations, applied only to entities, and ignore or skim over the heuristic character of these methods, the reliability arguments, and the importance of the independence of different means of access. See also Chapter 8 and Wimsatt (1976a, 1980a, and 1980b). Elaborated by psychologist Donald Campbell (as "triangulation") and mathematical ecologist Richard Levins, robustness also resonates with physicist Richard Feynman (1967, chapter 2) and Clark Glymour's (1980) work on bootstrapping. More recently, the late Sylvia Culp (1994, 1995) and J. D. Trout (1995, 1998) have exploited it in diverse studies of scientific methodology.
2. *Calibration* of procedures, instruments, and even theories is centrally important to both discovery and justification. Without calibration, we don't know how theories apply in the real world, how different parts of theories articulate with each other, how to scale our measurements, or how to tell reliable measurements from artifacts. Without it, we cannot pursue most forms of justification or discovery, so it clearly deserves equal time. Abuses of calibra-

tion lead to ad hoc curve fitting. Knowing how to do it is critical to assessing claims for realism and instrumentalism. A fuller analysis of calibration would use concepts of robustness (Chapter 4) and pattern-matching (Chapter 6 and Wimsatt, 1991).

3. A fifth and sixth property have since been added. All are discussed in Appendix A.
4. Giere (1988) parses models differently. I treat what he would call the application of a model. He sees truth and falsity as too crude for the relations between models and phenomena. (Models may be true in some respects, false in others, and in most respects only approximately true, where that is relativized to the purposes at hand.) I agree. But scientists talk this way, and don't detach models from their applications as freely as Giere's semantic view of theories suggests.
5. For complementary historical accounts, see Allen (1978), Darden (1991), Mainschein (1991), and Kohler (1994).
6. I accept her views on models and idealizations, but reject her anti-realistic instrumentalism for theories. They are instruments to be sure, but not "nothing but instruments." Their role as instruments is consistent with realism.
7. Software by Schank and Wimsatt (1993), and a 230-page manual and text (Wimsatt and Schank, 1993) are available for Macintosh as part of the BioQuest library of strategic simulations, published by Academic Press, and now free online at www.bioquest.org.

4. Robustness, Reliability, and Overdetermination

This chapter is the product of seeds planted around 1970 by the writings and ideas of Donald Campbell and Richard Levins. Since then these germinating ideas have received nurturance and selective pruning by many individuals. Particularly helpful were discussions with Bill Bechtel, Aaron Ben Ze'ev, Bob McCauley, and Bob Richardson. Sandy Zabell and Steve Stigler provided valuable guidance through the statistical literature on robustness that, while related, was less central than I had hoped. The editors of this volume have made creative suggestions improving the paper. This work was supported by the National Science Foundation under Grant NSF SOC78–07310.

5. Heuristics and the Study of Human Behavior

I thank Donald Fiske, Jim Griesemer, and Leigh Star for useful comments on an earlier draft of this chapter. This work was supported by the Systems Development Foundation, Grant no. 6357.

6. False Models as Means to Truer Theories

I thank James Crow and Bruce Walsh for useful feedback at the symposium where this was presented, particularly for discussions and references on modern theories and models of interference phenomena. Nils Roll-Hansen and another anonymous referee made suggestions on organization and content, and Matt Nitecki helped to clarify and streamline sticky prose. Janice Spofford and Ed Garber helped me find data on non-standard mating systems. I also thank a decade's students in my biology class, "Genetics in an Evolutionary Perspective," who have acted as sounding boards for some of these ideas and guinea pigs for the computer labs that grew out of them. Bill Bechtel, Jim Griesemer, Ron Laymon, Bob Richardson, Sahotra Sarkar, Leigh Star, Leigh van Valen, and Mike Wade all gave useful comments at earlier stages. Even though her *How the Laws of Physics Lie* is mostly in service of a different conclusion, Nancy Cartwright's beautiful study of model building in physics provides many further examples of the sort discussed here, demonstrating that this analysis is not limited to biology. Support from the National Science Foundation and the Systems Development Foundation, for time and computing equipment, made parts of the task substantially easier.

7. Robustness and Entrenchment

1. This is done by duplicating components with identical functions (Von Neumann, 1956) or functional multiplexing with multiple functions realized by and sharing multiple components (Winograd and Cowan, 1963), a feature also characteristic of connectionist networks.
2. Near-decomposability or modularity is a plausible fourth fundamental architectural principle and also interacts with GE. By cutting down on interactions, modularity characteristically decreases the GE of system parts (Schank and Wimsatt, 2000), but there is a big *ceteris paribus* attached to this claim, and some interesting exceptions. This is currently a hot area of research in evolutionary developmental biology (Schlosser and Wagner, 2003), linguistics, and cognitive and evolutionary psychology (Fodor, 1983).
3. For an evolutionary engineering epistemologist, not only the object of study (be it an adaptive system), but also the instruments and the investigators are evolved and evolving adapted systems, and the system of investigation includes all three. Sensory, cognitive, and social systems have evolved to generate adaptation to and knowledge of an environment that is external at a given time. Our intimate involvement in responding to and acting upon our environments over evolutionary and ontogenetic time have formed and calibrated the architecture of our internal thought and knowledge. Problems of knowledge must be formulated in recognition of this fact. Further, we are sometimes not, individually, bearers of anything recognizable as knowledge, whereas drawing the boundaries outwards reveals ordered adaptive information organized across, usable by, and extractable from larger organized groups (Hutchins, 1995). How do we get stability in objects, properties, adaptations, and parts of knowledge-structures?
4. Campbell's "vicarious selectors" don't necessarily comprise or operate at distinct compositional levels, but they are distinctive selection processes which can operate quasi-independently.

5. We *do* sometimes have to live in the house while it is being rebuilt. But this only works because the conceptual organization of science, and of engineering practice, is usually robust, modular, and local, each of which reduces GE. Shaking (local) foundations usually doesn't bring the house down, and we still have a place to stand (on neighboring timbers) while we do it. For a sense of life at the critical edge (!) read Rhodes (1986) on revising theory and practice in the design, construction, and testing of the first nuclear reactor and atom bomb, or Feynman's view (Gleick, 1992) of the groping development of theory and computation at Los Alamos when they *had* to have accurate results without direct experiments.
6. My thanks to Cornell historian Carol Kaman for discussions and help in checking this legend. What happened is equivocal, but the story told here is one of the more plausible alternatives (Parsons, 1968).
7. Hull's exhaustively documented review, analysis, and discussion (1988, pp. 379–382) of "Planck's Platitude" (that older scientists are slower to accept new theories) among the many English scientists responding to Darwin shows how dangerous easy generalizations are in this area. Instead, he showed that it depended most strongly on "how much they had to lose," a direct confirmation of the GE account offered here.
8. The old-fashioned series-connected Christmas tree lights seem an exception here: if any one fails, the string goes out. In the newer parallel-connected ones, failure of each bulb affects only itself. But in both strings, a failure in the plug or wire leading from plug to string causes the entire string to fail—like a bulb in the series string—though in that case they are designed to be easily replaced, unlike the lead-wire or plug.
9. This only begins the job. At the very least we need to be able to appropriately handle distinctions between pragmatic and semantic components, and to wonder how different sorts of conventions are to be distinguished. It could also be objected that if "a" proposition moves from one category to another, it could not remain the same proposition. I don't think these problems are insurmountable. Thus Turner (1991) uses generative entrenchment to address the nature of the difference between literal and figurative meaning, and I will (in a forthcoming book on GE) deal much more with questions of abstraction and of meaning and scientific change with detailed examples.
10. Though if some of those downstream nodes are robust—having independent support from other sources—they will be influenced, and weakened, but will not fail without further perturbations.
11. Many relevant complexities are not discussed here. At the least, propositions require a Boolean network (see, e.g., Kauffman, 1993; Glymour, 1980) rather than a directed graph, but comparable moves can be made there (and for at least some more complex structures). Thus arguments here will remain qualitative and occasionally border on the metaphorical, but better moves are expositionally much more complex so that will have to do for now.
12. Larger ordered conceptual structures could fit here too: models or procedures. These structures can be "black boxed" and hierarchically aggregated.

8. Lewontin's Evidence (That There Isn't Any)

1. Popper's defenders have been in the habit of inventing numerically subscripted versions of that philosopher, with higher numbers indicating more sophisticated versions of his views—all in the service of preventing it from being falsified. The subscripts in the literature have so far gone no higher than three or four, however.
2. In saying (with deliberate and self-indulgent animism) that nature does not tolerate contradictions, I explicitly do *not* here wish to be denying the importance of the use of contradictions in argument (e.g., *reductio* proofs) and of efforts to avoid them in the elaboration of theory, conflict among contradictory theories, or "collisions" between "contradictory" forces, optima, constraints, desiderata, or design requirements in the development of adaptive structures generally—including organisms and theories as two special cases. Thus, this view is quite consistent with the claim that scientific investigation, problem solving, and theory competition are sometimes (or always!) dialectical processes. It raises more problems, I think, for scientific instrumentalism: once it is accepted that we can and do work, and work well with theories that contain contradictions, *it is not clear how the instrumentalist can expect, explain, or justify their removal!* There is, to my knowledge, only one form of formal contradiction that we do not generally attempt to remove—though we are forced to justify them, and may be pushed to understand their limitations more carefully. These are the approximations that link together various parts of any quantitative theory of any significant size and moment in the mathematical and natural sciences. For an important start on the analysis (and a catalogue of some of the variety) of these important tools, see Jeffrey Ramsey (1990b, 1992).
3. This was written in 1993. The list is obviously much richer and more developed now.
4. This is not quite right. It would make a substantial difference to evolutionary theory if more than 99 percent or less than 1 percent of the variation were adaptive, but these extremes are clearly ruled out, and quite major variations in between these figures (say, 90 percent versus 10 percent) wouldn't matter too much. There is enough uncertainty about other important factors that distinguishable positions should differ by orders of magnitude rather than closely weighted percentages. For example, how many distinguishable functional units of the genome are there? How many of them are under selection at any given time? How much of the neutrality we observe among allelic variants is itself a product of selective design elsewhere in the phenotype? More generally, we must know a lot more about the organization of developmental programs, their response to selection, and whether any of the forces causing systematic rearrangement of the genome somatically during development are available for evolutionary modification, and how frequently, before arguments like this can become productive.

5. For the status of experiments as demonstrations, or paradigm justifying or extending moves, see Allchin, 1992.
6. See Chapter 5 on heuristics and biases, and Chapter 6 on deliberate use of false models.
7. The history of conservative estimates shows that the estimates often fail to be conservative, particularly when the estimator has an interest in how the results come out—e.g., as with estimates of risk by an industry responsible for adopting safeguards commensurate with the risk. These can also be dangerously flawed because it is hard to estimate the effects of qualitatively different causes that have not been considered. (Witness Kelvin's "robust" estimates that there wasn't more than 100 million years available for evolution—flawed by his lack of knowledge of thermonuclear processes in the sun and radioactive heating in the earth [Burchfield, 1975].)
8. Haldane (1919), analyzed in detail in Wimsatt (1992).
9. See Schank (1991).
10. "Treasure your exceptions" is a homily at least as well attended to by the contextualist generalizer as it is by the naive falsificationist. To the latter it provides grounds to trash the relevant universal; to the former, it provides not only grounds to qualify the generalization but also it usually provides information about how to do so.
11. Glennan (1992) provides a strikingly different, and to my mind, the best available account of the nature of causation, causal mechanisms, and mechanistic explanation.
12. See Dembski (1991) for an elegant and informative essay on the nature of randomness. However, it will not bear the weight he has subsequently tried to put on it to support Intelligent Design.

III. Reductionism(s) in Practice

1. Less than half of my writings on reduction and reductionism are reprinted here. Omitted are 1971a, 1976a, 1979, 1980a, 1980b, 1981b, 1986b, 1992, and 2006a. The 1992 work is a historical case study of the development of mechanistic explanations of linkage phenomena in classical genetics, elaborating and lending further support to the account given here. The 1979 work is an extensive review, with examples and programmatic notes not found elsewhere. Bechtel and Richardson (1993), Schank (1991), Glennan (1992), Waters (1994), and Sarkar (1998) provide complementary perspectives.
2. Ian McHarg's famous *Design with Nature* (1969) is a source of particularly rich examples of multiple overlapping boundaries individuated using different criteria (*descriptive complexity* in this chapter). His city planning projects dealt with richly heterogeneous information integrated (in a precomputer age) via multiple overlays (one for each variable) on a map of the region. More recent GIS technology utilizing diverse variables from remote sensing images has enormously elaborated this approach. And active process control like that found initially in Paris and, increasingly, in many city cen-

ters makes multiple representations of activity patterns a basis for controlling it—as is now done widely for traffic flow in big cities.

3. I've never seen a whole body chart of this type—at least at a molar level (there are biochemical charts of all the major metabolic pathways). For partial charts, see Grodins' (1963) description of the cardiovascular regulatory system, or the hierarchical chart of interactions controlling ovulatory cycles of communally living rats in Schank and McClintock (1992).
4. Recognition of the critical role of membrane structure in the theory of oxidative phosphorylation could count as the first fundamental hybrid of anatomical and physiological theories—see Allchin (1991) and Bechtel and Richardson (1993).
5. Thus understanding relations between the levels of chemistry and molecular biology and between bonding properties of simple molecules and the structural character and behavior of macromolecules involves at least getting from the *primary structure*, or linear sequence, of an assembled protein, through the *secondary structures* (such as alpha-helices, beta-sheets, and others) into which these fold, to the *tertiary structure* or 3-D stereochemical configuration that yields the shape and active sites of the macromolecular machine (assembling a primary sequence requires a working cell, see Moss, 1992). Many proteins have still higher levels of relevant organization. Hemoglobin, for example, has at least two more since it is a tetramer, and (in sickling variants of the molecule) can form still larger super-crystals, which deform red blood cells and have still higher-level effects.
6. We might still discover new levels or principles forcing recognition of new levels where we had seen only an unclear mixture of size and dynamical scales. Ecologist C. S. Holling (1992) urges that new instances of scaling relations are found within ecosystems.
7. See Appendix C for more on these italicized terms.
8. The choices here are conjectural but justified: each of these disciplines is now reaching out to its neighbors in unprecedented ways, and seeing their respective problems as related. Problems of human development, the nature of cultural evolution, and its relation to biological evolution require this kind of massive cross-pollination and cooperation. Evolutionary psychology, while seriously threatened by oversimplification as it is now developing, is at least moving psychologists in useful directions, integrating both downwards with biology and upwards with the social sciences.
9. Various authors have drawn piecemeal on parts of this paper since its original publication in 1976, but often seem to miss its points. Chapters 8, 9, and this introduction are intended to motivate its different perspective. Hooker (1981a–1981c), Kitcher (1981), Cartwright (1983, 1989), Churchland (1986, 1995), and Waters (1990) have since offered views overlapping it in different respects, and Bechtel and Richardson (1993), McCauley (1996), Ramsey (1995), Sarkar (1992b, 1998), and Glennan (1992) provide extensions, related analyses, and further support.
10. Ramsey (1995) elaborates this kind of reduction with cases where a "less

good" theory is a limiting special case of a more general, fundamental, or better theory, but derived from it afterwards in a new way to *apply it* to a special case. The ontology of the special case could even be "created" by the reduction! He also exploits connections with engineering practice. Batterman (1995) finds new complexities that may affect both types of reduction: mathematical singularities blocking deduction, and hybrid intermediate theories (notably, the so-called semi-classical theory relating quantum mechanics and classical physics) capturing phenomena not generable from either parent theory.

11. There may be theories at the various levels, but not necessarily. The aim is usually to get lower-level mechanistic explanations of higher-level phenomena (an "inter-level theory"—Darden and Maull, 1977). With complex systems it is harder to find well-articulated theoretical structures true enough to the phenomena at two different levels for them both to be called theories. Statistical mechanics and classical thermodynamics, and classical and molecular genetics are exceptions here. That explanations generally are explanations of phenomena rather than of data or linguistic entities has since been argued by Bogen and Woodward (1992).
12. See also McCauley's excellent (1996) discussion of the issues surrounding theory co-evolution and eliminativism.
13. This possibility precipitated out of a conversation with Bob McCauley and Paul Churchland at a PSA reception in 1996.
14. Lindley Darden (1991), Bill Bechtel and Bob Richardson (1993), and Sahotra Sarkar (1998) urged this correction of the views expressed here and in Wimsatt (1976a). They point out the explicit target of a developing inter-level account is often the (lower-level) *localization* (and mechanistic explanation) of an upper-level property, object, or relationship, not an inter-level identificatory hypothesis. Localizations are indeed often the target, but since they require the identity of spatial properties, they guarantee the causally relevant properties, so the main feature of the analysis is not much changed, and still works.
15. *Mechanisms* in my account play a similar role to *capacities* in hers, with this difference: mechanisms key more naturally to the ways we give, modify, test, and elaborate mechanistic explanations; capacities seem chosen in part to deal with her focus on foundational questions at the quantum mechanical level.
16. Others have since used screening-off to secure the autonomy of upper-level phenomena, though somewhat differently. Brandon (1982) argues that phenotype screens off genotype in selective explanations of evolutionary change, though his example invites a confounding of temporal with multi-level effects in the production of screening off. While attractive in other respects, McClamrock's analysis (1995) cannot do the job here without the equivalent of an effective screening off relation. (In fairness, he does not intend it to do this. Perhaps he should.)
17. Glennan (1992) and Schank (1991) each use insights from object-oriented

programming (OOP) to characterize mechanistic explanation and analysis. This should be the wave of the future. (Mechanisms are potentially portable objects in the sense of OOP—with standardized inputs and outputs. Like engineered interchangeable parts, we carry them around and install, replace, or adapt them where they fit. Our sense of mechanism as an object and an articulated set of objects [mechanisms are often composed of articulated processes] owes much to our engineering past. Our models of mechanisms do also, reflected in the OOP notion of an "object.") Other recent analyses of inter-level explanations drawing strongly on scientific practice, e.g., Darden (1991) or Bechtel and Richardson (1993), complement the approach taken here (see also Machamer, Darden, and Craver, 2000).

18. Why isn't aggregativity a kind of multiple realizability? It is—an unusually strong form, but so strong and implying such a homogeneity of the parts' actions as to deny the *multiplicity* of realizations! With aggregativity, there's no tendency to separate system property from sum of parts properties; they are simply identified.
19. Cowan was chair of the very successful new Department of Theoretical Biology, descendant of the historic Committee on Mathematical Biology, at Chicago. President John Wilson merged the departments, and the better-funded area of biophysics took over. The two departments' members had little in common. The theoretical biologists soon found appointments in other places in the university where Waring blenders were not used. And, of course, the biophysicists weren't *uninterested* in biological organization: disrupting cells and tissues made it easier for them to isolate the macromolecules whose organization they were interested in, and they avoided methods that would disrupt the macromolecules they were interested in.
20. How do we recognize when such a framework is only being used as a curve-fitting device? (Instrumentalists cannot even ask this one, since to them *all* theories are curve-fitting devices.) We need criteria for distinguishing realistic theories from curve-fitting ones in a world where there are both kinds of entities, and hybrids between them. Ramsey (1990a, 1997) discusses the "semi-empirical method" in chemical kinetics—a provocative case of such a hybrid—and I (Wimsatt, 1992) present a revealing juxtaposition of the two in Haldane's (1919) self-conscious simultaneous use of causal and predictive non-causal models of linkage and mapping functions.
21. Giere (1988, p. 106) argues that distinguishing two dimensions of approximation solves many puzzles. A model is similar in some *respects* but not in others to the thing modeled, and accurate to different *degrees* in each of these respects. We get consilience if we treat the four conditions as his "respects," and orderings within conditions as his "degrees."
22. The kind of ex post facto reification discussed in the last several paragraphs is central to Wittgenstein's (1962) concern in *Remarks on the Foundations of Mathematics* when discussing how we follow rules. Before the fact, it appears that we have a choice in how we apply a rule, but after the fact it appears that we *had* to go in the chosen direction. This important psychological phenom-

enon arises when we assimilate the new application to our understanding of the rule. This assimilation is central to the perceived sense of continuity we have in many of our activities. As an epistemological question—how much and why we had to continue as we did—the discussion of generative entrenchment in Chapter 7 elaborates why and how some epistemic actions may become effectively irreversible after they are performed, and the consequences of this fact. Finally, the lovely chapter by Hoffrege and Hertwig on hindsight bias as a product of adaptive processes (in Gigerenzer, Todd, and the ABC Research Group, 1999) brings this whole topic successfully within the scope of heuristic inference.

9. Complexity and Organization

Parts of this chapter are based on my doctoral dissertation (Wimsatt, 1971a) and on work done during the tenure of a Woodrow Wilson Dissertation Fellowship at the University of Pittsburgh and a post-doctoral fellowship with The Committee on Evolutionary Biology at the University of Chicago, supported by the Hinds Fund for Studies in Evolution. I gratefully acknowledge their support.

1. This also seems to be true in physics in active research areas such as (but not limited to) meteorology and magnetohydrodynamics, though those arguments for complexity based on evolutionary phenomena have no obvious application in these areas.
2. The problem of how to delineate the domain of a scientific theory has received little attention. See Shapere (1974) for an examination of these issues.
3. Discussions of theoretical pluralism have arisen in the context of scientific change, but implications of this pluralism (including the hotly debated problems of translation and meaning variance) have not been investigated for *simultaneously held* partially overlapping theories and models. Richard Levins' views on the nature and use of theories and models in biology (Levins 1966; 1968, chapter 1) are a notable exception, and Stuart Kauffman's (1971) views on the plurality of ways of seeing or describing systems are also suggestive. It is tempting to dismiss this as a kind of pre-paradigm or multi-paradigm science, but this is to ignore the fact that many of the scientists must then be viewed as simultaneously (or alternatively) using several of the paradigms at any given time.
4. See Bergmann (1957, pp. 93–96) for discussion of the assumptions of closure and completeness of a system.
5. On this, see also Bishop Berkeley (1709, paragraphs 41–51, especially 48–50). There should be obvious application of these remarks to the problem of interrelating information from different sensory modalities in the construction of "external objects." Why, and when, do we objectify?
6. It is thus ironic that Goodman's (1966) calculus of individuals or something quite like it (pp. 46–61) seems admirably suited to the analysis of what I describe as *descriptive complexity*. I thank Leonard Linsky for this reference.

7. Some of the most interesting contributions in this area are Campbell's (1958, 1959, 1973). Campbell's most relevant point in the present context is an attention to the boundaries of systems. He argues, in part, that we consider objects the more real and substantial the more there is a coincidence of boundaries on different criteria of individuation. He applies this, among other things, to the individuation of social units (1958) and the order in which we learn different kinds of concepts (1973). For more analyses of the importance of the boundaries of systems on different criteria, see Platt (1969) and Simon (1962, 1996). On the localization of functions and questions concerning whether functions are objects, see Gregory (1961, 1962) and Wimsatt (1971a, 1972, and the last section below).
8. These two factors represent in part an attempt to give a conceptual basis and motivation for some of the ideas expressed by Richard Levins (1970a, 1973, and personal communications).
9. Kauffman (1971) seeks to argue that these perspectives may not be reducible, one to another, so he emphasizes the possibility of non-isomorphism. Spatial coincidence of the parts' boundaries under two decompositions implies spatial isomorphism of the decompositions but not conversely. It is coincidence, not isomorphism, that is important here.
10. It is important to unambiguously identify and individuate the different theoretical perspectives appropriate to a system. *Any* system is trivially descriptively simple relative to just a single T_i and $K(T)_i$, so two or more theoretical perspectives and correlative decompositions must be considered. But if criteria for individuating these are in doubt, then someone might claim that a system purportedly classified as descriptively simple relative to a set of perspectives (like the granite case discussed below) is not, because the perspectives are not in fact distinct, but are parts of the *same* perspective.
11. For an ordered pair of such parts, the mapping relations 1–1, 1-many, 1-part, 1-many part, many-1, part-1, and many part-1 are exhaustive. Other considerations are important, such as whether the mappings are continuous. Discontinuities in mappings will increase the complexity of description of a system.
12. A number of measures are possible, and different ones would be preferable under different circumstances. For illustrative purposes, ϵ_c can be thought of as the maximum strength of intersystemic interactions divided by the minimum strength of intrasystemic interactions. This ranges between 0 and 1, approaching 0 as intersystemic interactions become negligible.
13. I am *not* an anti-reductionist. There is not even any reason to believe that these theories are incorrect for what we intuitively regard as complex systems. On the other hand, there is no reason to believe that they will describe other systems as simply as the systems they were generated to explain. Adequate theories of complex systems may require new categories if they are to be described and analyzed more simply. Statistical mechanics owes its acceptability to just such considerations.
14. The view criticized here is Levins' reading of Simon, and this line of attack is

basically an elaboration and adaptation of a point Levins has made many times in conversation. Levins' interpretation seems not to be too idiosyncratic: Howard Pattee (1970) also reports Simon's analysis as showing that hierarchically organized structures are nearly decomposable (see p. 136n10). Nonetheless, it may not be a fair reading of Simon. The evidence is at least equivocal: In addition to the "stable subassemblies" argument, this point of view is also suggested by Simon's discussion of near-decomposability in physical systems (pp. 103–104) and the first part (pp. 114–115) of his discussion of genetic control of development in terms of hierarchically organized computer programs. (The latter part, pp. 116–117, appears to suggest that Simon is aware of the issues raised here.)

15. Some of the most interesting cases of this are to be found in so-called cooperative phenomena in polymers. Thus, it has been suggested that hemoglobin (a tetramer) might have evolved from its monomeric precursor because interactions among the four subunits facilitate binding of the hemes (which carry oxygen). See Jukes (1966, chapter 5) for further details.

16. This argument would yield firmer grounds than the randomness assumption for a belief that hierarchically aggregating systems will become interactionally complex through co-adaptation. In fact, the argument here seems to be just the other side of the argument made by von Neumann and Morgenstern (1964, pp. 11–12) that a game-theoretic interaction is not a simple maximum or minimum problem. Their point is that with two or more individuals acting, there are two (or more, as appropriate) functions, each of which some individual is trying to maximize, subject to the constraints of the actions of the others, and there is no overall function that tends to an extremal value.

Here we begin with independent systems with their own (relatively) independent optima. When they aggregate and begin to coevolve, their individual survival becomes dependent upon the "collective welfare," with something more akin to a single optimum. I assume, as von Neumann and Morgenstern do, that it is overwhelmingly improbable that the conditions for individual and collective optima coincide. The failure of this coincidence results in selection for interactional complexity.

Indeed, the situation is more complicated than indicated here, where it is supposed that either individual optima or the group optimum dominates absolutely. In many cases (see Lewontin, 1970) selection can operate independently upon units at two or more levels of the *same* hierarchy, in concert or in opposition. In these cases, we have co-action of forces, and a compromise among conflicting optima is inevitable.

17. I have attempted detailed analysis of functional organization 1n chapters 6 and 7 of Wimsatt (1971a). Some of the conceptual complexities inherent in this problem are discussed in Wimsatt (1972).

18. Campbell has argued (in conversation) that even for systems that I have chosen to call complex in one of the two above ways, there is still probably a coincidence of boundaries for the vast majority of properties. This seems

reasonable: if there were an arbitrary thicket of overlapping boundaries we probably would not be able to pick out *any* systems. Furthermore, the use of modular construction techniques (both as new as third generation computers and as old as multi-cellularity) would appear to imply the coincidence of at least *many* boundaries at the boundaries of the modules.

Simon has recently similarly argued that these remarks on complexity are a second-order approximation to modify the first-order approximation of describing systems as nearly decomposable (personal communication). I agree with both of these remarks, though not with the implication one might draw that second-order effects are, in the relevant sense, always ignorable relative to first-order effects. Whether they are or not depends upon which phenomena you are interested in. Weak hydrogen bonding is, chemically speaking, energetically negligible. It happens, however, to be crucial to the proper functional behavior of biologically important macro-molecules.

Another interesting and powerful line was suggested by Levins (1973) when he argued that an initially arbitrarily complex system will tend toward greater simplification, and perhaps this is implicit in Campbell and Simon's arguments. Thus bounded (by co-evolution) away from aggregative simplicity and (by the need to have parts of the system responding semi-autonomously to selection pressures) away from total interactional complexity, it would appear that there must be a relatively stable intermediate level of complexity. Whether this would involve an intuitively and immediately recognizable degree of modularity is an open and important question. Might it be, for example, that there would be different modes of modularity or near-decomposability for systems that arise by aggregation of stable subassemblies (à la Simon) and for those that rise by specialization and differentiation of subparts of a single system (à la Levins)? This is not an academic question. A recent challenge to Margulis' (1971) aggregative account of the origin of eukaryotic cells is offered by Raff and Mahler (1972), who suggest that eukaryotic cells evolved by specialization. It would be extremely useful to have criteria for adjudicating this and other similar disputes.

19. Jaegwon Kim (1971, pp. 329–334) discusses views suggesting that problems with spatial localization of function have been a problem in this context and seems himself to suppose that events of this type must have a precise location.
20. Obviously, talk about the location of mental events in this sense is not intended to apply to those cases where we most frequently *do* talk about location—cases such as feeling an "itch in the leg" or a "pain in the tooth." These kinds of cases are probably best explicated as Margaret Boden does (1970, pp. 207–209), as events that are occurrences in an "internal model" representing states of the organism. As such the events can refer to occurrences at the locations in question without themselves occurring at those locations. See also Globus (1972).
21. Keith Gunderson (1970, p. 303) has suggested that one of the problems with conceptualizing the self arises from the feeling that the concept of the self

requires that one entity be in *two* places at the same time (in one place as observ*er* and in another place as observ*ed*). It seems reasonable to suggest that the belief that such spatial paradoxes bedevil the mental realm is or was influential in denying its spatiality.

22. I think that this will be the judgment of the future even though I admit that these points about functional organization and localization go but one small part of the way there. I believe that there are three classes of paradoxical phenomena to be handled in attributing spatiality to the mental realm: those, respectively, associated primarily with first person knowledge, with third person knowledge, and with interactions between first and third person knowledge. The terms *first person knowledge* and *third person knowledge* must first be reanalyzed in terms of asymmetries among and limitations on the locational information given by the various sensory modalities. The functional localization problems discussed here are then seen to be pure third person problems, whereas most of the interesting problems discussed by philosophers are seen to be one of the other two types. Interestingly, there is a sense in which first person knowledge of the mental realm *is* non-spatial, though this fact turns out to be of no comfort for those who would wish to use it against materialism. This argument builds substantially upon key points raised by Gunderson (1970) and Globus (1972). [Author's note, 2005: I would now argue that the importance of second person perspectives—particularly second person plural—have been insufficiently recognized in looking at the impact of social and cultural issues in the analysis of the mental.]

10. The Ontology of Complex Systems

I would like to thank Irene Appelbaum, Bill Bechtel, Chuck Dyke, Stuart Glennan, Sergio Martinez, Alirio Rosales, Jeff Schank, Bob Ware, and Barbara Wimsatt for discussion and commentary on matters both substantial and stylistic; the talented (and sadly now deceased) Sylvia Culp for very useful last-minute input; and Bob Ware for his tolerance as an editor.

1. For a philosophical response to "the new messiness," see for example John Dupre's provocatively titled, *The Disorder of Things* (1993). But while Dupre and I both urge major surgery on our ontologies, methodologies, and epistemological assumptions, and make movements in many of the same directions, I believe that my surgery is both ultimately more conservative, particularly in defending a liberalized and non-eliminative descendant of classical mechanistic materialism, and is also more in accord with actual scientific practice. (Our differences on the former but I think not on the latter point may be in part ideological or rhetorical rather than substantive.) Dupre could urge in return that I haven't paid sufficient attention to the social determinants and aspects of our practice. To this I plead guilty, though I think the view argued here can both deal with and in part explain those complexities.

2. Simple mathematical models of this and other reliability calculations are given in Chapter 5. When probabilities of being correct (or of introducing error) are bounded between zero and one, then serial dependencies always reduce reliability and parallel redundancies always increase it.
3. The apparent robustness of the "result" that group selection could not be causally efficacious is one such case where the various supposedly independent considerations supporting this conclusion turned out not to be independent after all. Similarly, the once highly touted validity of IQ scales is seriously compromised by the fact that agreement with older tests was used as a criterion for the inclusion of questions in newer tests, so the tests—even as composed with entirely different questions—failed to be causally or probabilistically independent in the relevant sense (see Wimsatt, 1980b, and Chapter 5 in this volume). A more challenging case is provided by the bacterial mesosome, as discussed by Nicholas Rasmussen (1993). This entity was once thought to be a new kind of cellular organelle, but is now widely regarded as an artifact of preparation methods. Although the mesosome appeared under a variety of treatments (thus showing some robustness), they are classified as artifacts (as Rasmussen presented it) in part because crucial properties (such as the size and number found) appeared to vary with the preparation methods in ways that were inappropriate for cellular bodies. These are ultimately failures of robustness, though mediated in part by theory: objects of the sort detectable through these means ought not to behave like that—therefore they are artifactual. Culp (1994) analyzed this case in more detail, tracking the dispute longer, with additional evidence. Contra Rasmussen, she argues that lack of robustness was, as it should have been, ultimately the downfall of the mesosome. She points out that positive support for it as a natural entity stopped accumulating, while robust support for it as an artifact of preparation methods continued to increase. By the middle 1980s one had more techniques and a virtual recipe book for how to produce or avoid mesosomes. There was very strong evidence that they were invaginations in the cytoplasmic membrane produced by preparation-induced contractions of the nucleosome, and their production was facilitated by damage to the cytoplasmic membrane and inhibited by breaking connections between the membrane and the nucleosome, which would tend to produce invaginations. The complexity of this case might suggest possible circularity for uses of robustness as a criterion for reality, but given different degrees of robustness and independence, and given specific knowledge about how different means of access may break down, and under what conditions, I think that any circularities present are not vicious. See further discussion below.
4. Thus, in fascinating work in the late 1960s on the interactions of visual and tactual modalities, Rock and Harris (1967) found a complex conditional dependency in which sense we trusted when both were used. When no disparity in judgment between the two was noticed, vision was taken over touch—a judgment justified evolutionarily by the fact that we can make higher-

resolution and more accurate discriminations (for shape, pattern, texture, and the like) with vision than with any of the other spatial sensory modalities. When a disparity between the judgment of the senses is noticed, however, touch is taken over vision—again a reflection of the fact that vision is more subject to systematic distortion than touch (witness the "bent stick" illusion). The accuracy and reliability of our different scientific instruments are related and interdigitated in ways that are at least as complex as this case, and we have to learn which to trust, under what conditions, and why.

5. A serious obstacle to both activities occurs if we cannot discover how the means of access (our instruments) work and how they can be biased or break down. In reviewing work in the 30 years since their classic paper recommending a variant of this methodology in the social sciences, Fiske and Campbell conclude (Fiske, 1992) that its limited success there is due to the more complex processes affecting measurement in the human sciences, and the lack of an adequate "theory of the instrument." Crucial to a theory of the instrument (such things as questionnaires, interviews, participant observation, and the like) is knowing how it interacts with the object. This is essential to ecognizing whether a result is a property or product of the object, the instrument, the testing situation, or some complex relation among some or all of the above.
6. As Sergio Martinez (1992) points out, it was also often argued that primary qualities were aggregative, which would further increase the tendency to locate the "real" properties at the lowest level of aggregation—a close cousin to foundationalism. Martinez points to interesting historical connections between ideas of realism, robustness, and aggregativity. See Wimsatt (1986a) and Chapter 12 for further discussion of aggregativity.
7. We will see examples of this later in the level-relativity of explanation: between-level phenomena are always referred up or down in level for explanation—illustrated using the example of Brownian motion.
8. I do not think this results in the demise of deductive arguments, or of philosophy or in the fusion of philosophy and science, though there are more activities that could be viewed as either or both. It does suggest a broader critique of philosophical methodologies that urges formulation of each argument in deductive form, come hell what may, and heaps scorn on other styles of argument without concern for what is the appropriate argument for the context. A valid but unsound argument, or even an argument that is both valid and sound, but whose conditions of soundness are extremely restrictive, may sometimes wisely be replaced by a broader but less fragile argument, or by a number of mutually supporting though individually weaker arguments. We are similarly sometimes ill-served by the search for exceptionless laws. In both of these, we have been misled by our concerns with logic. This is not to deny that there are contexts for which deductive arguments are the tool of choice—but only to deny that they have a priority or preferability over all other tools in all contexts, or have a particular foundational significance. Robustness also gives qualitatively plausible ways of

dealing with structures that contain local contradictions and approximations—two features common to rich scientific theories that are difficult to handle naturally with deductivist tools (see also Wimsatt, 2006b).

9. For those for whom this matters, I have deliberately picked cases where the replicator is also an interactor.
10. They are represented in a compositional manner—and may be compositionally related to other mentalistic entities. This does not imply, however, that they or any mental objects would map to physical objects (as opposed to physical configurations, stable dynamical patterns, or whatever) in a successful material theory of the mind—any more than we would expect the objects of "object-oriented programming" to do so. In fact, there is reason (suggested in part by examples like this) to believe that they wouldn't. See Wimsatt, 1976a, part III, for further discussion.
11. Levels are probably most frequently discussed in conjunction with accounts of hierarchical organization, of which there is an enormous literature, much of it suggestive and useful for present purposes. While many of the systems with multiple levels are hierarchical in character, I don't wish to couple levels talk to hierarchies, since I will also be interested in situations where the conditions required to define hierarchies are violated. For more on hierarchical organization, see Pattee, 1973, and for a more recent work that draws particularly broadly on the literature, see Salthe, 1985.
12. There is a significant epistemological parallel here with the research program of Stuart Kauffman's *The Origins of Order* (1993), where Kauffman seeks to argue that many adaptive "emergent" properties of systems emerge relatively directly as ensemble properties of classes of systems. Since they are in effect high entropy properties of such systems, we get them "from the physics" for free, and we don't have to invoke special selective processes at higher levels to explain them. Indeed, Kauffman's research program turns out to be a particularly revealing special case of this, which merits substantial further study for its methodological lessons.
13. Philosophical constructs like Wilfred Sellars' "manifest image" (Sellars, 1962) beg to be analyzed in terms of levels of organization, though the mixture of psychological and physical properties in Sellars' construct render the analysis not straightforward, and unlike the "folk psychology" that some writers have derived from Sellars' "manifest image," levels are not eliminable through conceptual revolutions. (Of course, folk psychology may not be either!)
14. For effects to go up a level, a lower-level process must be a deviation-amplifying process (at least under those conditions) and, more generally, if it is an adaptation, it must be such under a suitably broad range of conditions. This suggests that a fuller exposition of this condition may require reference to the concepts of canalization and of deterministic chaos. One important (and perhaps the only systematic) way for an event to have effects that go down in level, is through a selection process.
15. The fortuitous term *level leakage* I owe to Stuart Glennan, whose inventive

and highly original (1992) work on causation and mechanism provides important further explication of both notions that support the central role that a mechanistic perspective plays in this account.

The ways in which we exploit *level leakage* to gain access to other levels became much clearer to me through my involvement in 1979–81 in helping to program a custom ROM module for the Hewlett-Packard HP-41C programmable calculator. The calculator was designed to be programmed in RPN (Reverse Polish Notation), a sort of assembly-level language that allowed direct manipulation of program instructions, numbers, and alphabetical characters in a controlled region of the calculator's memory, preventing access to other regions of memory used by the calculator's "system software," on the other side of a "curtain." A "bug" in the definition of some of the keyboard functions on some early calculators gave unintended ways of creating new "synthetic instructions," which made it possible to move the curtain and directly manipulate the contents of control registers behind the curtain on all HP-41 calculators, whether they had that bug or not. This led to a new machine-specific discipline, called *synthetic programming*, which gave the synthetic programmer control over many things on the HP-41 that Hewlett-Packard engineers never intended (e.g., individual elements in the LCD display, and individual pixels in the printer output, and the ability to do all sorts of bit manipulations to compose new kinds of instructions). Synthetic programming thus gave new capabilities, and sometimes striking increases in efficiency, speed, or power. On the down side, it also gave new and dangerous ways of crashing the calculator, and exploiting this new resource required much greater knowledge of the details of the machine, such as the Hexidecimal code for all machine instructions, and greater knowledge of how they worked, and how they interacted with the hardware. See the *PPC ROM User's Manual* (1981) for the section on the history and description of "Synthetic Programming" and also the 1982 book of that same title by its main developer, William Wickes.

16. *Identity theories* are now out of fashion. *Token identities* seem too weak since they claim nothing more than spatio-temporal coincidence, and say nothing about how the upper-level phenomena are explained by the lower-level characterization. *Type identities* are subject to a myriad of possible counterexamples, both of the ordinary garden variety derived by considering how two different people with two different and presumably differently wired brains can think the same thought, and also of the more Procrustean variety preferred by philosophers (stimulated by images of the mental life of the population of China or a cerebral Martian plant, both of which are supposed to have the same functional architecture as you or I). I find the first kind of case much more convincing and important than the two of the latter, because we have an existence proof in the first case that it is indeed possible, whereas it is not clear what if anything follows from such unconstrained "thought experiments" as are imagined in the latter two. Type theorists prefer talk of instantiation rather than identity. On the other hand, more sci-

entifically motivated accounts rooted in biology such as that of Darden (1991) or Bechtel and Richardson (1993) favor talk of functional localization rather than identity. I am sympathetic with the latter kind of approach, but still feel the need for a kind of identity that falls somewhere in between type and token identities in its logical characteristics. For a realistic account of scientific theorizing, we want a kind of context-bound type identity; not one that is expected to be valid across all possible worlds, but neither one bound rigidly to this particular single instance. It should be valid, possibly with qualifications, across the range of contexts we extend our theories and mechanisms to, including, if we are lucky, some quite unlike those with which we started. It is a contextual generalization of uncertain but not too narrow scope, where the properties of the upper level thing are explained, *ceteris paribus,* by the operation of the spatio-temporally coincident lower-level causal machine. The generalization has exceptions, as do all generalizations relative to their lower-level instantiations. (Donald Davidson [1970] was too parochial in boasting about the anomalousness of the mental—it's anomalies at each level of organization, all of the way down.)

17. It is tempting to think that this can't be true for natural kinds, but I think rather that we haven't paid enough attention to the right kinds of more complex cases—and along with this (though not here) to ask more carefully what functions we want our concept of a *natural kind* to serve. Thus consider ecological definitions of species types—where the ecological niche defines the type of organism that it fills—a kind of functional equivalence. There are both marsupial and placental "squirrels" and "dogs"—though the distance to their common ancestors are great, and the two marsupials (or the two placentals) are closer relatives by evolutionary genealogy (and DNA sequence) than the two squirrels or the two dogs. Nor will it do to say that terms *squirrel* and *dog* only refer to categories of folk psychology and can't be natural kinds, for these terms enter into many regularities of behavior, and have genuine predictive and explanatory import—indeed, probably more for most aspects of their behavior than any characterization derived from their genealogy or distances in a DNA sequence space. To say that natural kinds must have definitions in terms of intrinsic rather than functional properties would beg all of the most important questions. See also the last part of the discussion of Brownian motion earlier in this chapter for further consideration of what makes intrinsic properties important and their apparent absence from Brownian motion particles.
18. It obviously would not be for a functionally defined entity.
19. For a systematic discussion of the importance of size in the biological realm, see Schmidt-Nielsen's (1984) fine book *Scaling: Why Is Animal Size So Important?*
20. This criterion is different from the compositional one, but has related presuppositions, and like any statistical property of collections, is further indirect evidence of the importance of the compositional criterion. It isn't necessary because system properties—indeed, most of the interesting ones—needn't be

purely additive or aggregative functions of the properties of the parts, like an average is. The property of aggregativity—or its denial—is crucially connected with an important concept of emergence in which the higher-level properties depend upon how the parts are strung together. The dimensions of aggregativity turn out to be very useful tools in describing the modes of organization of complex hierarchically organized systems (see Chapter 12 and Wimsatt, 1986a).

21. There is an interesting and suggestive relation here between "average" and "stereotype" (an abstraction depicting an average or distorted average) and level, where in this case level is broadened to include not only quasi-compositional level, but also social status or power relations. Different compositional or quasi-compositional levels are involved when a member or representative of a corporation relates stereotypically to an individual—say a customer—while the individual is forced to relate to the corporation according to *its* individual characteristics. The differential behavior of members of different (upper and lower) classes toward one another is legendary, and the stuff of novels. If the stereotypic or stereotyped object relates to the agent as an individual, this puts that individual at a higher level. If both individuals are stereotyping each other, they are just cogs in their respective institutional or social machines, "or acting out their roles." There is an oft-observed confusion between average and stereotype or between statistical norm and some sort of evaluative norm in common thought, but it is interesting how this maps onto the compositional, power, or status levels distinctions.

22. Lest one believe that *any* reduction in dimensionality should lead to a simplification in the observed behavior, I suggest a visit to Edwin Abbott's classic *Flatland* (1884), in which two-dimensional creatures see very complex and confusing behavior of three-dimensional objects passing through their two-dimensional world. In fact, the predictive accuracy of a model is usually increased by increasing the dimensionality of the model—which is part of what makes levels such remarkable beasts. See also the discussion of Lewontin's "dimensionality" argument in the units of selection controversy in Wimsatt (1980b).

23. Walter Fontana (1991) and Leo Buss (Fontana and Buss, 1994) worked on simulations of the evolution of life using symbolic biochemistries based upon expressions and rules of the λ-calculus—expressions that can combine and operate on one another. They found the spontaneous evolution of a subset of expressions that occupied a small (i.e., lower-dimensional) subspace of the space of possible λ-expressions, and whose interactions were governed by a grammar. Neither the expressions nor the rules for their interaction were built into their simulation at the start. What they have in effect done is generated the evolution of a level of organization, complete with entities and laws, in a different (and in this case, abstract) material, which has the important properties described here of a level of organization. This suggests that compositional levels of organization may in fact be an extremely general property of spontaneously evolved complex systems.

24. This suggests that ontological changes of centrally located entities at a given level should differentially affect other entities and properties at that level most strongly, and be more weakly connected with other changes at other levels. This demands further thought.
25. Of course, with intentional agents, categories in theories can acquire a causal role in the generation of behavior, and if the behavior involves the production of material systems, such categories or decisions using them can result in the generation or creation of physical, biological, psychological, social, and cultural order. But in this way, theories become *parts* of the physical world as well as lenses through which it is viewed. The interests and needs of human agents can become materialized in similar fashion, becoming instantiated through hardware and software technology, our choice of research projects, and of how they are to be pursued, producing (in Stuart Glennan's fortuitous words) "changes both in the lens and in the picture it presents" (personal communication). In this way, the picture I urge combines elements of a constructivism in a broader-based realism, although it may be extremely hard (and perhaps pointless) to tease the aspects of construction and realism apart. Nonetheless, it is plausible to assert that theories will become more causally efficacious in that world to the extent that theoretical categories map accurately onto natural categories in the world—or onto cost-benefit approximations to them.
26. But see Waismann (1951) for a rich and perceptive paper on levels written from the linguistic perspective, which (particularly in his accounts of the limitations of inter-level translation) makes many points I would agree with.
27. Indeed, I would argue that almost all robust entities are at levels, for reasons given in the next section. Here, as elsewhere, when I use terms like *most,* or *almost all,* I do not assume that the entities are counted, or even countable. (They could fail to be countable to a fallibilist either by being of a nondenumerable infinity in number, or more paradoxically, by being finite, but not orderable in any compact rule-governed way, so that the only way to tell would be by doing an exhaustive survey of all cases.) In this or in most other such cases, when I say *most,* I refer to the proportion among the cases sampled, on the assumption that they are representative—a judgment subject to the normal array of availability biases discussed by Tversky and Kahneman (1974).
28. It does of course in the human realm, and for parallel reasons: counterpredictive purposive agents are the stuff of game theory, and not surprisingly, the one place game theory has found a home outside of the realm of human behavior is in evolutionary biology, for which, see Maynard-Smith, 1982.
29. Matching levels involves most obviously matching size scale and frequency dynamics parameters so that the main desiderata of the niche that the organism has chosen can be fulfilled. This will mean evolving with the behavior of the other organisms that are evolutionary factors as constraints—something that may sometimes call for matching size and frequency (most often

for conspecifics), and sometimes for *mis*-matching one or the other or both (most often with prey or predators).

30. The still much shorter half-life of elementary particles should make it clear that the lifetime required for an object to count as a good object is level-relative. With this observation, we can then formulate the problem more precisely as that the lifetime of the "clusters" is not appropriate to—it is too short for—entities of that size scale. See item (n) below.
31. The converse does not follow—that anything given a functional definition is necessarily between levels. Philosopher Chuck Dyke argues (personal communication) that if we accept functional definitions of objects, there is nothing wrong with speaking of Brownian motion particles and their colliding assemblages as constituting a level. This suggests either that functional definitions alone are not enough, or alternatively (perhaps suggested by the mental realm) that we need a more strongly connected set of interlocking functional definitions or a stronger notion of function (Wimsatt, 2002) to be willing to reify a level on that basis alone.
32. Ramsey (2000) argues to a strikingly similar conclusion.
33. The usefulness of this kind of measure depends upon the widespread scale-invariance of these fractions for different reactions over different size deviations from equilibrium—producing a recognizable exponential approach to the equilibrium state. The best known example is the half-life of different radioisotopes—the time it takes for half of their nuclei to decay (there is a different measurable constant for different isotopes). There are other measures (the infamous LD-50—the dose that kills half of the relevant type of test organisms) that don't have this exponential character: two LD-50s will probably kill all or nearly all of the organisms, not three quarters of them.
34. This is not quite true. I give conditions—characterized both formally and informally—under which going to the lower level is explanatory and conditions under which it is not in Chapter 11—see especially the definition of "effective screening off" in the appendix.
35. At still higher (e.g., planetary, solar, or cosmological) levels, this process seems to be reversed for several complex reasons: many of the still higher level processes are driven by bulk or average processes at lower levels and thus tend to produce regular behavior, but some represent divergent processes producing chaotic irregularities whose effects we cannot characterize except in terms of fractal, and thus scale-independent patterns. Perhaps the absence of differential selection processes and predator-prey and parasitism networks which ferret out any useable order while simultaneously generating designed unpredictabilities (see section h) allows a less complicated dynamics, or perhaps it is equally or even more complicated, and we simply do not yet have the tools to see it.
36. This is a qualitative remark. There is of course no reason to suppose that there should be this kind of relation.
37. Hugh R. Wilson, personal communication.
38. This is perhaps just a first approximation: I am unhappy with the implicit

claim that ecological niches are confined to a single level. See item i above. Perhaps "predominately to a single level."

39. The terms *descriptive complexity* and *interactional complexity* refer to the complexity of mappings of object boundaries from one perspective to another, and to the strength and structure of causal interactions between variables and parts in different perspectives. See Chapter 9.
40. The path here is fraught with error and tempting but dangerous inferences. I think that the tendencies are real, but there are exceptions to everything in the rest of this paragraph. The kinds of qualifications about finding genuinely aggregative properties in nature urged in Chapter 12 apply here. At present, it is important to regard this claim as a statement about discovery heuristics or psychological tendencies. I think that more can be said which is sound, but a lot more careful exploration is required first.
41. Published in Cartwright (1994).

11. Reductive Explanation

Most of this chapter was written as a visiting research fellow in Humanities, Science, and Technology at Cornell University. I wish to thank the program and especially Max Black and Stuart Brown for their support.

1. On *in principle* translatability, see Boyd (1972) for a masterful discussion and doubts of a more general and pervasive nature.
2. Boyd (1973, 1980), and especially Kauffman (1970).
3. It is naturally important to distinguish between disputes over details of particular mechanisms from objections (e.g., like those of Haldane, 1914, or Elsasser, 1966) that challenge the adequacy of an entire approach.
4. See Boyd (1980), and Wimsatt (1976a, parts II and III). Boyd locates the primary difficulty in verificationism and acceptance of the Humean account of causation, but as a realist, would also agree with the views advanced here.
5. Ruse (1976) retreated from his earlier attack on the formal model and attempt to characterize "informal reduction." I am more in sympathy with his earlier views.
6. This line of criticism was initiated by Nickles (1973). See also Wimsatt (1976a, part II).
7. All of us believe that some reconstruction is necessary. Hull and I appear to believe that less is necessary (me) or appropriate (both of us) than Schaffner or Ruse. See Hull's (1976) discussion, which does *not* mention the specific alternative discussed here: reconstruction of reduction as an efficient end-directed activity.
8. This is explicit in Kim's (1964) analysis of the deductive-nomological (or D-N) model for explanation and prediction. Although Kim advances this as a defense of the D-N model, by suggesting that the differences are pragmatic and epistemological rather than structural, I am using it as an attack on the formal model, by suggesting that the structural similarities are more superficial than the functional differences.

   Nickles (1973) individuates two types of reduction on both functional and structural grounds, but concentrates on what I call "successional" or "intra-level" reductions. He largely accepts the formal model for the other kind (which is most relevant here) and does not draw the close links between functional and structural characterizations that can be made for each of the two types. Schaffner's (1974b) argument for the peripherality of (formal) reduction in the development of molecular biology invokes Bayesian arguments for choosing scientific research strategies. This presupposes a purposive account of scientific activity, but he has not attempted a functional analysis of reduction or of other related activities.

9. Schaffner (1976, pp. 626–628) treats Nickles' "reduction$_2$" and the correlative notion of a transformation as a competitor to his condition of strong analogy, and criticizes it, uncharitably, I think, for being too open ended because (says Nickles) there is no general way to characterize what kinds of transformations should be allowable and what should not. Schaffner claims that the notion of "strong analogy" can be applied with general agreement (3 out of 4, at least—*pace* Hull!) for genetics and thus, though it is unanalyzed and primitive, it is at least testable. But surely Nickles could claim as much for the notion of an allowable transformation. I suspect that there would be general agreement in any given case on what transformations would be allowable in constructing a reduction$_2$. Nickles doubted he could find something that I doubt exists: a general theory-independent criterion to determine allowable transformations. This is impossible for the same reason as a theory independent notion of "strong analogy": what transformations are allowable (or even interesting) and what features of an analogy are salient depend upon usually quite general and important features of theory in that area. And on these, there would usually be general agreement. Further, the notion of a transformation is mathematically an extremely powerful and suggestive one, and is less tied down to intuitive notions of similarity than analogy. Consider three examples that are very different in terms of allowable transformations, but for which there would be agreement on allowable transformations. First, Minsky and Papert's (1969) applications of linear transformations to the analysis of the data-manipulating capabilities of certain classes of neural networks; second, the "law of similitude" and its use in building scale models of ships and aircraft for testing in wind tunnels and towing tanks; and third, the continuous deformations allowable in the applications of conformal mapping to two-dimensional airfoil theory (see Prandtl and Tietjens, 1957) and in D'Arcy Thompson's application of his (1961) theory of transformations to problems of development and allometric growth. Indeed, none of these has been seen as involving anything like reduction, though Nickles' analysis suggests the possibility of seeing them in a new light.

10. Nickles (1976) gives a more complete account of theory succession and elaboration. His work suggests and may require modifications to the account of successional reduction adumbrated here, but seems to lend further support to the general functionalist approach. His account shows some features of

both intra- and inter-level reduction. This is to be expected in the analysis of any multi-level historical case, which should involve both components of change. The ways his new account differs from his earlier one or from the view advanced here would give no comfort to advocates of the standard model.

11. Ruse (1971) suggests that reducibility is a similarity relation, but gives different reasons (which I do not accept) for saying so.
12. In his paper Ruse (1976) gives up this view and attacks Hull for holding it. Here I agree with Ruse (and Schaffner) though in virtually all other respects, I agree with Hull.
13. For the earliest statement of a related view, see Simon (1996, chapter 4). See also Bronowski (1970). For my general approval of and some dissatisfactions with Simon's view, see Chapter 9 in this volume; for a thorough discussion of levels, see Wimsatt (1976a), part III, or Chapter 10 in this volume.
14. Thus, e.g., in his first presentation of his general reduction paradigm, Schaffner (1967) made provision for upper-level modifications or corrections, but not for lower-level ones, a matter corrected later (Schaffner, 1969).
15. This picture resembles that drawn by Friedrich Waismann in his penetrating essays, "Verifiability" (1951) and "Language Strata" (1953), though he put more weight on the language and less on the underlying structure of the world than I would. Waismann suggests that different language strata might not fit exactly, but would permit nearly exact translations at some points and none or only rough and partial ones at others. This is true for the languages that best describe phenomena and entities at different levels of organization.
16. Traits that are erratically variable are unusually difficult to select for in most cases, so one might argue that it would be highly unlikely that they would be included as part of a *functional* mechanism. But all or virtually all mechanisms that are of interest in biological organisms are functional. Thus highly variable things would not likely be included as parts of biological mechanisms. Ecologist G. E. Hutchinson (1964) used this elegantly to argue that certain trace materials probably could not be utilized by organisms to perform any characteristic functions because they were present in amounts of less than about 10^4 atoms per cell. Hutchinson suggested this as a rough stochastic threshold below which fluctuation phenomena made their presence too unreliable to be used by selection in any biological processes. Unfortunately, this reasoning does not apply symmetrically to allow one to assume (as Dinman, 1972, does) that lower concentrations of trace elements could not *disrupt* functional processes.
17. This realism may look superficially very much like a kind of instrumentalism, because our perceptual apparatus, senses, cognitive apparatus, and theories are treated as instruments designed by biological, psychological, and social selection processes according to cost-benefit constraints that naturally introduce biases. But the biases are taken seriously as deviations from a

correct portrayal of the real world. We regard the biases of the senses, theories, etc., as leading to *false* judgments that we try to correct when appropriate. That a good theory is a useful *instrument* for getting around in the world is a product of the fact that it contains a good deal of *truth*. This is no form of instrumentalism.

18. I doubt that this is the *best* way to argue for this conclusion. Judgments as to where to look for an explanation of a phenomenon are made on other grounds that determine whether a standard causal, micro-level, or functional explanation is appropriate. The judgments of relative likelihood follow from these in any given case. But, roughly, I think that the likelihoods are assumed to be as they are in the text, and the matter is clearly worth further study.
19. If there were a single micro-variable partitioning the macroscopic reference class into exceptions and non-exceptions to the macro-law, this micro-variable would give the relevant lower-level–type descriptions for a reduction. The force of Hull's complaint concerning the complexity of reduction functions is that there is not even a small number of such variables. The force of ergodic theory is to suggest that the same problem affects statistical mechanics, but that the number of "pathological" states involved is so small (of measure 0) that we nonetheless treat it as a reduction (see Sklar, 1973). The number of pathological states in the case of genetics is *not* likely to be of measure 0, however.
20. I am playing loose with Mendel's terminology (and assumptions) in this description, but the respects in which it is thereby distorted do not affect the present argument. See Chapter 6.
21. I am here considering a single break, so the complications of interference and multiple crossing-over do not arise. But even this ignores the complication that breakage strength may vary along the chromosome. All of these factors were recognized and discussed by the Morgan school. See Chapter 6 and Wimsatt (1992).
22. Underestimates in the number of genes was a crucial factor in overestimating their size. Further progress raised questions about the classical model, such as Muller's doubts that the unit of mutation was the same as the unit of recombination.
23. For various reasons it becomes experimentally more difficult to handle a large number of markers in a given experiment. The largest number ever followed at once to my knowledge was 6, by Muller, and that for a very special kind of test of the linearity hypothesis. (See Muller, 1920, especially Table II, for discussion of why smaller numbers of marker genes were usually followed, and Wimsatt, 1992, for a detailed analysis of Muller's methodology.)
24. This may exaggerate the difference. Evidence is accumulating in *Neurospora* (a bread mold widely used in genetic experiments) that there is a strong or even an absolute bias against intra-genic recombination at a molecular level. This is a product of site specificities in where the "nickases" (enzymes that nick open the DNA to allow recombination) will act. If this phenomenon is

veridical and generalizable, then the beads-on-a-string view of the genome is inappropriate only for suggesting a macro-mechanical metaphor rather than a chemical or a micro-mechanical one (see Whitehouse, 1973, pp. 367–369, for a relevant discussion). I thank Thomas Kass for helpful discussion of this and other related points.

12. Emergence as Non-Aggregativity and the Biases of Reductionisms

Prepared originally for a festschrift for Richard Levins, which did not appear. My thoughts here owe a lot to two writers—Richard Levins and Ernest Nagel. Some distinctions here are noted by Nagel in his classic, "Wholes, Sums and Organic Unities" (later published in Nagel, 1961)—though we diverge in what we made of them. The immediate stimulus for this analysis was Levins' germinal essay "The Limits of Complexity" (1973, from 1971 draft) and his distinction between aggregate, engineering, and evolved systems, which also influenced Chapter 9 of this volume. Stuart Glennan, Peter Taylor, and Bill Bechtel gave especially useful comments on prior versions. See acknowledgements in Wimsatt (1986b), dating from 1971 notes.

1. Herbert Simon's essay on complexity (1996, chapter 7) adopts basically this view of emergence. For more examples see Kauffman (1992).
2. Wimsatt (1976a, 1976b, 1979). For similar views, see Sarkar (1992b); Waters (1990); Bechtel and Richardson (1993). Some characterized by Waters (1990) as anti-reductionists are better viewed as opposing the standard philosophical analysis of reduction. (I'm not willing to cede ownership of the term to philosophers.) What most scientists view as reductionism is a species of explanatory mechanistic materialism that may not even be properly regarded as involving laws or theories, as those terms are traditionally understood by philosophers (see the discussion of "particularistic mechanism" in Wimsatt, 1992).
3. For a complementary account of the relation between causation and mechanistic explanation, see Glennan (1992).
4. This proposal—by Stuart Glennan—is worth more elaboration than it finds here.
5. There is a curious mismatch here: such philosophers also talk about "supervenience"—a relation between a system property and lower-level realizations of that property in which there is a many-one relationship from micro-states to that higher-level property, using micro-state variables that are solely *intra*-systemic. Now called "narrow supervenience," this is contrasted with "wide supervenience," where the many-one mappings depend also on the context of the system (see, e.g., Rosenberg, 1978, on fitness). Philosophers of mind often see wide supervenience as irrelevant to their problems. This is a mistake. Reductionist biases (Wimsatt, 1980b) should often lead us to think that narrow supervenience will do when wide supervenience is required. Thus we slip into regarding fitness as a property of organisms (or even of

genes), rather than as a relation between organism and environment. Moss (1992) catalogues ways in which we have attributed properties to genes that are properly regarded as characteristics of the cellular milieu or larger entities. McClamrock (1995) argues that our Cartesian heritage has mislead us—seeing consciousness as an organismally or computationally "inner" property, rather than a fundamentally relational set of properties of a developed, socialized, embodied brain (Wimsatt, 1976a, gives an early statement of this view). But there are problems with accepting that there are any cases of supervenience (as defined) in science: I have argued that Levin's heuristic analogue—the idea of a "sufficient parameter"—better fits all claimed cases, is more operationally usable, and better fits with actual scientific practice than supervenience (chapters 5 and 10 of this volume).

6. As urged by Sarkar (1992b), this corrects earlier analyses (Wimsatt, 1976a) where I called an account reductionist if there were a mechanism at any level or combination of them that explained the phenomena—even if it included mechanisms at a higher level or external to the system in question. In drawing the distinction, I would now require explicitly relativizing an account to two (or more) reference levels: (1) a bottom level—the lowest level at which specific parts of a mechanism must be invoked to explain the phenomena (so an explanation of "position effect" requires characterizing an operon as a genetic control structure and how it works, but not the atomic physics accounting for the relevant electron orbitals that yield the binding which occurs); and (2) a top level—drawing the system boundaries broadly enough so that all relevant parts of the mechanisms involved are included. (Still broader things may sometimes be required, but taken for granted, as features of solar and planetary dynamics may be crucial to evolutionary processes on a variety of different time scales.) This "multi-level reductionist analysis" picks out the appropriate levels for objects, processes, and phenomena, and explicates their relations to complete the explanatory task with no further mystery.
7. Not much hangs on this. These are relevant and important criteria, but may not be completely independent, and thus fail to be separately necessary (in all combinations). They seem sufficient, but I'd welcome additional sensible requirements: each gives new handles on tough cases or special circumstances. This pragmatic attitude, first urged upon me 35 years ago by Dick Levins, makes sense—classification of a property as aggregative emerges in most cases as highly conditional and qualified, and has no or little foundational or architectonic significance. But it may have lots of practical import: by choosing decompositions of the system to maximize their fit, these criteria can help us choose good boundaries around objects and parts of objects, acting as a tool of discovery in theory formulation and construction.
8. Some philosophers distinguish synthetic identities from realizations or instantiations. It doesn't make a difference here, but see Chapter 10, note 30.
9. Addition, multiplication, and other operations (e.g., logical disjunction) could be appropriate in different contexts. The context of the parts proper-

ties, the system property, the question being asked, the purposes of the investigation, and the relevant applicable theories can all play important roles in such judgments.

10. Sergio Martinez (1992) brought this most strikingly to my attention in the context of a historical discussion of and argument for the robustness of primary qualities. He points out that in the seventeenth century something very much like aggregativity was used, together with robustness, as a criterion for natural kinds.
11. These approximations are endemic in the formal sciences, and in all attempts I know to build mathematical models of phenomena. Ramsey (1990a, 1992) gives an analysis of the role of approximations in theory construction and justification.
12. Deviation from multiplicative proportions in multi-locus systems is called "linkage disequilibrium" (see Crow and Kimura, 1970, or any population genetics text). Linkage equilibrium values are equilibrial because they are a state of highest entropy—a maximally mixed state. Linkage disequilibrium is misnamed because it is a product of both linkage and gametic packaging. This and similar cases discussed in Wimsatt (1981b, pp. 152–164) provide the basis for higher-level "segregation analogues" defining higher levels of genetic organization. Systems capable of linkage disequilibrium have more complicated dynamics than one always in linkage equilibrium. Here, as elsewhere, a system that behaves aggregatively is simpler. Complexities involved in going from relative locations of genes along the chromosome to their linkage distance—which further qualify the remarks here—are discussed in Wimsatt (1992).
13. This complexity arises only for multi-locus genotypes, and is the single largest difference between single- and multi-locus theories. Moreover, under special circumstances, linkage can sometimes be ignored, and multi-locus problems can be treated as an aggregate of single-locus problems (at or near linkage equilibrium, and when selection forces, migration rates, or other things that could displace the population significantly are kept small). This illustrates again situations of conditional and approximate aggregativity, or "quasi-independence" (Lewontin, 1978).
14. We should have two different terms for the effects of sufficiently close location in the same chromosome ($0 > r > .5$) and chromosomal association in the gametes. I propose *chromosomal* and *gametic* linkage, though the latter term misleadingly suggests that the association is solely a product of gametic packaging. It is a product of this packaging *in the context of a haploid gamete/diploid genotype/two-sex system.* If, for example, there were n sexes each contributing a monoploid gamete to make an n-ploid genotype, with arbitrary recombination and independent assortment among the n homologous chromosomes of each type, then $(0 > r > (n-1)/n)$, and the asymptotic equilibrium value, now $(n-1)/n$, would be approached successively faster for larger n. It would behave more like the one-generation equilibration of the single locus case as n became large enough for the individual n-adic

genotype to be potentially a good sample of the chromosomal and gametic variability in the population.

15. Signs in the four squares of Figure 12.3 indicate where recombination contributes to (+) or retards (−) approach to equilibrium, measured for a population starting with AABB and aabb homozygous genotypes. Starting with double heterozygotes AaBb would reverse the signs. These two pairs of squares are always opposite in sign, except at equilibrium.
16. In the more general quantitative version of this diagram, for arbitrary gametic and recombination frequencies, gametic frequencies p, q, s, and t (of gametes A−−B, A−−b, a−−B, and a−−b) are entered along row and column margins, and cross-multiplied to give zygotic frequencies. Then r of those in the squares along the diagonal undergo recombination.
17. It is not clear whether the squares that produce heterozygotes or those that produce homozygotes should be the analogous ones (both occur in equal frequency). The segregants—homozygotes—are the new combinations produced in a cross among heterozygotes, and thus analogous to the new combinations one can find along the reverse diagonal in the two-locus case. But it is the heterozygotes that involve dissimilar combinations and (thus) occur along the reverse diagonal. The structural differences between alternative alleles and alternative loci prevent formulation of an exactly analogous case.
18. This is the story emerging from classical genetics. Molecular genetics complicates it with new levels of modularity—intron, exon, and transposon—in between base pair and codon and the whole gene, and mechanisms that explain deletion, insertion, inversion, duplication, and transposition. These don't seriously change the picture, but further demonstrate that various design processes exist for increasing the apparent aggregativity in meiosis. The design of mitosis and meiosis for reliable transmission and assortment of hereditary factors is one of the most elegant adaptations in the history of evolution, and their functional analyses through the work of Roux (1883), Weismann (1892), Boveri (1902), Sutton (1903), and others is a high point in the history of biology. The editing and rearrangement of the genome in somatic lineages to mediate gene expression in development make this point in yet another way: the properties of the genome are not an aggregate of the properties of its parts. The arrangement of the genes in the genome DOES matter for development (see Shapiro, 1992, for more details).
19. Figure 12.3 used two-strand models of recombination to emphasize the parallels between recombination and independent assortment in preserving larger structures. More realistic models from classical cytogenetics (ca. 1928–1934) use four-strand models of mechanisms (and failures) of normal recombination to explain the kinds of chromosomal rearrangements taken as givens by population geneticists in describing species differences and giving cytological explanations of inversions, some cases of meiotic drive, and other mutations of large fitness effect.
20. Mechanisms of öogenesis are actually inconsistent with this story: only 1 out of the 4 haploid genomic products normally survives in females. Mecha-

nisms of spermatogenesis are more like it, but there is no recombination in *Drosophila* males. But all kinds of reduced or fragmented products occur naturally in low frequency, and the loss of three-fourths of the genetic material in öogenesis re-emphasizes the non-conservative character of gamete formation.

21. "Abnormal" gametes are considered when the occurrence of deletions is systematic, and one is evaluating their effects, as with population genetic studies of hybridization between species differing in inversions or chromosome numbers, or their proposed use in insect control programs because of the populational consequences of such hybridization (Whitten et al., 1974). But these studies presuppose the non-aggregativity of the effects, and count the deleted gametes only for the fitness reductions these produce for the parents.
22. Sober's widely advertised objections to the additivity criterion (1981, 1984b; Sober and Lewontin, 1982) are not compelling. His criticisms misfire: his "counterexample" to my analysis only *tries* to meet one of two necessary conditions for a unit of selection, so it fails to apply, and his other claims against it fail (Wimsatt, 1981b, pp. 147–152). Sober does not cite this paper. As Lloyd (1989, 1994) points out, he fails to test *his* analysis, which fails the same "counterexamples" he proposes to my analysis! Finally, perhaps most relevant here, he does not address how the "additivity" criterion is actually used by scientists—who do so in ways that would not be subject to his criticisms (Griesemer and Wade, 1988) even if they were sound. I don't actually think his analysis is wrong-headed, though the "additivity" approach is more powerful and captures central features of the theory that his does not. His account provides correct intuitions about the level at which causes act if there is group selection, but—unlike the "additivity" account—gives no operational criteria for applying it. Brandon's (1982) critique is more interesting: if the units of selection debate is taken to be about individuating the relevant genetic units for selection or evolution, it neglects the fact that selection generally acts directly on phenotypes (or "interactors") rather than genotypes (or "replicators"), except insofar as the material entities picked out as genotypes are also interactors (which sometimes occurs). I showed above that additive models didn't make fitness an aggregative property of genic fitnesses. Those complexities are reflections of Brandon's point. See also Sarkar (1994) for a more general and compelling critique of additivity—which still, however, does not vitiate the points of Griesemer and Wade (1988)—nor the usefulness of the additivity approach in most situations, especially if we recognize that additivity is context dependent, and (I'll say it again) *not* aggregative.
23. But Levin's "fine grained" adaptive function—unlike expected utility—is what would be called the utility of a mixture (of environments). Expected utility is glossed as the utility of a lottery among (exclusive) alternatives with given probabilities, but the above definitions of fitness (and of mean Darwinian fitness) are the utilities of mixed ensembles of subenvironments in

proportions given by the probabilities. (Similarly for the mixed populations of different genotypes in the common definitions of mean Darwinian fitness.) As von Neumann and Morgenstern note (1964, chapter 1), it is like comparing the utility of a lottery in which you get coffee *or* cream with probabilities p and $1-p$, to the utility of getting (for sure) a mixture of coffee and cream in proportions p and $1-p$. These are obviously not the same. Thus Levins' "adaptive function" is an *empirical* hypothesis about how to model certain biological situations, whereas the definition of expected utility is an *analytical* (though useful) exercise in probability theory.

24. Were the p_i's integers rather than probabilities, this would *be* the multiplicative law.
25. This last qualifier and similar phrases are dead giveaways that approximations are being used—commonly to treat non-linear relationships as linear. They are warning flags that all bets are off regarding the aggregativity of described relationships. The models are aggregative all right, but the phenomena aren't.
26. Actually, the derivations of both functions implicitly assume that the order of the subenvironments doesn't matter, but it is clearer for the coarse-grained adaptive function (see Strobeck, 1975).
27. See discussions of perceptual grain in predation and "contest" vs. "scramble" competition in Wimsatt (1980a).
28. I know of no close precedent for Lewontin's table, though there have been implicit recognitions of the relations between dimensionality and the combinatorial properties of genetics from the beginning. Mendel in 1866 pointed out that for two segregating factors per locus, and with simple dominance, the number of gametes should scale as 2^n, the number of distinguishable phenotypes as 2^n, and the number of genotypes (treating Aa and identical with aA) as 3^n. This awareness of dimensionality is next reflected in Sutton's famous 1903 paper, in which he shows in a table for from 1 to 18 chromosomes how the number of possible combinations increases, making it plausible that the combinatorial explosion of chromosomal combinations had the right kinds of properties to explain the enormous variety of organic life.
29. My table differs from Lewontin's in arrangement, in the inclusion of additional values, and corrects a minor error. However, the conception of it is unchanged from the original.
30. This does not of course imply that mating or attraction is a random affair, but only that there is no statistical correlation between who mates with who that is detectable at the genotypic level of description.
31. If this relative context-independence facilitates elaboration of simpler theory and models using these properties, they will also tend to become generatively entrenched (Chapter 7), and statements using them to become more resistant to falsification. This adds a veneer of quasi-tautology, definitional truth, or a priori status that will exacerbate the tendency to see such properties as natural kinds.

32. In Chapter 11 I discuss the purported meaning and testability of these in-principle claims and show that in empirical science (as opposed to in mathematics) they are better seen as corollaries of identity (or localization) claims that don't make impossible demands on our computational capabilities than as claims to knowledge we cannot have (Wimsatt, 1976b).
33. See Allen's (1978) biography of Morgan; Maienschein's (1991) study of Wilson, Conklin, Morgan; and Harrison; Darden (1991) on early theory construction in genetics, and Wimsatt (1987, 1992) on the early history of linkage mapping.

IV. Engineering an Evolutionary View of Science

1. In addition to Simon's classic, many crucial works of this period had a strong engineering flavor: the second edition of Wiener's *Cybernetics* (1961); the reprint of Lotka's 1924 classic, *Elements of Physical Biology*; and Ashby's influential books, *Design for a Brain* (1952) and his important but often overlooked *An Introduction to Cybernetics* (1956). Rosenblatt (1962) founded modern connectionism: his strategies combined themes for modeling learning and development in neural nets, and his "ensemble" modeling gave a powerful new approach explored independently for gene-control networks by Kauffman (1969, 1993) and now common through the growth of Monte-Carlo simulation. Von Foerster's symposia on self-organizing systems between 1959 and 1962 included many whose views later became more widely known—Simon, Campbell, Holland, Minsky, and Rosenblatt. This broad fusion narrowed significantly with the rise of artificial intelligence (AI) in the mid-1960s, reemerging with connectionism and Artificial Life (AL) in the 1980s. Lewontin's classic (1970) generalization of the idea of a unit of selection also contributed to an emerging general account of evolution.
2. This crucial essay was delayed in publication for six years—Popper could not bear the idea that contributors to "his" Schilpp volume had anything critical to say about his "Critical Philosophy." I told Campbell in 1970 that he was giving Popper far too much credit for ideas that were his own. Years later he told me he had thought that if Popper adopted it as his own, it would promote the theory among philosophers. Popper lauded cultural evolution ("Our theories can die in our stead"), but really, he meant *other* people's theories: he fought for the immortality of all he put his name on, and tried to put his name on as much as possible. Campbell's kind of cultural evolutionary altruism was beyond his comprehension.
3. That the order in which alternatives were considered could affect choice was long seen as a mark of irrationality, so the pervasiveness of such phenomena (and their importance) was ignored. I first came across it in a more positive light in 1968, and discuss it in Wimsatt (2006c).
4. See the classic of city planning by McHarg (1969). On visualization, see Gregory (1970), Tufte (1983, 1988, 1996), and Ferguson (1992). Gould's works celebrate the intersection of adaptive design and contingency in biology and

culture from *Ever Since Darwin* (1992) forward, while Petroski (1985, 1994) draws analogous insights from engineering.

13. Epilogue

1. My colleague Howard Stein felt that the rhetoric of this chapter was unfair to logical positivism—at least to Carnap, Hempel, and others who openly disavowed telling scientists how to do their work. I agreed, but had trouble finding overt statements to change. I had conveyed an anti-positivist message mainly by innuendo, and they had done the reverse—not necessarily intentionally: (1) Some (e.g., Popper and Ayer), acting like missionaries, did make such recommendations, and had broad influence. (2) In favoring logical reconstruction, precise definition, and clarity, the writings of all positivists communicated an attitude toward such work that constituted a de facto recommendation for how to do science—especially in the human sciences, where debates about foundations and methodology were active. (3) Comparisons of scientific practice to positivist idealizations were easily seen as critical of those practices. Hempel's essay on "The Function of General Laws in History" (1942) suggested that narrative explanations were not explanations but "explanation sketches," which presupposed general laws even when they didn't mention them. So it might seem natural to search for such laws and make them explicit, thereby "improving" the explanations. (Perhaps with enough detail and precision we could promote them to full-blooded deductive-nomological explanations!) This is absurd, but in context, the crisp idealizations often had this normative force. As such, they were not good norms to try to follow. The more sensitive positivists would not have wanted their models to be taken as *ex cathedra* pronouncements. But this is not to say that they weren't normative.
2. Giere's *Explaining Science* (1988) systematically defends a satisficing account of the pursuit and acceptance of models in scientific research, and Gigerenzer, Todd, and the ABC Research Group (1999) document satisficing's superior performance in naturalistic circumstances.
3. This was a subtheme of many writers influenced by Wittgenstein, including Ryle, Hanson, and Toulmin. It has reappeared among sociologists, historians, and psychologists of science who have gone beyond theory and justification to study experiments, laboratory procedures, instruments, the generation, treatment, transformation, and visualization of data, heuristics, and problem-solving strategies; and how our social structures support, condition, and inform cognition and scientific investigation.
4. To my knowledge, this marvelous play on words was coined by theoretical biologist Robert Rosen, tweaking unnecessary formalisms in his own discipline, in a book review (in *Science*) around 1970.
5. Feynman (1967), chapter 2. Feynman was hostile to philosophy, but still a most provocative and philosophical methodologist. Perhaps philosophers that he met were foundationalists or formalists—his "Euclideans." Physicist

Max Dresden told me that arguing with a foundationalist (recounted in Callebaut, 1993, p. 74) motivating his marvelous essay "Remarks on Fundamentality and Complexity." Feynman clearly identified with his other ideal type—the ever-pragmatic "Babylonians," and the spirit of Dresden's remarks place him there too. I discuss Feynman's views in Chapter 4.

6. Engineering physics (EP) was a natural home for forming my later theoretical prejudices. Physicists in the war efforts to develop radar and the atomic bomb found collaborating engineers inadequately trained in basic science. The strengths of American experimental physics derived in part from the practical experience physicists got during the depression in designing and building their own equipment (Schweber, 1992; Galison, 1997), so they vowed to create a hybrid discipline. Cornell's was the first such of similar programs elsewhere—the inspiration of theoretical physicist Hans Bethe, whom I was privileged to meet several times before I got out of high school. Bethe's influence explains why EP not only demanded as much physics as a physics major and a lot of engineering, but also an unusually large number of liberal arts electives, which it managed by going to a 5-year program and asking all students to take a 30% heavier load.
7. Resler tells me that my memory of this part of the story—use of the computer—was faulty, though he confirmed the rest (personal communication, March 1993). He never made use of the computer until much later, and only then, as he says "In self defense, to illustrate the errors people have made misusing canned programs that they don't understand." So much for eyewitness testimony! But this phenomenology of computer use in complex simulations does fit lots of real research, and many stories I have heard since. Resler's specialty was finding interesting approximations and ways of tweaking the equations to simplify them, and in his hands it was an art. After all, it is a feat of some ingenuity to find ways of eliminating 18 out of 22 of the variables from the equations, and still have them tell you something of any use!
8. The relevance of limits are not exhausted here. The existence of singularities, indicating the failures of limits to exist or of series to converge produce some of the most interesting and challenging mathematical behavior in attempts to relate different theories. They are critical to a fuller understanding of what is required for reduction (Batterman, 2002).
9. The lineage of today's older computer manufacturers is lost when their names become unexpanded acronyms. NCR stood for the National Cash Register company, more recently bought by IBM (International Business Machines).
10. A whole genre of books and techniques common through the late nineteenth century up to about 30 years ago has simply vanished—the handbooks, tables, and graphical methods for the solution of a variety of numerical problems and equations. Lipka (1918) is representative. An advertisement in the front of his book declares it one of six like books (for various applications) by Lipka, a professor at MIT. He teaches about the slide rule, and how

to construct charts for the solution of second- and third-degree equations, as well as periodic functions, and nomographs or alignment charts for the solution of multiple simultaneous equations. The book came with two removable scales—one for logarithms (for making your own slide rules) and a second (given the labors of Newton's method!) for finding square roots. You could tell 'serious' engineering students by the K&E log-log-duplex-decitrig slide rules at their belts (that's Keuffel and Esser—makers of beautiful things in ivory and mahogany, and the computers of every generation of engineers until the mid-1970s). Petroski (1985) is a nostalgic and critical lament for the slide rule and what we have lost since we exchanged it for calculator and computer: the skill of doing order of magnitude calculations and other kinds of informal reckoning have declined precipitously. These skills are no less necessary than they once were for checking results. There are unfortunately fewer people who realize it.

11. *Slop* is a technical term. Every circular hole has to be made with a larger diameter than that of a shaft, which has to rotate in it—large enough to offset uncertainties in the actual manufactured diameter in that shaft, its actual location, and the actual location of the hole. This excess means that in each linkage one of the two links will usually start to move before its bearing surface contacts that of the other link, producing slop; delay; and with flexion of the parts, a net flexibility or lack of rigidity; and uncertainty in the location in the parts of the linkage. No two parts or linkages are exactly alike, and the aim of manufacturing at each stage of assembly is to produce a sufficient number that work sufficiently well (enter cost-benefit considerations). Some of these uncertainties may be taken up with flexion of the components, but the manufacturing tolerances of all components and their effects on friction and wear are learned through experience—sometimes bitter experience. There are optimal amounts of "slop" and tolerances for parts that differ from case to case and problem to problem, and the design, slop, and tolerances must be "solved" together. Unnecessarily long linkages accumulate too much slop. I remember loosely adjusted linkages in which a displacement at one end disappeared before it got to the other. Unnecessarily high tolerances to limit slop increase cost, time of manufacture, and rejection rates for parts during manufacture. Insufficiently high tolerances increase assembly time, rejection rates of parts during assembly, and later maintenance problems and component failures.
12. Variations from machine to machine induced by variations in parts in a combinatorial explosion of possible assemblages would make it pointless to try to get an exact solution. See also the preceding note and discussion below.
13. When I asked to make such a movie, I found that one already existed. Others had taken similar paths for related problems, and needed the same transformation for their solutions. That the same movie could act as data input to multiple design tasks is another aspect of the near-decomposability or modularity of the design problem noted in the text, and also indicates the importance of that particular data as pivotal in the design. The mixture of obser-

vation and theory here makes it an instance of a "semi-empirical method." See Ramsey (1997); Humphreys (1991).

14. Aside from the inevitable slop in the linkage, I assumed that displacement of the cam (and ribbon) was just a function of the position of the driveshaft and the geometry of the linkage. I was treating the links as rigid bodies—ignoring effects of acceleration (force) and "jerk" (see below) on them. Why ignore them here when they are so important below? The answer shows the simultaneously contextual and realistic justification of approximations (see also Ramsey, 1990a, 1997; Vincenti, 1990). The carbon ribbon was the weak "link" in the chain. One thus had to consider the jerk in the analysis of *ribbon* breakage, but could safely ignore it in analyzing behavior of the metal links (no links had broken!). If necessary, one *could* have gotten rigid-body accelerations from the paper dolly simulations, measured elastic deformation (of individual links or the whole linkage) under static loads, and estimated the force necessary to accelerate the tape roll under appropriate ranges of conditions, thus providing estimates of elastic deformation in the linkage. But no one would do this without observed breakage or bending of links. The deep principle here is: "If it ain't broke, don't fix it." In fact, don't even look at it.
15. It was a Hewlett-Packard HP-41C, on which I did everything until I got my first computer in 1982. I bought my first programmable calculator, an HP-25, in 1975. Engineers have always tended to carry their calculating tools with them. Hewlett-Packard's first scientific calculator, the HP-35 in 1972, was called (by them) "The electronic slide rule." At $395, it was 15 times as much as the K&E, and sadly, I had no use for it! The programmable HP-25 I bought three years later was immeasurably more powerful: it could do programmed iterative calculations at about 50 steps a second and cost half as much. (I didn't have a use for *it* either, but the stimulus was just too great! I told my wife I needed it for my courses, and set out to *develop* a use for it.) With it I wrote simulations for virtually every simple model in population genetics and mathematical ecology, and discovered the beauties of chaos (in April 1976). Simulation and model-building became a way of life for me.
16. This fact is strikingly obvious in Giere's (1988) discussion. There are exceptions in large projects, such as the wartime Manhattan and RadLab (radar) projects, and high energy physics today, where different members of the research team specialize in different areas and problems. Even so they have to be able to communicate with one another—a task that has lead to the invention of new specialties and new training curricula. Thus the invention of engineering physics as a discipline. See Galison's (1997) discussion of the RadLab experience.
17. See Paul and Kimmelman (1988) for the crucial role that applied research played in the theoretical development of genetics. Provine's (1986) definitive biography of Sewall Wright showed how applied problems formed Wright's conception of the proper aims and scope of theory.
18. Kadanoff more recently told me that *he* worked for Schlumberger too! And

on whether this is respectable high church physics, his eyes sparkled mischievously: "Well, that depends upon who you talk to!" Condensed-matter physicists have the same problems with particle physicists as evolutionary biologists have with the human genome project machine. It is called "little science–big science" and the battle over reductionism and foundationalism is often as much about money and influence as anything else.

19. The ironies here are obvious from the biographies of many of the giants of modern theoretical physics. Hans Bethe, Enrico Fermi, and Richard Feynman were renowned as wizards of rapid methods for getting good approximate answers, and often challenged one another and colleagues to contests. Those who needed reliable but not necessarily exact answers invented many applied and heuristic mathematical methods during World War II. Their percolation into theoretical physics after the war (and the growth of the technology for doing it—the computer) revolutionized that discipline, and ultimately all of the mathematical sciences. For a flavor of these influences, see Gleick's (1992) moving and illuminating biography of Feynman (e.g., pp. 175–183). Feynman often comments on the uselessness of the solvable limiting cases and idealizations for real-world prediction. This recalls Chapter 6.
20. Perhaps the most general methodological principle for the study of mechanisms, or more generally, functionally organized systems—one independently discovered many times—is: "Learn how it works by studying how it breaks down!" It is implicit in Glymour's (1980) "bootstrapping" approach to theory testing. I urged him to explore connections with fault localization in integrated circuits—which he did in 1983 (see Glymour, 1983, section 5, "Fault detection in logic circuits"). Darden (1991) also exploits "fault localization" in her account of theory testing and modification. There are further riches here to be mined.
21. Lakatos (1976) characterizes different classes of counterexamples in mathematics, and Star and Gerson (1986) have a rich classification and treatment of kinds of anomalies in science.

Bibliography

Abbott, A. (1995). "Things of Boundaries." *Social Research,* 62: 857–882.

———. (2001). "Chaos of Disciplines." In *Self-Similar Social Structures.* Chicago: University of Chicago Press.

Abbott, E. (1884). *Flatland,* 6th ed. New York: Dover, 1953.

Alexander, C. (1964). *Notes on the Synthesis of Form.* Cambridge, MA: Harvard University Press.

Allchin, D. (1991). "Resolving Disagreement in Science: The Ox-Phos Controversy, 1961–1977." Ph.D. diss., Committee on the Conceptual Foundations of Science, University of Chicago.

———. (1992). "How Do You Falsify a Question? Crucial Tests v. Crucial Demonstrations." In *PSA-1992,* vol. 1, ed. D. Hull, M. Forbes, and K. Okruhlik. East Lansing, MI: Philosophy of Science Association, pp. 74–88.

Allen, G. E. (1978). *Thomas Hunt Morgan: The Man and His Science.* Princeton, NJ: Princeton University Press.

Aloimonos, Y., and A. Rosenfeld. (1991). "Computer Vision." *Science,* 253: 1249–1254.

Ando, A., F. M. Fisher, and H. A. Simon. (1963). *Essays on the Structure of Social Science Models.* Cambridge, MA: MIT Press.

Arthur, B. (1994). *Increasing Returns and Path-Dependence in the Economy.* Ann Arbor: University of Michigan Press.

Arthur, W. (1982). "A Developmental Approach to the Problem of Evolutionary Rates." *Biological Journal of the Linnean Society,* 18: 243–261.

———. (1984). *Mechanisms of Morphological Evolution.* New York: Wiley.

Ashby, W. R. (1952). *Design for a Brain.* New York: Wiley.

———. (1956). *An Introduction to Cybernetics.* New York: Chapman and Hall. Reprint [London: Chapman & Hall, University Paperbacks, 1964].

Bailey, N. T. J. (1961). *Introduction to the Mathematical Theory of Genetic Linkage*. Oxford: Oxford University Press.

Barlow, R. E., and F. Proschan. (1975). *Statistical Theory of Reliability and Life Testing: Probability Models*. New York: Wiley.

Bateson, W., and R. C. Punnett. (1911). "On Gametic Series Involving Reduplication of Certain Terms." *Journal of Genetics,* 1: 239–302.

Bateson, W., E. Saunders, and R. C. Punnett. (1906). "Report III: Experimental Studies in the Physiology of Sex." *Reports to the Evolution Committee of the Royal Society,* 3: 1–53.

Batterman, R. (1995). "Theories between Theories: Asymptotic Limiting Intertheoretic Relations." *Synthese,* 103: 171–201.

———. (2000). "Multiple Realizability and Universality." *The British Journal for the Philosophy of Science,* 51: 115–145.

———. (2002). *The Devil in the Details: Asymptotic Reasoning in Explanation, Reduction, and Emergence*. New York: Oxford University Press.

Bechtel, W., and J. Mundale. (1999). "Multiple Realizability Revisited: Linking Cognitive and Neural States." *Philosophy of Science,* 66: 175–207.

Bechtel, W., and R. C. Richardson. (1993). *Discovering Complexity: Decomposition and Localization as Strategies in Scientific Research*. Princeton, NJ: Princeton University Press.

Bergmann, Gustav. (1957). *Philosophy of Science*. Madison: University of Wisconsin Press.

Berkeley, Bishop George. (1709). *An Essay towards a New Theory of Vision*. Dublin: Printed by A. Rhames for J. Pepyat.

Beurton, P. J., R. Falk, and H-J. Rheinberger. (2000). *The Concept of the Gene in Development and Evolution*. Cambridge: Cambridge University Press.

Boden, M. (1970). "Intentionality in Physical Systems." *Philosophy of Science,* 37: 200–214.

———. (1972). *Purposive Explanation in Psychology*. Cambridge, MA: Harvard University Press.

Bogen, J., and J. Woodward. (1988). "Saving the Phenomena." *The Philosophical Review,* 97(3): 302–352.

———. (1992). "Observations, Theories and the Evolution of the Human Spirit." *Philosophy of Science,* 59: 590–611.

Boveri, T. (1902). "On Multipolar Mitosis as a Means of Analysis of the Cell Nucleus." English translation of "Über mehrpolige Mitosen als Mittel zur Analyse des Zellkerns," *Verhandlungen der physikalish-medizinischen Gesellschaft zu Würzburg,* 35: 67–90. Reprinted in *Foundations of Experimental Embryology,* 2nd ed., ed. B. H. Willier and J. M. Oppenheim. New York: Macmillan, 1974.

Boyd, Richard M. (1972). "Determinism, Laws, and Predictability in Principle." *Philosophy of Science,* 39: 431–450.

———. (1973). "Realism, Underdetermination, and a Causal Theory of Evidence." *Nous,* 7(1): 1–12.

———. (1980). "Materialism without Reductionism: What Physicalism Does

Not Entail." In *Readings in Philosophy of Psychology,* vol. 1, ed. N. Block. Cambridge, MA: Harvard University Press, pp. 67–106.

Boyd, Robert, and P. Richerson. (1985). *Culture and the Evolutionary Process.* Chicago: University of Chicago Press.

Brandon, R. (1982). "The Levels of Selection." In *PSA-1982,* vol. 1, ed. P. Asquith and T. Nickles. Dordrecht: Reidel, pp. 315–323.

———. (1999). "The Units of Selection Revisited: The Modules of Selection." *Biology and Philosophy,* 14(2): 167–180.

Bridges, C. B. (1917). "An Intrinsic Difficulty for the Variable Force Hypothesis of Crossing Over." *American Naturalist,* 51: 370–373.

Bronowski, Jakob. (1970). "New Concepts in the Evolution of Complexity: Stratified Stability and Unbounded Plans." *Synthese,* 21: 228–246.

Brush, Stephen. (1974). "Should the History of Science Be Rated X?" *Science,* 183: 1164–1172.

Burchfield, J. (1975). *Lord Kelvin and the Age of the Earth.* Chicago: University of Chicago Press.

Callebaut, W. (1993). *Taking the Naturalistic Turn, or How Real Philosophy of Science Is Done.* Chicago: University of Chicago Press.

Campbell, D. T. (1958). "Common Fate, Similarity, and Other Indices of the Status of Aggregates of Persons as Social Entities." *Behavioral Science,* 3: 14–25.

———. (1959). "Methodological Suggestions from a Comparative Psychology of Knowledge Processes." *Inquiry,* 2: 152–182.

———. (1966). "Pattern Matching as an Essential in Distal Knowing." In *The Psychology of Egon Brunswik,* ed. K. R. Hammond. New York: Holt, Rinehart and Winston, pp. 81–106.

———. (1969a). "Definitional versus Multiple Operationalism." Reprinted in *Methodology and Epistemology for Social Science: Selected Papers,* ed. E. Samuel Overman. Chicago: University of Chicago Press, 1988.

———. (1969b). "Prospective: Artifact and Control." In *Artifact in Behavioral Research,* ed. R. Rosenthal and R. Rosnow. New York: Academic Press, pp. 351–382.

———. (1973). "Ostensive Instances and Entitativity in Language Learning." In *Unity through Diversity,* part 2, ed. W. Gray and N. D. Rizzo. New York: Gordon and Breach, pp. 1043–1057.

———. (1974a). "Evolutionary Epistemology." In *The Philosophy of Karl Popper,* vol. 2, ed. P. A. Schilpp. La Salle, IL: Open Court, pp. 412–463.

———. (1974b). " 'Downwards Causation' in Hierarchically Organized Biological Systems." In *Studies in the Philosophy of Biology,* ed. F. J. Ayala and T. Dobzhansky. Berkeley: University of California Press, pp. 179–186.

———. (1977). "Descriptive Epistemology: Psychological, Sociological, and Evolutionary." Unpublished William James Lectures, given at Harvard University.

Campbell, D. T., and D. W. Fiske. (1959). "Convergent and Discriminant Validation by the Multitrait-Multimethod Matrix." *Psychological Bulletin,* 56: 81–105.

Carlson, Elof A. (1966). *The Gene: A Critical History*. Philadelphia: Saunders.
Cartwright, N. (1983). *How the Laws of Physics Lie*. London: Oxford University Press.
———. (1989). *Nature's Capacities and Their Measurement*. New York: Oxford University Press.
———. (1994). "Fundamentalism vs. the Patchwork of Laws." *Proceedings of the Aristotelean Society,* 93: 279–292. Reprinted in *The Philosophy of Science,* ed. with introduction by D. Papineau. Oxford: Oxford University Press, 1996, pp. 314–326.
Castle, W. E. (1919a). "Is the Arrangement of the Genes in the Chromosome Linear?" *Proceedings of the National Academy of Science,* 5: 25–32.
———. (1919b). "The Linkage System of Eight Sex-Linked Characters of *Drosophila virilis*." *Proceedings of the National Academy of Science,* 5: 32–36.
———. (1919c). "Are Genes Linear or Non-linear in Arrangement?" *Proceedings of the National Academy of Science,* 5: 500–506.
Castle, W. E., and J. Phillips. (1914). *Piebald Rats and Selection*. Washington, DC: Carnegie Institute of Washington.
Causey, R. L. (1972). "Attribute-Identities in Microreductions." *Journal of Philosophy,* 69: 407–422.
Churchland, P. S. (1986). *Neurophilosophy: Towards a Unified Science of the Mind-Brain*. Cambridge, MA: MIT Press.
———. (1995). *The Engine of Reason, the Seat of the Soul*. Cambridge, MA: MIT Press.
Clark, A. (1997). *Being There: Putting Brain, Body, and World Together Again*. Cambridge, MA: MIT Press.
Clark, A., and J. Fujimura (eds.). (1992). *The Right Tools for the Job*. Princeton, NJ: Princeton University Press.
Constant, Edward W. (1980). *The Origins of the Turbojet Revolution*. Baltimore: Johns Hopkins University Press.
Cook, T. D., and D. T. Campbell. (1979). *Quasi-Experimentation: Design and Analysis for Field Settings*. Chicago: Rand McNally.
Cowie, F. (1999). *What's Within? Nativism Reconsidered*. New York: Oxford University Press.
Craver, C. F. (2007). *Explaining the Brain: Mechanisms and the Mosaic Unity of Neuroscience*. Oxford: Oxford University Press.
Cronbach, L. J., and P. E. Meehl. (1955). "Construct Validity in Psychological Tests." *Psychological Bulletin,* 52: 281–302.
Crow, J. (1987). "Neutral Models of Molecular Evolution." In *Neutral Models in Biology,* ed. M. Nitecki and A. Hoffman. Oxford: Oxford University Press.
Crow, J. F., and M. Kimura. (1970). *An Introduction to Population Genetics Theory*. New York: Harper & Row.
Culp, S. (1994). "Defending Robustness: The Bacterial Mesosome as a Test Case." In *PSA-1994,* vol. 1, ed. D. Hull, M. Forbes, and R. M. Burian. East Lansing, MI: Philosophy of Science Association, pp. 46–57.

———. (1995). "Objectivity in Experimental Inquiry: Breaking Data-Technique Circles." *Philosophy of Science,* 62: 438–458.

D'Andrade, R. (1986). "Three Scientific World Views and the Covering Law Model." In *Metatheory in Social Science: Pluralisms and Subjectivities,* ed. D. Fiske and R. Shweder. Chicago: University of Chicago Press, pp. 19–41.

Darden, L. (1974). "Reasoning in Scientific Change: The Field of Genetics at Its Beginnings." Ph.D. diss., Committee on the Conceptual Foundations of Science, University of Chicago.

———. (1991). *Theory Change in Science: Strategies from Mendelian Genetics.* New York: Oxford University Press.

Darden, L., and N. Maull. (1977). "Interfield Theories." *Philosophy of Science,* 44: 43–64.

Daston, L. (1992). "Objectivity and the Escape from Perspective." *Social Studies of Science,* 22: 597–618.

———. (1995). "How Nature Became the Other: Anthropomorphism and Anthropocentrism in Early Modern Science." In *Biology as Society, Society as Biology: Metaphors, Yearbook for the Sociology of Science,* vol. 18, ed. Sabine Maasen, Everett Mendelsohn, and Peter Weingart. Dordrecht: Kluwer, pp. 37–56.

Daston, L., and P. Galison. (1992). "The Image of Objectivity." *Representations,* 40: 81–128.

Davidson, D. (1970). "Mental Events." In *Essays on Actions and Events.* Oxford: Oxford University Press.

———. (1973–1974). "The Very Idea of a Conceptual Scheme." Presidential Address. *Proceedings and Addresses of the American Philosophical Association,* 47: 5–20.

Dawkins, R. (1976). *The Selfish Gene.* Oxford: Oxford University Press.

Dembski, W. (1991). "Randomness by Design." *Nous,* 25: 75–106.

Dennett, D. C. (1971). "Intentional Systems." *Journal of Philosophy,* 68: 87–106. Reprinted in *Brainstorms: Philosophical Essays on Mind and Psychology.* Cambridge, MA: MIT Press, 1979.

———. (1979). *Brainstorms: Philosophical Essays on Mind and Psychology.* Cambridge, MA: MIT Press.

———. (1991). *Consciousness Explained.* Boston: Little, Brown, and Co.

———. (1995). *Darwin's Dangerous Idea: Evolution and the Meanings of Life.* New York: Simon and Schuster.

Diamond, J. (1997). *Guns, Germs and Steel.* New York: W. W. Norton.

Dinman, Bertram D. (1972). "The 'Non-Concept' of 'No-Threshold' Chemicals in the Environment." *Science,* 175: 495–497.

Downes, S. M. (1992). "The Importance of Models in Theorizing: A Deflationary Semantic View." In *PSA-1992,* vol. 1, ed. D. Hull, M. Forbes, and K. Okruhlik. East Lansing, MI: Philosophy of Science Association, pp. 142–153.

Dresden, M. (1974). "Reflections on 'Fundamentality and Complexity.'" In *Physical Reality and Mathematical Description,* ed. C. P. Enz and J. Mehra. Dordrecht: Reidel, pp. 133–166.

Dupre, John. (1993). *The Disorder of Things: Metaphysical Foundations of the Disunity of Science.* Cambridge, MA: Harvard University Press.

Dyke, Charles. (1988). *The Evolutionary Dynamics of Complex Systems: A Study in Biosocial Complexity.* New York: Oxford University Press.

Ehrlich, P. (2000). *Human Natures: Genes, Cultures and the Human Prospect.* New York: Penguin.

Elsasser, Walter M. (1966). *Atom and Organism: A New Approach to Theoretical Biology.* Princeton, NJ: Princeton University Press.

Felsenstein, J. (1979). "A Mathematically Tractable Family of Genetic Mapping Functions with Different Amounts of Interference." *Genetics,* 91: 769–775.

Ferguson, Eugene S. (1992). *Engineering and the Mind's Eye.* Cambridge, MA: MIT Press.

Festinger, L. (1957). *A Theory of Cognitive Dissonance.* Evanston, IL: Row, Peterson.

Feynman, R. P. (1967). *The Character of Physical Law.* Cambridge, MA: MIT Press.

Fisher, R. A. (1930). *The Genetical Theory of Natural Selection.* New York: Oxford University Press.

Fiske, D. M. (1992). "Citations Do Not Solve Problems." *Psychological Bulletin,* 112: 393–395.

Fiske, D. W., and R. Shweder (eds.). (1986). *Metatheory in Social Science: Pluralisms and Subjectivities.* Chicago: University of Chicago Press.

Fodor, Jerry. (1965). "Functional Explanation in Psychology." In *Philosophy in America,* ed. Max Black. London: Allen and Unwin, pp. 161–179.

———. (1968). *Psychological Explanation: An Introduction to the Philosophy of Psychology.* New York: Random House.

———. (1974). "Special Sciences (Or: The Disunity of Science as a Working Hypothesis)." *Synthese,* 28: 97–115.

———. (1983). *The Modularity of Mind.* Cambridge, MA: Bradford/MIT Press.

Fontana, W. (1991). "Algorithmic Chemistry." In *Artificial Life II,* ed. C. G. Langton, C. Taylor, J. D. Farmer, and S. Rasmussen. Redwood City, CA: Addison-Wesley, pp. 211–254.

Fontana, W., and L. Buss. (1994). "The Arrival of the Fittest: Towards a Theory of Biological Organization." *Bulletin of Mathematical Biology,* 56: 1–64.

———. (1996). "The Barrier of Objects: From Dynamical Systems to Bounded Organizations." In *Boundaries and Barriers,* ed. J. Casti and A. Karlqvist. Redwood City, CA: Addison-Wesley, pp. 56–116.

Frank, Robert H. (1988). *Passions within Reason: The Strategic Role of the Emotions.* New York: W. W. Norton.

Gaddis, J. L. (2002). *The Landscape of History: How Historians Map the Past.* New York: Oxford University Press.

Galison, P. (1997). *Image and Logic: A Material Culture of Microphysics.* Chicago: University of Chicago Press.

Garber, D. (1986). "Learning from the Past." *Synthese,* 67: 91–114.

Gasking, D. A. T. (1955). "Causation and Recipes." *Mind,* new series, 64: 479–487.

Giere, R. (1988). *Explaining Science: A Cognitive Approach.* Chicago: University of Chicago Press.

Gigerenzer, G. (1993). "The Bounded Rationality of Probabilistic Mental Models." In *Rationality: Psychological and Philosophical Perspectives,* ed. K. I. Manktelow and D. E. Over. London: Routledge, pp. 284–313.

Gigerenzer, G., and D. G. Goldstein. (1996). "Reasoning the Fast and Frugal Way: Models of Bounded Rationality." *Psychological Review,* 103: 650–669.

Gigerenzer, G., P. M. Todd, and the ABC Research Group. (1999). *Simple Heuristics That Make Us Smart.* New York: Oxford University Press.

Glassman, R. B. (1978). "The Logic of the Lesion Experiment and Its Role in the Neural Sciences." In *Recovery from Brain Damage: Research and Theory,* ed. S. Finger. New York: Plenum, pp. 3–31.

Glassmann, R. B., and W. C. Wimsatt. (1984). "Evolutionary Advantages and Limitations of Early Plasticity." In *Early Brain Damage,* vol. 1, ed. R. Almli and S. Finger. New York: Academic Press, pp. 35–58.

Gleick, J. (1992). *Genius: The Life and Science of Richard Feynman.* New York: Pantheon.

Glen, William (ed.). (1994). *The Mass-Extinction Debates: How Science Works in a Crisis.* Palo Alto, CA: Stanford University Press.

Glennan, S. S. (1992). "Mechanisms, Models, and Causation." Ph.D. diss., University of Chicago.

———. (1996). "Mechanisms and the Nature of Causation." *Erkenntnis,* 44: 49–71.

———. (1997). "Probable Causes and the Distinction between Subjective and Objective Chance." *Nous,* 31: 496–519.

Globus, Gordon G. (1972) "Biological Foundations of the Psychoneural Identity Hypothesis." *Philosophy of Science,* 39: 291–301.

Glymour, Clark. (1975). "Relevant Evidence." *Journal of Philosophy,* 72: 403–425.

———. (1980). *Theory and Evidence.* Princeton, NJ: Princeton University Press.

———. (1983). "On Testing and Evidence." In *Testing Scientific Theories,* ed. J. Earman. Minneapolis: University of Minnesota Press.

Goldschmidt, R. (1917). "Crossing-over ohne Chiasmatypie?" *Genetics,* 2: 82–95.

Goldstein, D. G., and G. Gigerenzer. (1999). "The Recognition Heuristic: How Ignorance Makes Us Smart." In *Simple Heuristics That Make Us Smart.* G. Gigerenzer, P. M. Todd, and the ABC Research Group. New York: Oxford University Press, pp. 37–58.

Goodman, Nelson. (1966). *The Structure of Appearance,* 2nd ed. Indianapolis: Bobbs-Merrill.

Gould, S. J. (1977). *Ontogeny and Phylogeny.* Cambridge, MA: Harvard University Press.

———. (1980). *The Panda's Thumb*. New York: Norton.

———. (1992). *Ever since Darwin: Reflections on Natural History*. New York: Norton.

Gould, S. J., and E. Vrba. (1982). "Exaptation: A Missing Term in the Science of Form." *Paleobiology*, 8: 4–15.

Grant, R., and P. Grant. (1989). *Evolutionary Dynamics of a Natural Population*. Chicago: University of Chicago Press.

Gregory, R. L. (1958). "Models and the Localization of Function in the Central Nervous System." *National Physical Laboratory Symposium*, 10: 671–681. Reprinted in *Key Papers: Cybernetics*, ed. C. R. Evans and A. D. J. Robertson. London: Butterworths, pp. 91–102.

———. (1961). "The Brain as an Engineering Problem." In *Current Problems in Animal Behavior*, ed. W. H. Thorpe and O. L. Zangwill. London: Cambridge University Press.

———. (1962). "The Logic of the Localization of Function in the Nervous System." In *Biological Prototypes and Synthetic Systems*, vol. 1, ed. R. Bernard and B. Kare. New York: Plenum Press.

———. (1969). "On How So Little Information Controls So Much Behavior." In *Toward a Theoretical Biology*, vol. 2., ed. C. H. Waddington. Edinburgh, Scotland: University of Edinburgh Press, pp. 236–247.

———. (1970). *The Intelligent Eye*. New York: McGraw-Hill.

Griesemer, J. (1983). "Communication and Scientific Change: An Analysis of Conceptual Maps in the Macroevolution Controversy." Ph.D. diss., Committee on the Conceptual Foundations of Science, University of Chicago.

Griesemer, J., and M. Wade. (1988). "Laboratory Models, Causal Explanation and Group Selection." *Biology and Philosophy*, 3: 67–96.

Griesemer, J. R., and W. C. Wimsatt. (1989). "Picturing Weismannism: A Case Study in Conceptual Evolution." In *What Philosophy of Biology Is: Essays Dedicated to David Hull*, ed. M. Ruse. Dordrecht: Martinus Nijhoff, pp. 75–137.

Griffiths, P. E. (1996). "Darwinism, Process Structuralism and Natural Kinds." *Philosophy of Science*, 63(3) (supplement): S1–S9.

Griffiths, P. E., and R. D. Gray. (1994). "Developmental Systems and Evolutionary Explanation." *Journal of Philosophy*, 91: 277–304.

Grodins, F. S. (1963). *Control Theory and Biological Systems*. New York: Columbia University Press.

Gunderson, K. (1970). "Asymmetries and Mind-Body Perplexities." In *Minnesota Studies in the Philosophy of Science*, vol. 4, ed. M. Radner and S. Winokur. Minneapolis: University of Minnesota Press, pp. 273–309.

Hacking, I. (1983). *Representing and Intervening: Introductory Topics in the Philosophy of Natural Science*. Cambridge, MA: Cambridge University Press.

Haldane, J. B. S. (1919). "The Combination of Linkage Values, and the Calculation of Distance between the Loci of Linked Factors." *Journal of Genetics*, 8: 299–309.

———. (1927). "On Being the Right Size." In *On Being the Right Size and Other*

Essays [of J. B. S. Haldane], ed. John Maynard Smith. Oxford: Oxford University Press, pp. 1–8.

Haldane, J. S. (1914). *Mechanism, Life and Personality: An Examination of the Mechanistic Theory of Life and Mind.* London: J. Murray.

Harper, D. (1987). *Working Knowledge: Skill and Community in a Small Shop.* Chicago: University of Chicago Press.

Hempel, C. (1942). "The Function of General Laws in History." *Journal of Philosophy,* 39: 35–48.

Holling, C. S. (1992). "Cross-Scale Morphology, Geometry, and Dynamics of Ecosystems." *Ecological Monographs,* 62(4): 447–502.

Hooker, C. A. (1981a). "Towards a General Theory of Reduction. Part I: Historical and Scientific Setting." *Dialogue,* 20: 38–59.

———. (1981b). "Towards a General Theory of Reduction. Part II: Identity in Reduction." *Dialogue,* 20: 201–236.

———. (1981c). "Towards a General Theory of Reduction, Part III: Cross-Categorial Reduction." *Dialogue,* 20: 496–529.

Hubby, J. L., and R. C. Lewontin. (1966). "A Molecular Approach to the Study of Genic Heterozygosity in Natural Populations I. The Number of Alleles at Different Loci in *Drosophila pseudoobscura.*" *Genetics,* 54(2): 595–609.

Hull, David L. (1972). "Reduction in Genetics Biology or Philosophy?" *Philosophy of Science,* 39: 491–499.

———. (1974). *Philosophy of Biological Science.* Englewood Cliffs, NJ: Prentice-Hall.

———. (1976). "Informal Aspects of Theory Reduction." In *PSA-1974,* ed. A. C. Michalos, C. A. Hooker, G. Pearce, and R. S. Cohen. Dordrecht: Reidel, pp. 653–670.

———. (1988). *Science as a Process: An Evolutionary Account of the Social and Conceptual Development of Science.* Chicago: University of Chicago Press.

Humphreys, P. (1991). "Computer Simulations." In *PSA-1990,* vol. 2, ed. A. Fine, M. Forbes, and L. Wessels. East Lansing, MI: Philosophy of Science Association, pp. 497–506.

Hutchins, E. (1995). *Cognition in the Wild.* Cambridge, MA: MIT Press.

Hutchinson, G. E. (1964). "The Influence of the Environment." *Proceedings of the National Academy of Sciences,* 51: 930–934.

Jackson, W. (1994). *Becoming Native to This Place.* Lexington: University of Kentucky Press.

Janis, I. L., and L. Mann. (1977). *Decision Making: A Psychological Analysis of Conflict, Choice, and Commitment.* New York: The Free Press.

Jeans, James. (1940). *An Introduction to the Kinetic Theory of Gases.* Cambridge, MA: Cambridge University Press.

Jukes, Thomas H. (1966). *Molecules and Evolution.* New York: Columbia University Press.

Kahneman, D., P. Slovic, and A. Tversky. (1982). *Decision under Uncertainty: Heuristics and Biases.* London: Cambridge University Press.

Kaufmann, A. (1968). *The Science of Decision Making: An Introduction to Praxeology*. New York: McGraw-Hill.

Kauffman, S. A. (1969). "Metabolic Stability and Epigenesis in Randomly Constructed Genetic Networks." *Journal for Theoretical Biology*, 22: 437–467.

———. (1970). "Articulation of Parts Explanation in Biology and the Rational Search for Them." In *PSA-1970*, ed. R. Buck and R. Cohen. Dordrecht: Reidel, pp. 257–272.

———. (1971). "Gene Regulation Networks: A Theory of Their Global Structure and Behavior." In *Current Topics in Development Biology*, vol. 6, ed. A. Moscona and A. Monroy. New York: Academic Press, pp. 145–182.

———. (1985). "Self-Organization, Selective Adaptation and Its Limits: A New Pattern of Inference in Evolution and Development." In *Evolution at a Crossroads*, ed. D. J. Depew and H. Weber. Cambridge, MA: MIT Press, pp. 169–207.

———. (1993). *The Origins of Order: Self-Organization and Selection in Evolution*. London: Oxford University Press.

Kim, Jaegwon. (1964). "Inference, Explanation, and Prediction." *Journal of Philosophy*, 61: 360–368.

———. (1966). "On the Psycho-Physical Identity Thesis." *American Philosophical Quarterly*, 3: 225–235.

———. (1971). "Materialism and the Criteria of the Mental." *Synthese*, 22: 323–345.

———. (1978). "Supervenience and Nomological Incommensurables." *American Philosophical Quarterly*, 15: 149–156.

Kimura, M. (1983). *The Neutral Theory of Molecular Evolution*. Cambridge: Cambridge University Press.

King, R. C. (1974). *Bacteria, Bacteriophages and Fungi—Handbook of Genetics*, vol. 1. New York: Plenum.

Kitcher, P. (1981). "Explanatory Unification." *Philosophy of Science*, 48: 507–531.

Kohler, R. (1994). *Lords of the Fly: Drosophila Genetics and the Experimental Life*. Chicago: University of Chicago Press.

Kornfeld, W., and C. Hewitt. (1981). "The Scientific Community Metaphor." *IEEE Transactions on Systems, Man, and Cybernetics*, SMC-11(1): 24–33.

Kosambi, D. D. (1944). "The Estimation of Map Distances from Recombination Values." *Annals of Eugenics*, 12: 172–175.

Kuhn, T. S. (1970). *The Structure of Scientific Revolutions*, 2nd ed. Chicago: University of Chicago Press.

Lakatos, I. (1970). "Falsification and the Methodology of Scientific Research Programmes." In *Criticism and the Growth of Knowledge*, ed. I. Lakatos and A. Musgrave. Cambridge: Cambridge University Press, pp. 91–196.

———. (1976). *Proofs and Refutations: The Logic of Mathematical Discovery*. Cambridge: Cambridge University Press.

Lashley, K. (1951). "The Problem of Serial Order in Behavior." In *Cerebral*

Mechanisms in Behavior, ed. L. Jeffress. New York: John Wiley, pp. 112–136; discussion pp. 136–146.

Latour, B. (1987). *Science in Action: How to Follow Scientists and Engineers through Society*. Cambridge, MA: Harvard University Press.

Latour, B., and S. Woolgar. (1979). *Laboratory Life: The Social Construction of Scientific Facts*. Beverly Hills, CA: Sage.

Laudan, L. (1971). "William Whewell on the Consilience of Inductions." *The Monist*, 55: 368–391.

Lenat, D. B. (1982). "The Nature of Heuristics." *Artificial Intelligence*, 19: 189–249.

Levins, R. (1966). "The Strategy of Model Building in Population Biology." *American Scientist*, 54: 421–431.

———. (1968). *Evolution in Changing Environments: Some Theoretical Explorations*. Princeton, NJ: Princeton University Press.

———. (1970a). "Complex Systems." In *Towards a Theoretical Biology*, vol. 3, ed. C. H. Waddington. Chicago: Aldine Publishing Company, pp. 73–88.

———. (1970b). "Extinction." In "Some Mathematical Questions in Biology: Lectures on Mathematics in the Life Sciences." *American Mathematical Society*, 2: 75–108.

———. (1973). "The Limits of Complexity." In *Hierarchy Theory: The Challenge of Complex Systems*, ed. H. Pattee. London: Braziller, pp. 73–88.

———. (1974). "The Qualitative Analysis of Partially Specified Systems." *Annals of the New York Academy of Sciences*, 231: 123–138.

Levins, R., and R. C. Lewontin. (1985). *The Dialectical Biologist*. Cambridge, MA: Harvard University Press.

Lewontin, R. (1966). "Is Nature Probable or Capricious?" *BioScience*, 16: 25–27.

———. (1970). "The Units of Selection." *Annual Review of Ecology and Systematics*, 1: 1–18.

———. (1974). *The Genetic Basis of Evolutionary Change*. New York: Columbia University Press.

———. (1978). "Adaptation." *Scientific American*, 239(3): 157–169.

———. (1991). "Facts and the Factitious in Natural Sciences." *Critical Inquiry*, 18(1): 140–153.

Lighthall, F., and S. D. Allan. (1989). *Local Realities, Local Adaptations*. London: The Falmer Press.

Lipka, J. (1918). *Graphical and Mechanical Computation*. New York: John Wiley and Sons.

Lloyd, E. (1988). *The Structure and Confirmation of Evolutionary Theory*. New York: Greenwood Press. Reissued (with new preface), Princeton, NJ: Princeton University Press, 1994.

———. (1989). "A Structural Approach to Defining Units of Selection." *Philosophy of Science*, 56(3): 395–418.

———. (1995). "Objectivity and the Double Standard for Feminist Epistemologies." *Synthese*, 104: 351–381.

Lorenz, K. Z. (1965). *Evolution and Modification of Behavior*. Chicago: University of Chicago Press.

Lotka, A. J. (1956). *Elements of Mathematical Biology*. New York: Dover Books. Reprint of *Elements of Physical Biology*, 1924.

Luce, R. D., and H. Raiffa. (1957). *Games and Decisions: Introduction and Critical Survey*. New York: Wiley.

Machamer, P., L. Darden, and C. Craver. (2000). "Thinking about Mechanisms." *Philosophy of Science*, 57: 1–25.

Maienschein, J. (1991). *Transforming Traditions in American Biology, 1880–1915*. Baltimore: Johns Hopkins University Press.

Malcolm, Norman. (1964). "Scientific Materialism and the Identity Theory." *Dialogue*, 3: 15–125.

———. (1971). *Problems of Mind: Descartes to Wittgenstein*. New York: Harper and Row.

Mandelbrot, Benoit. (1982). *The Fractal Geometry of Nature*, rev. ed. New York: W. H. Freeman.

Marey, G. E. (1895). *Movement*. Trans. E. Pritchard. New York: Appleton.

Margenau, H. (1950). *The Nature of Physical Reality*. New York: McGraw-Hill.

Margulis, L. (1971). "The Origin of Plant and Animal Cells." *The American Scientist*, 59: 230–235.

Marr, David. (1982). *Vision: A Computational Investigation into the Human Representation and Processing of Visual Information*. New York: W. H. Freeman.

Martinez, Sergio. (1992). "Objetividad Contextual y Robustez." *Dianoia* (Annual of the Instituto de Investigaciones Filosoficas, Universidad Nacional Autonoma de Mexico).

Maull, N. (1974). "Progress in Modern Biology: An Alternative to Reduction." Ph.D. diss., Committee on the Conceptual Foundations of Science, University of Chicago.

———. (1976). "Reconstructed Science as Philosophical Evidence." In *PSA-1974*, ed. A. C. Michalos, C. A. Hooker, G. Pearce, and R. S. Cohen. Dordrecht: Reidel, pp. 119–129.

———. (1977). "Unifying Science without Reduction." *Studies in History and Philosophy of Sciences*, 8: 143–162. Reprinted in *Conceptual Issues in Evolutionary Biology: An Anthology*, ed. Elliot Sober. Cambridge, MA: MIT Press, 1983.

May, R. M. (1974). "Biological Populations with Non-overlapping Generations: Stable Points, Stable Cycles, and Chaos." *Science*, 186(4164): 645–647.

Maynard-Smith, John. (1982). *Evolution and the Theory of Games*. Cambridge, MA: Cambridge University Press.

Mayr, E. (1982). *The Growth of Biological Thought: Diversity, Evolution, and Inheritance*. Cambridge, MA: Belknap Press.

Mayr, E., and W. Provine (eds.). (1980). *The Evolutionary Synthesis: Perspectives on the Unification of Biology*. Cambridge, MA: Harvard University Press.

McCauley, R. N. (1986). "Problem Solving in Science and the Competence Approach to Theorizing in Linguistics." *Journal for the Theory of Social Behaviour,* 16: 299–312.

———. (1996). "Explanatory Pluralism and the Co-evolution of Theories in Science." In *The Churchlands and Their Critics,* ed. R. N. McCauley. Cambridge, MA: Blackwell, pp. 17–47.

McClamrock, R. M. (1991). "Marr's Three Levels: A Re-evaluation." *Minds and Machines,* 1: 185–196.

———. (1995). *Existential Cognition: Computational Minds in the World.* Chicago: University of Chicago Press.

McClelland, D. D. (1973). "Testing for Competence Rather than for 'Intelligence.'" *American Psychologist,* 28(1): 1–14.

McClintock, M. K. (1971). "Menstrual Synchrony and Suppression." *Nature,* 229: 244–245.

McHarg, Ian. (1969). *Design with Nature.* Garden City, NY: Natural History Press.

Medvedev, Grigori. (1991). *The Truth about Chernobyl.* Trans. David MacRae. New York: Basic Books.

Mendel, G. (1866/1956). *Experiments in Plant Hybridization.* Cambridge, MA: Harvard University Press (translation by William Bateson, 1902).

Mertz, D. B., and D. MacCauley. (1980). "The Domain of Laboratory Ecology." *Synthese,* 43: 95–110. Reprinted in *Conceptual Issues in Ecology,* ed. E. Saarinen. Dordrecht: Reidel, 1982, pp. 229–244.

Mikkelson, G. (1997). "Stretched Lines, Averted Leaps, and Excluded Competition: A Theory of Scientific Counterfactuals." *Philosophy of Science,* 63(3) (supplement): S194–S201.

Mill, J. S. (1843). *A System of Logic.* London: Longmans, Green and Co.

Minsky, Marvin, and Seymour Papert. (1969). *Perceptrons: An Introduction to Computational Geometry.* Cambridge, MA: MIT Press.

Moore, G. E. (1903). *Principia Ethica.* Cambridge: Cambridge University Press.

Moore, J. A. (1972). *Heredity and Development,* 2nd ed. New York: Oxford University Press.

Morgan, T. H. (1909). "What Are 'Factors' in Mendelian Explanations?" *American Breeders Association Reports,* 5: 365–368.

———. (1910a). "Chromosomes and Heredity." *American Naturalist,* 44: 449–496.

———. (1910b). "Sex-Limited Inheritance in *Drosophila.*" *Science,* 32: 120–122.

———. (1911). "Chromosomes and Associative Inheritance." *Science,* 34: 636–638.

Morowitz, Harold J. (1992). *The Beginnings of Cellular Life: Metabolism Recapitulates Biogenesis.* New Haven, CT: Yale University Press.

Moss, L. (1992). "A Kernel of Truth? On the Reality of the Genetic Program." In *PSA-1992,* vol. 1, ed. D. Hull, M. Forbes, and K. Okruhlik. East Lansing, MI: Philosophy of Science Association, pp. 335–348.

Muller, H. J. (1916a–1916d). “The Mechanism of Crossing Over I–IV.” *American Naturalist,* 50: 193–221, 284–305, 350–366, 421–434.

———. (1920). “Are the Factors of Heredity Arranged in a Line?” *American Naturalist,* 54: 97–121.

Mynatt, C. R., M. E. Doherty, and R. D. Tweney. (1977). “Confirmation Bias in a Simulated Research Environment: An Experimental Study of Scientific Inference.” *Quarterly Journal of Experimental Psychology,* 29: 85–95.

Nagel, Ernest. (1961). *The Structure of Science: Problems in the Logic of Scientific Explanation.* New York: Harcourt, Brace & World.

Nagel, T. (1974). “What Is It Like to Be a Bat?” *The Philosophical Review,* 83: 435–450.

Newell, A., J. C. Shaw, and H. A. Simon. (1957). “Empirical Explorations of the Logic Theory Machine: A Case Study in Heuristic.” *Proceedings of the Western Joint Computer Conference,* Institute of Radio Engineers, pp. 218–230. Reprinted in *Computers and Thought,* ed. E. A. Feigenbaum and J. Feldman. New York: McGraw-Hill, 1963, pp. 109–133.

Nickles, Thomas. (1973). “Two Concepts of Intertheoretic Reduction.” *Journal of Philosophy,* 7: 181–201.

———. (1976). “Theory Generalization, Problem Reduction, and the Unity of Science.” In *PSA-1974,* ed. A. C. Michalos, C. A. Hooker, G. Pearce, and R. S. Cohen. Dordrecht: Reidel, pp. 31–74.

——— (ed.). (1980a). *Scientific Discovery, Logic, and Rationality.* Dordrecht: Reidel.

——— (ed.). (1980b). *Scientific Discovery: Case Studies.* Dordrecht: Reidel.

———. (1981). “What Is a Problem that We May Solve It?” *Synthese,* 47: 85–118.

———. (2002). “Normal Science: From Logic to Case-Based and Model-Based Reasoning.” In *Thomas Kuhn,* ed. T. Nickles. Cambridge, MA: Cambridge University Press.

Nisbett, R., and L. Ross. (1980). *Human Inference: Strategies and Shortcomings of Social Judgment.* Englewood Cliffs, NJ: Prentice-Hall.

Nitecki, M., and A. Hoffman. (1987). *Neutral Models in Biology.* Oxford: Oxford University Press.

Norman, D. (1993). *Things That Make Us Smart: Defending Human Attributes in the Age of the Machine.* New York: Addison-Wesley.

North, Douglass. (1990). *Institutions, Institutional Change, and Economic Performance.* Cambridge: Cambridge University Press.

Nozick, R. (1993). *The Nature of Rationality.* Princeton, NJ: Princeton University Press.

Olby, R. (1974). *The Path to the Double Helix.* Seattle: University of Washington Press.

Odenbaugh, J. (2001). “Searching for Patterns, Hunting for Causes: A Philosophical Examination of Mathematical Modeling in Theoretical Ecology.” Ph.D. diss., University of Calgary.

Oster, G., and E. O. Wilson. (1979). *Caste and Ecology in the Social Insects.* Princeton, NJ: Princeton University Press.

Oyama, S. (1985). *The Ontogeny of Information: Developmental Systems and Evolution.* New York: Cambridge University Press.
Oyama, S., R. Gray, and P. Griffiths (eds.). (2001). *Cycles of Contingency: Developmental Systems and Evolution.* Cambridge, MA: MIT Press.
Painter, T. S. (1934). "A New Method for the Plotting of Chromosome Aberrations and the Plotting of Chromosome Maps in *Drosophila melanogaster.*" *Genetics,* 19: 175–188.
Parsons, K. C. (1968). *The Cornell Campus: A History of its Planning and Development.* Ithaca, NY: Cornell University Press.
Pattee, Howard. (1970). "The Problem of Biological Hierarchy." In *Towards a Theoretical Biology,* vol. 3, *Drafts,* ed. C. H. Waddington. Chicago: Aldine, pp. 117–136.
——— (ed.). (1973). *Hierarchy Theory: The Challenge of Complex Systems.* New York: Braziller.
Paul, D., and B. Kimmelman. (1988). "Mendel in America: Theory and Practice, 1900–1919." In *The American Development of Biology,* ed. R. Rainger, K. Benson, and J. Maienschein. Philadelphia: University of Pennsylvania Press, pp. 281–310.
Peirce, C. S. (1936). "Some Consequences of Four Incapacities." In *Collected Papers of Charles Sanders Peirce,* vol. 5, ed. C. Hartshorne and P. Weiss. Cambridge, MA: Harvard University Press. (Originally published in 1868.)
Peterson, Ivars. (1993). *Newton's Clock: Chaos in the Solar System.* San Francisco: Freeman.
———. (1995). *Fatal Defect: Chasing Killer Computer Bugs.* New York: Random House.
Petroski, Henry. (1985). *To Engineer Is Human: The Role of Failure in Successful Design.* New York: St. Martin's Press.
———. (1992). "From Slide-Rule to Computer: Forgetting How It Used to Be Done." In *To Engineer Is Human: The Role of Failure in Engineering Design.* New York: Alfred Knopf, pp. 189–203.
———. (1994). *Design Paradigms: Case Histories of Error and Judgment in Engineering.* Cambridge: Cambridge University Press.
Platt, John R. (1969). "Some Theorems on Boundaries in Hierarchial Systems." In *Hierarchical Structures,* ed. L. L. Whyte, Albert G. G. Wilson, and Donna Wilson. New York: American Elsevier, pp. 201–213.
Plutinski, A. (2006). "Strategies of Model Building in Population Genetics." *Philosophy of Science,* 73(5): in press.
Polya, Georg. (1954). *Patterns of Plausible Inference.* Princeton, NJ: Princeton University Press.
Prandtl, Ludwig, and O. G. Tietjens. (1957). *Fundamentals of Hydro- and Aeromechanics.* New York: Dover. (Originally published in 1934.)
Provine, W. (1986). *Sewall Wright and Evolutionary Biology.* Chicago: University of Chicago Press.
Purcell, R. (1977). "Life at Low Reynolds Number." *American Journal of Physics,* 45: 3–11.

Putnam, H. (1962). "The Analytic and the Synthetic." In *Minnesota Studies in the Philosophy of Science,* vol. 3, ed. H. Feigl and G. Maxwell. Minneapolis: University of Minnesota Press, pp. 358–397.

———. (1967). "The Mental Life of Some Machines." In *Intentionality Minds and Perception,* ed. H. Castaneda. Detroit: Wayne State University Press.

Quine, W. V. O. (1953). "Two Dogmas of Empiricism." In *From a Logical Point of View.* Cambridge, MA: Harvard University Press.

———. (1960). *Word and Object.* Cambridge, MA: MIT Press.

Raff, Rudolf A. (1996). *The Shape of Life: Genes, Development, and the Evolution of Animal Form.* Chicago: University of Chicago Press.

Raff, R. A., and H. R. Mahler. (1972). "The Nonsymbiotic Origin of Mitochondria." *Science,* 177: 575–582.

Ramsey, J. (1990a). "Metastable States of Belief: The Justification of Approximative Procedures in Reaction Kinetics, 1923–1947." Ph.D. diss., Committee on the Conceptual Foundations of Science, University of Chicago.

———. (1990b). "Beyond Numerical and Causal Accuracy: Expanding the Set of Justificational Criteria." In *PSA-1990,* vol. 1, ed. A. Fine and M. Forbes. East Lansing, MI: Philosophy of Science Association, pp. 485–499.

———. (1992). "Towards an Expanded Epistemology for Approximations." In *PSA-1992,* vol. 1, ed. D. Hull, M. Forbes, and K. Okruhlik. East Lansing, MI: Philosophy of Science Association, pp. 154–164.

———. (1995). "Construction by Reduction." *Philosophy of Science,* 62: 1–20.

———. (1997). "Between the Fundamental and the Phenomenological: The Challenge of 'Semi-Empirical' Methods." *Philosophy of Science,* 64(4): 627–653.

———. (2000). "Realism, Essentialism and Intrinsic Properties: The Case of Molecular Shape." In *Of Minds and Molecules: Philosophical Perspectives on Chemistry,* ed. N. Bhushan and S. Rosenfeld. Oxford: Oxford University Press, pp. 117–128.

Rasmussen, N. (1987). "A New Model of Developmental Constraints as Applied to the *Drosophila* System." *Journal for Theoretical Biology,* 127: 271–299.

———. (1993). "Facts, Artifacts, and Mesosomes: Practicing Epistemology with the Electron Microscope." *Studies in History and Philosophy of Science,* 24: 227–265.

Raup, D. (1987a). *The Nemesis Affair: A Story of the Death of Dinosaurs and the Ways of Science.* New York: Norton.

———. (1987b). "Neutral Models in Paleontology." In *Neutral Models in Biology,* ed. M. Nitecki and A. Hoffman. Oxford: Oxford University Press, pp. 121–132.

———. (1991). *Extinction: Bad Genes or Bad Luck?* New York: Norton.

Raup, D., S. Gould, T. Schopf, and D. Simberloff. (1973). "Stochastic Models of Phylogeny and the Evolution of Diversity." *Journal of Geology,* 81: 525–542.

Reichenbach, H. (1958). *The Philosophy of Space and Time.* New York: Dover.

Rhodes, Richard. (1986). *The Making of the Atomic Bomb.* New York: Simon and Schuster.

Rock, I., and C. S. Harris. (1967). "Vision and Touch." *Scientific American,* 216: 96–104.

Rosenberg, A. (1978). "The Supervenience of Biological Concepts." *Philosophy of Science,* 45: 368–386.

Rosenblatt, Frank. (1962). *Principles of Neurodynamics: Perceptrons and the Theory of Brain Mechanisms.* Washington DC: Spartan Books.

Roth, Nancy Maull. (1974). "Progress in Modern Biology: An Alternative to Reduction." Ph.D. diss., Committee on the Conceptual Foundations of Science, University of Chicago.

Roux, W. (1883). *Über die Bedeutung der Kerntheilungsfiguren* (On the Significance of Nuclear Division Figures: A Hypothetical Discussion). Reprinted in *The Chromosome Theory of Inheritance,* ed. B. Voeller. New York: Appleton-Century-Crofts, 1968.

Ruse, Michael. (1971). "Reduction, Replacement, and Molecular Biology." *Dialectica,* 25: 39–72.

———. (1973). *The Philosophy of Biology.* London: Hutchinson University Library.

———. (1976). "Reduction in Genetics." In *PSA-1974,* ed. A. C. Michalos, C. A. Hooker, G. Pearce, and R. S. Cohen. Dordrecht: Reidel, pp. 632–652.

Russell, B. (1918/1985). *The Philosophy of Logical Atomism.* Ed. with introduction by D. Pears. LaSalle, IL: Open Court.

Salmon, Wesley C. (1971). *Statistical Explanation and Statistical Relevance.* Pittsburgh: University of Pittsburgh Press.

Salthe, Stanley. (1985). *Evolving Hierarchical Systems: Their Structure and Representation.* New York: Columbia University Press.

Sarkar, S. (1989). "Reductionism and Molecular Biology: A Reappraisal." Ph.D. diss., University of Chicago.

——— (ed.). (1992a). *Fisher, Haldane, Muller, and Wright: Founders of the Modern Mathematical Theory of Evolution.* Dordrecht: Martinus Nijhoff.

———. (1992b). "Models of Reduction and Categories of Reductionism." *Synthese,* 91: 167–194.

———. (1994). "The Selection of Alleles and the Additivity of Variance." In *PSA-1994,* vol. 1, ed. R. M. Burian, M. Forbes, and A. Fine. East Lansing, MI: Philosophy of Science Association, pp. 3–12.

———. (1998). *Genetics and Reductionism: A Primer.* London: Cambridge University Press.

Schaffer, W. M. (1984). "Stretching and Folding in Lynx Fur Returns: Evidence for a Strange Attractor in Nature?" *American Naturalist,* 124: 798–820.

Schaffner, Kenneth F. (1967). "Approaches to Reduction." *Philosophy of Science,* 34: 137–147.

———. (1969). "The Watson-Crick Model and Reductionism." *British Journal for the Philosophy of Science,* 20: 325–348.

———. (1974a). "Logic of Discovery and Justification in Regulatory Genetics." *Studies in History and Philosophy of Science,* 4: 349–385.

———. (1974b). "The Peripherality of Reductionism in the Development of Molecular Biology." *Journal of the History of Biology,* 7: 111–139.

———. (1976). "Reductionism in Biology: Prospects and Problems." In *PSA-1974,* ed. A. C. Michalos, C. A. Hooker, G. Pearce, and R. S. Cohen. Dordrecht: Reidel, pp. 613–632.

———. (1993). *Discovery and Explanation in Biology and Medicine.* Chicago: University of Chicago Press.

Schank, J. C. (1991). "The Integrative Role of Model Building and Computer Simulation in Experimental Biology." Ph.D. diss., Committee on the Conceptual Foundations of Science, University of Chicago.

Schank, J. C., and M. K. McClintock. (1992). "A Coupled-Oscillator Model of Ovarian-Cycle Synchrony among Female Rats." *Journal of Theoretical Biology,* 157: 317–362.

Schank, J. C., and W. C. Wimsatt. (1988). "Generative Entrenchment and Evolution." In *PSA-1986,* vol. 2, ed. A. Fine and P. K. Machamer. East Lansing, MI: Philosophy of Science Association, pp. 33–60.

———. (1993). *Modelling 2.0.* Simulation software to run on the Macintosh family for teaching model building and its critical analysis. A BioQUEST Strategic Simulation. Available online at http://www.bioquest.org/index.php.

———. (2000). "Evolvability: Modularity and Generative Entrenchment." In *Thinking about Evolution: Historical, Philosophical and Political Perspectives,* ed. R. Singh, C. Krimbas, D. Paul, and J. Beatty (Festschrift for Richard Lewontin, vol. 2). Cambridge: Cambridge University Press.

Schlosser, G., and G. Wagner. (2004). *Modularity in Evolution and Development.* Chicago: University of Chicago Press.

Schmidt-Nielsen, Knut. (1984). *Scaling: Why Is Animal Size So Important?* Cambridge: Cambridge University Press.

Schoener, T. (1989). "The Ecological Niche." In *Ecological Concepts: The Contribution of Ecology to an Understanding of the Natural World,* ed. J. M. Cherrett. London: Blackwell.

Schweber, S. S. (1992). "Big Science in Context: Cornell and MIT." In *Big Science: The Growth of Large-Scale Research,* ed. P. Galison and B. Hevly. Stanford: Stanford University Press, pp. 149–189.

Sellars, Wilfrid. (1962). "Philosophy and the Scientific Image of Man." In *Frontiers of Science and Philosophy,* ed. Robert Colodny. Pittsburgh: University of Pittsburgh Press, pp. 35–78.

Senchuk, Dennis. (1991). *Against Instinct: From Biology to Philosophical Psychology.* Philadelphia: Temple University Press.

Shaffer, Jerome. (1961). "Could Mental States Be Brain Processes?" *Journal of Philosophy,* 58: 813–822.

———. (1965). "Recent Work on the Mind-Body Problem." *American Philosophical Quarterly,* 2: 81–104.

Shannon, C., and J. McCarthy (eds.). (1956). *Automata Studies.* Princeton, NJ: Princeton University Press.

Shapere, Dudley. (1974). "Scientific Theories and Their Domains." In *The Struc-*

ture of Scientific Theories, ed. Frederick Suppe. Urbana: University of Illinois Press.

Shapiro, J. (1992). "Natural Genetic Engineering in Evolution." *Genetica,* 86: 99–111.

Shimony, A. (1970). "Statistical Inference." In *The Nature and Function of Scientific Theories,* ed. R. G. Colodny. Pittsburgh: University of Pittsburgh Press, pp. 79–172.

———. (1971). "Perception from an Evolutionary Point of View." *Journal of Philosophy,* 68: 571–583.

Shweder, R. A. (1977). "Likeness and Likelihood in Everyday Thought: Magical Thinking in Judgments about Personality." *Current Anthropology,* 18: 637–648. Reply to discussion: 652–658.

———. (1979a). "Rethinking Culture and Personality Theory, Part I." *Ethos,* 7: 255–278.

———. (1979b). "Rethinking Culture and Personality Theory, Part II." *Ethos,* 7: 279–311.

———. (1980a). "Rethinking Culture and Personality Theory, Part III." *Ethos,* 8: 60–94.

——— (ed.). (1980b). "Fallible Judgment in Behavioral Research." In *New Directions for Methodology of Social and Behavioral Science.* San Francisco: Jossey-Bass, pp. 37–58.

Simon, H. A. (1955). "A Behavioral Model of Rational Choice." *Quarterly Journal of Economics,* 69: 99–118.

———. (1957). *Administrative Behavior: A Study of Decision-Making Processes in Administrative Organization,* 2nd ed. New York: Macmillan.

———. (1962). "The Architecture of Complexity." *Proceedings of the American Philosophical Society,* 106(6): 467–482.

———. (1973). "On the Structure of Ill-Structured Problems." *Artificial Intelligence,* 4: 181–201.

———. (1979). *Models of Thought.* New Haven, CT: Yale University Press.

———. (1996). *The Sciences of the Artificial,* 3rd ed. Cambridge, MA: MIT Press. (Originally published in 1969.)

Sklar, Lawrence. (1967). "Types of Inter-Theoretic Reduction." *British Journal for the Philosophy of Science,* 18: 109–124.

———. (1973). "Statistical Explanation and Ergodic Theory." *Philosophy of Science,* 40: 194–212.

Sober, E. (1981). "Holism, Individualism, and the Units of Selection." In *PSA-1980,* vol. 2, ed. P. D. Asquith and R. N. Giere. East Lansing, MI: Philosophy of Science Association, pp. 93–101.

——— (ed.). (1984a). *Conceptual Issues in Evolutionary Biology: An Anthology.* Cambridge, MA: MIT Press.

———. (1984b). *The Nature of Selection: Evolutionary Theory in Philosophical Focus.* Cambridge, MA: MIT Press.

Sober, E., and R. Lewontin. (1982). "Artifact, Cause, and Genic Selection." *Philosophy of Science,* 49(2): 157–180.

Solomon, Miriam. (1996). "Situatedness and Specificity." Paper presented at the American Philosophical Association Eastern Division meetings, December.

Sperber, Dan. (1996). *Explaining Culture: A Naturalistic Approach,* London: Blackwell.

Sperry, R. W. (1976). "Mental Phenomena as Causal Determinants." In *Consciousness and the Brain: A Scientific and Philosophical Inquiry,* ed. Gordon G. Globus, Grover Maxwell, and Irwin Savodnik. New York: Plenum, p. 167.

Stadler, L. J. (1954). "The Gene." *Science,* 120: 811–819.

Star, S. L. (1983a). "Scientific Theories as Going Concerns: The Development of the Localizationist Perspective in Neurophysiology, 1870–1906." Ph.D. diss., University of California, Berkeley.

———. (1983b). "Simplification in Scientific Work: An Example from Neuroscience Research." *Social Studies of Science,* 13: 205–228.

Star, S. L., and E. Gerson. (1986). "The Management and Dynamics of Anomalies in Scientific Work." *Sociological Quarterly,* 28: 147–169.

Stein, N., and C. Glenn. (1979). "An Analysis of Story Comprehension in Elementary School Children." In *Advances in Discourse Processes: New Directions in Discourse Processing,* vol. 2, ed. R. O. Freedle. Norwood, NJ: Ablex, pp. 53–120.

Stigler, S. (1987). "Testing Hypotheses or Fitting Models: Another Look at Mass Extinctions." In *Neutral Models in Biology,* ed. M. Nitecki and A. Hoffman. Oxford: Oxford University Press.

Strobeck, K. (1975). "Selection in a Fine-Grained Environment." *American Naturalist,* 109: 419–426.

Strong, John C., in collaboration with H. V. Neher, A. E. Whitford, C. H. Cartwright, and R. Hayward. (1938). *Procedures in Experimental Physics.* Englewood Cliffs, NJ: Prentice-Hall.

Sturtevant, A. H. (1913). "The Linear Arrangement of Six Sex-Linked Factors in *Drosophila,* as Shown by Their Mode of Association." *Journal of Experimental Zoology,* 14: 43–59.

———. (1917). "Crossing-Over without Chiasmatype?" *Genetics,* 2: 301–304.

Sturtevant, A. H., and G. W. Beadle. (1939). *An Introduction to Genetics.* Philadelphia: W. B. Saunders.

Sturtevant, A. H., C. B. Bridges, and T. H. Morgan. (1919). "The Spatial Relation of Genes." *Proceedings of the National Academy of Science,* 5: 168–173.

Sutton, W. (1903). "The Chromosomes in Heredity." *Biological Bulletin of the Marine Biological Laboratory at Woods Hole,* 4: 231–248.

Taylor, P. J. (1985). "Construction and Turnover of Multi-species Communities: A Critique of Approaches to Ecological Complexity." Ph.D. diss., Harvard University.

Thompson, D'Arcy W. (1961). *On Growth and Form,* abridged ed., introduction by J. T. Bonner. Cambridge: Cambridge University Press. (Originally published in 1917.)

Thompson, E., E. Rosch, and F. Varela. (1991). *The Embodied Mind: Cognitive Science and Human Experience.* Cambridge, MA: MIT Press.

Thompson, Evan. (1992). "Novel Colors." *Philosophical Studies,* 68: 321–349.

Trabasso, T., and N. Stein. (1997). "Narrating, Representing, and Remembering Event Sequences." In *Developmental Spans in Event Comprehension and Representation,* ed. P. van den Broek, P. Bauer, and T. Bourg. Hillsdale, NJ: LEA, pp. 237–270.

Tribe, L. (1972). "Policy Science: Analysis or Ideology?" *Philosophy and Public Affairs,* 2: 66–110.

Trout, J. D. (1995). "Diverse Tests on an Independent World." *Studies in History and Philosophy of Science,* 26: 407–429.

———. (1998). *Measuring the Intentional World: Realism, Naturalism, and Quantitative Methods in the Behavioral Sciences.* Oxford: Oxford University Press.

Tufte, Edward R. (1983). *The Visual Display of Quantitative Information.* Cheshire, CT: Graphics Press.

———. (1988). *Envisioning Information.* Cheshire, CT: Graphics Press.

———. (1996). *Visual Explanations.* Cheshire, CT: Graphics Press.

Turner, M. (1991). *Reading Minds: The Study of English in the Age of Cognitive Science.* Princeton, NJ: Princeton University Press.

Tversky, A., and D. Kahneman. (1974). "Judgment under Uncertainty: Heuristics and Biases." *Science,* 185: 1124–1131.

Tweney, R., M. Doherty, and C. Mynatt (eds.). (1981). *On Scientific Thinking.* New York: Columbia University Press.

Tyson, C. (1994). *New Foundations for Scientific Social and Behavioral Research: The Heuristic Paradigm.* New York: Allyn and Bacon.

Valenstein, E. (1973). *Brain Control: A Critical Examination of Brain Stimulation and Psychosurgery.* New York: Wiley.

Van Fraasen, B. (1990). *Laws and Symmetry.* Oxford: Oxford University Press.

Van Valen, Leigh. (1973). "A New Evolutionary Law." *Evolutionary Theory,* 1: 1–30.

Varela, F. J., E. Thompson, and E. Rosch. (1991). *The Embodied Mind: Cognitive Science and Human Experience.* Cambridge, MA: MIT Press.

Vincenti, Walter. (1990). *What Engineers Know and How They Know It: Analytical Studies from Aeronautical History.* Baltimore: Johns Hopkins University Press.

Von Baer, K. (1828). *Über Entwickelungsgeschichte der Thiere,* vol. 1, pp. 221–224. In *Scientific Memoirs, Selections from Foreign Academies of Science, and from Foreign Journals: Natural History,* trans. A. Henfrey and T. Huxley. London: Taylor and Francis, 1853.

Von Neumann, J. (1956). "Probabilistic Logic and the Synthesis of Reliable Organisms from Unreliable Components." In *Automata Studies,* ed. C. E. Shannon and J. McCarthy. Princeton, NJ: Princeton University Press, pp. 43–98.

Von Neumann, John, and Oskar Morgenstern. (1964). *Theory of Games and Economic Behavior.* New York: Wiley. (Originally published in 1944.)

Von Üexkull, Jacob. (1934). "A Stroll through the Worlds of Animals and Men: A Picture Book of Invisible Worlds." Reprinted in *Instinctive Behavior: The*

Development of a Modern Concept, ed. Claire H. Schiller. New York: International Universities Press, 1957, pp. 5–80.

Wade, M. J. (1978). "A Critical Review of the Models of Group Selection." *Quarterly Review of Biology,* 53: 101–114.

———. (1996). "Adaptation in Subdivided Populations: Kin Selection and Interdemic Selection." In *Adaptation,* ed. M. Rose and G. Lauder. New York: Academic Press, pp. 381–405.

Wade, M. J., and C. J. Goodnight. (1998). "Perspective: The Theories of Fisher and Wright in the Context of Metapopulations: When Nature Does Many Small Experiments." *Evolution,* 52: 1537–1553.

Wagner, A. (2005). *Robustness and Evolvability in Living Systems.* Princeton, NJ: Princeton University Press.

Wagner, G. P., and L. Altenberg. (1996). "Complex Adaptations and the Evolution of Evolvability." *Evolution,* 50: 967–976.

Waismann, F. (1951). "Verifiability." In *Logic and Language,* first series, ed. A. G. N. Flew. Oxford: Blackwell, pp. 139–168.

———. (1953). "Language Strata." In *Logic and Language,* second series, ed. A. G. N. Flew. Oxford: Blackwell, pp. 11–31.

Waters, K. (1990). "Why the Anti-reductionist Consensus Won't Survive the Case of Classical Mendelian Genetics." In *PSA-1990,* ed. A. Fine, M. Forbes, and L. Wessels. East Lansing, MI: Philosophy of Science Association, pp. 125–139.

———. (1994). "Genes Made Molecular." *Philosophy of Science,* 61: 163–185.

Weisberg, M. (2006a). "Robustness Analysis." *Philosophy of Science,* 73(5): in press.

——— (ed.). (2006b). Special issue of *Biology and Philosophy* on Richard Levins, in press.

Weismann, A. (1892). *Das Keimplasma, Eine theorie der Vererbung.* Jena: Gustav Fischer. English translation W. Parker and H. Ronnfeldt, *The Germ-Plasm, A Theory of Heredity.* New York: Charles Scribner's Sons, 1893.

Whitehouse, H. L. K. (1973). *Towards an Understanding of the Mechanisms of Heredity,* 3rd rev. ed. New York: St. Martin's Press.

Whitten, M. J. et al. (1974). "Chromosome Rearrangements for the Control of Insect Pests." *Science,* 176: 875–880.

Wiener, N. (1961). *Cybernetics: Or, Control and Communication in the Animal and the Machine,* 2nd ed. New York: MIT Press.

Williams, G. C. (1966). *Adaptation and Natural Selection: A Critical Review of Some Current Evolutionary Thought.* Princeton, NJ: Princeton University Press.

Williams, Michael. (1996). *Unnatural Doubts: Epistemological Realism and the Basis of Skepticism.* Princeton, NJ: Princeton University Press.

Wimsatt, W. C. (1971a). "The Concepual Foundations of Functional Analysis." Ph.D. diss., University of Pittsburgh.

———. (1971b). "Self-Organization, Selection, and Dissipative Structures" (comments on a paper by Aharon Katchalsky). *Zygon,* 6: 269–274.

———. (1971c). "Some Problems with the Concept of Feedback." In *PSA-1970,* ed. R. C. Buck and R. S. Cohen. Dordrecht: Reidel, pp. 241–256.

———. (1972). "Teleology and the Logical Structure of Function Statements." *Studies in History and Philosophy of Science,* 3: 1–80.

———. (1974). "Complexity and Organization." In *PSA-1972,* ed. K. F. Schaffner and R. S. Cohen. Dordrecht: Reidel, pp. 67–86.

———. (1976a). "Reductionism, Levels of Organization, and the Mind-Body Problem." In *Consciousness and the Brain,* ed. G. G. Globus, G. Maxwell, and I. Savodnik. New York: Plenum, pp. 199–267.

———. (1976b). "Reductive Explanation: A Functional Account." In *PSA-1974,* ed. A. C. Michalos, C. A. Hooker, G. Pearce, and R. S. Cohen. Dordrecht: Reidel, pp. 671–710.

———. (1979). "Reduction and Reductionism." In *Current Research in Philosophy of Science,* ed. P. D. Asquith and H. Kyburg Jr. East Lansing, MI: Philosophy of Science Association, pp. 352–377.

———. (1980a). "Randomness and Perceived-Randomness in Evolutionary Biology." *Synthese,* 43: 287–329.

———. (1980b). "Reductionistic Research Strategies and Their Biases in the Units of Selection Controversy." In *Scientific Discovery,* vol. 2, *Case Studies,* ed. T. Nickles. Dordrecht: Reidel, pp. 213–259.

———. (1981a). "Robustness, Reliability, and Overdetermination." In *Scientific Inquiry and the Social Sciences,* ed. M. Brewer and B. Collins. San Francisco: Jossey-Bass, pp. 124–163.

———. (1981b). "Units of Selection and the Structure of the Multi-level Genome." In *PSA-1980,* vol. 2., ed. P. D. Asquith and R. N. Giere. East Lansing, MI: Philosophy of Science Association, pp. 122–183.

———. (1985). "Heuristics and the Study of Human Behavior." In *Metatheory in Social Science: Pluralisms and Subjectivities,* ed. D. W. Fiske and R. Shweder. Chicago: University of Chicago Press, pp. 293–314.

———. (1986a). "Developmental Constraints, Generative Entrenchment, and the Innate-Acquired Distinction." In *Integrating Scientific Disciplines,* ed. W. Bechtel. Dordrecht: Martinus Nijhoff, pp. 185–208.

———. (1986b). "Forms of Aggregativity." In *Human Nature and Natural Knowledge,* ed. A. Donagan, N. Perovich, and M. Wedin. Dordrecht: Reidel, pp. 259–293.

———. (1987). "False Models as Means to Truer Theories." In *Neutral Models in Biology,* ed. M. Nitecki and A. Hoffman. London: Oxford University Press, pp. 23–55.

———. (1991). "Taming the Dimensions—Visualizations in Science." In *PSA-1990,* vol. 2, ed. M. Forbes, L. Wessels, and A. Fine. East Lansing, MI: Philosophy of Science Association, pp. 111–135.

———. (1992). "Golden Generalities and Co-opted Anomalies: Haldane vs. Muller and the *Drosophila* Group on the Theory and Practice of Linkage Mapping." In *Fisher, Haldane, Muller, and Wright: Founders of the Modern

Mathematical Theory of Evolution, ed. S. Sarkar. Dordrecht: Martinus Nijhoff, pp. 107–166.

———. (1994a). "The Ontology of Complex Systems: Levels, Perspectives, and Causal Thickets." In *Biology and Society: Reflections on Methodology,* ed. M. Matthen and R. Ware. *Canadian Journal of Philosophy,* supplementary vol. 20: 207–274.

———. (1994b). "Lewontin's Evidence (That There Isn't Any!)." In *Questions of Evidence,* ed. J. Chandler, A. Davidson, and H. Haroutunian. Chicago: University of Chicago Press, pp. 492–503.

———. (1997a). "Functional Organization, Functional Analogy, and Functional Inference." *Evolution and Cognition,* 3: 2–32.

———. (1997b). "Aggregativity: Reductive Heuristics for Finding Emergence." *Philosophy of Science,* 64(4): S372–S384.

———. (1999a). "Genes, Memes, and Cultural Inheritance." *Biology and Philosophy,* 14: 279–310.

———. (1999b). "Generativity, Entrenchment, Evolution, and Innateness." In *Biology Meets Psychology: Philosophical Essays,* ed. V. Hardcastle. Cambridge, MA: MIT Press, pp. 139–179.

———. (2000). "Emergence as Non-Aggregativity and the Biases of Reductionism(s)." *Foundations of Science,* 5: 269–297.

———. (2001). "Generative Entrenchment and the Developmental Systems Approach to Evolutionary Processes." In *Cycles of Contingency: Developmental Systems and Evolution,* ed. S. Oyama, R. Gray, and P. Griffiths. Cambridge, MA: MIT Press, pp. 219–237.

———. (2002a). "Functional Organization, Functional Inference, and Functional Analogy." In *Function,* ed. A. Ariew, R. Cummins, and R. Perlman. New York: Oxford University Press, pp. 174–221.

———. (2002b). "False Models as Means to Truer Theories: Blending Inheritance in Biological vs. Cultural Evolution." *Philosophy of Science,* 69(3): S12–S24.

———. (2003). "Evolution, Entrenchment, and Innateness." In *Reductionism and the Development of Knowledge,* ed. T. Brown and L. Smith. Mahwah, NJ: Lawrence Erlbaum, pp. 53–81.

———. (2006a). "Reductionism and Its Heuristics: Making Methodological Reductionism Honest." *Synthese,* 151: 445–475.

———. (2006b). "Inconsistencies, Optimization and Satisficing—Steps towards a Philosophy for Limited Beings: Commentary on Russell Hardin." In *Is There Value in Inconsistency?* ed. C. Engel and L. Daston. Baden-Baden: Nomos Verlagsgesellschaft, pp. 201–220.

———. (2006c). "Optimization, Consistency, and Kluged Adaptations: Can Maximization Survive? Comment on Kachalnik, et al." In *Is There Value in Inconsistency?* ed. C. Engel and L. Daston. Baden-Baden: Nomos Verlagsgesellschaft, pp. 309–420.

———. (2007a). "On Building Reliable Pictures with Unreliable Data: An Evolutionary and Developmental Coda for the New Systems Biology." In *Systems*

Biology: Philosophical Foundations, ed. F. C. Boogerd, F. J. Bruggeman, J-H. S. Hofmeyer, and H. V. Westerhoff. Amsterdam: Reed-Elsevier.

———. (2007b). "Echoes of Haeckel? Reentrenching Development in Evolution." In *From Heredity and Development to Evo-Devo,* ed. J. Maienschein and M. Laubichler. Cambridge, MA: MIT Press, pp. 309–356.

Wimsatt, W. C., and J. R. Griesemer. (2007). "Re-Producing Entrenchments to Scaffold Culture: The Central Role of Development in Cultural Evolution." In *Integrating Evolution and Development,* ed. R. Sansom and R. Brandon. Cambridge, MA: MIT Press.

Wimsatt, W. C., and J. C. Schank. (1988). "Two Constraints on the Evolution of Complex Adaptations and the Means for Their Avoidance." In *Evolutionary Progress,* ed. M. Nitecki. Chicago: University of Chicago Press.

———. (1993). *Modelling—A Primer, or: The Crafty Art of Making, Modifying, Extending, Transforming, Tweaking, Bending, Disassembling, Questioning, and Breaking Models.* A 230-page lab manual/textbook to accompany Schank and Wimsatt, 1993. New York: Academic Press. Now available online at http://www.bioquest.org/index.php.

———. (2004). "Generative Entrenchment, Modularity and Evolvability: When Genic Selection Meets the Whole Organism." In *Modularity in Evolution and Development,* ed. G. Schlosser and G. Wagner. Chicago: University of Chicago Press, pp. 359–394.

Winograd, S., and J. Cowan. (1963). *Reliable Computation in the Presence of Noise.* Cambridge, MA: MIT Press.

Wittgenstein, L. (1962). *Remarks on the Foundations of Mathematics.* London: Blackwell.

Wright, L. (1973). "Functions." *The Philosophical Review,* 82: 139–168.

Wright, S. (1977). *Evolution and the Genetics of Populations,* vol. 3, *Experimental Results and Evolutionary Deductions.* Chicago: University of Chicago Press.

Credits

Some of the chapters in this book originally appeared elsewhere, as noted below. Thanks are due the various journals and publishers for permission to reprint.

Chapter 4: "Robustness, Reliability, and Overdetermination," in *Scientific Inquiry and the Social Sciences,* ed. M. Brewer and B. Collins (San Francisco: Jossey-Bass, 1981), pp. 124–163.

Chapter 5: "Heuristics and the Study of Human Behavior," in *Metatheory in Social Science: Pluralisms and Subjectivities,* ed. D. W. Fiske and R. Shweder (Chicago: University of Chicago Press, 1985), pp. 293–314.

Chapter 6: "False Models as Means to Truer Theories," in *Neutral Models in Biology,* ed. M. Nitecki and A. Hoffman (London: Oxford University Press, 1987), pp. 23–55.

Chapter 8: "Lewontin's Evidence (That There Isn't Any!)," in *Questions of Evidence,* ed. J. Chandler, A. Davidson, and H. Haroutunian (Chicago: University of Chicago Press, 1994), pp. 492–503.

Chapter 9: "Complexity and Organization," in *PSA-1972,* ed. K. F. Schaffner and R. S. Cohen (Dordrecht: Reidel, 1974), pp. 67–86.

Chapter 10: "The Ontology of Complex Systems: Levels, Perspectives, and Causal Thickets," in *Biology and Society: Reflections on Methodology,* ed. M. Matthen and R. Ware, supplementary vol. 20 of the *Canadian Journal of Philosophy* (1994): 207–274.

Chapter 11: "Reductive Explanation: A Functional Account," in *PSA-1974,* ed. A. C. Michalos, C. A. Hooker, G. Pearce, and R. S. Cohen (Dordrecht: Reidel, 1976), pp. 671–710.

Chapter 12 is greatly expanded (by 70 percent) from an earlier paper of the same title: "Emergence as Non-Aggregativity and the Biases of Reductionism(s)," *Foundations of Science*, 5 (2000): 269–297.

Chapters 1, 2, 3, 7, and 13, the part introductions, and the appendixes are all published here for the first time.

Index

www.ingramcontent.com/pod-product-compliance
Ingram Content Group UK Ltd.
Pitfield, Milton Keynes, MK11 3LW, UK
UKHW010829150826
474047UK00007B/32/J

9 780674 015456